Eigenvalues and Completeness for Regular and Simply Irregular Two-Point Differential Operators

MEMOIRS
of the
American Mathematical Society

Number 911

Eigenvalues and Completeness for Regular and Simply Irregular Two-Point Differential Operators

John Locker

September 2008 • Volume 195 • Number 911 (second of 4 numbers) • ISSN 0065-9266

American Mathematical Society
Providence, Rhode Island

2000 *Mathematics Subject Classification.*
Primary 34L10, 34L20; Secondary 47E05.

Library of Congress Cataloging-in-Publication Data

Locker, John.
Eigenvalues and completeness for regular and simply irregular two-point differential operators / John Locker.
p. cm. — (Memoirs of the American Mathematical Society, ISSN 0065-9266 ; no. 911)
Includes bibliographical references and index.
ISBN 978-0-8218-4171-6 (alk. paper)
1. Eigenvalues. 2. Differential operators. I. Title.
QA193.L63 2008
515′.7242—dc22 2008020748

Memoirs of the American Mathematical Society

This journal is devoted entirely to research in pure and applied mathematics.

Subscription information. The 2008 subscription begins with volume 191 and consists of six mailings, each containing one or more numbers. Subscription prices for 2008 are US$675 list, US$540 institutional member. A late charge of 10% of the subscription price will be imposed on orders received from nonmembers after January 1 of the subscription year. Subscribers outside the United States and India must pay a postage surcharge of US$38; subscribers in India must pay a postage surcharge of US$43. Expedited delivery to destinations in North America US$53; elsewhere US$130. Each number may be ordered separately; *please specify number* when ordering an individual number. For prices and titles of recently released numbers, see the New Publications sections of the *Notices of the American Mathematical Society.*

Back number information. For back issues see the *AMS Catalog of Publications.*

Subscriptions and orders should be addressed to the American Mathematical Society, P. O. Box 845904, Boston, MA 02284-5904, USA. *All orders must be accompanied by payment.* Other correspondence should be addressed to 201 Charles Street, Providence, RI 02904-2294, USA.

Memoirs of the American Mathematical Society (ISSN 0065-9266) is published bimonthly (each volume consisting usually of more than one number) by the American Mathematical Society at 201 Charles Street, Providence, RI 02904-2294, USA. Periodicals postage paid at Providence, RI. Postmaster: Send address changes to Memoirs, American Mathematical Society, 201 Charles Street, Providence, RI 02904-2294, USA.

This publication is indexed in *Science Citation Index*®, *SciSearch*®, *Research Alert*®, *CompuMath Citation Index*®, *Current Contents*®/*Physical, Chemical & Earth Sciences.*
Printed in the United States of America.

♾ The paper used in this book is acid-free and falls within the guidelines established to ensure permanence and durability.
Visit the AMS home page at `http://www.ams.org/`

10 9 8 7 6 5 4 3 2 1 13 12 11 10 09 08

Contents

Abstract

In this monograph we develop the spectral theory for an nth order two-point differential operator L in the Hilbert space $L^2[0,1]$, where L is determined by an nth order formal differential operator ℓ having variable coefficients and by n linearly independent boundary values $B_1, \dots, B_n$. Using the Birkhoff approximate solutions of the differential equation $(\rho^n I - \ell)u = 0$, the differential operator L is classified as belonging to one of three possible classes: regular, simply irregular, or degenerate irregular. For the regular and simply irregular classes, we develop asymptotic expansions of solutions of the differential equation $(\rho^n I - \ell)u = 0$, construct the characteristic determinant and Green's function, characterize the eigenvalues and the corresponding algebraic multiplicities and ascents, and show that the generalized eigenfunctions of L are complete in $L^2[0,1]$. We also give examples of degenerate irregular differential operators illustrating some of the unusual features of this class.

Received by the editor December, 2004.

2000 *Mathematics Subject Classification.* Primary: 34L10, 34L20; Secondary: 47E05.

Key words and phrases. Birkhoff approximate solutions, regular differential operators, simply irregular differential operators, degenerate irregular differential operators, asymptotic expansions of solutions, characteristic determinant, Green's function, eigenvalues, completeness.

CHAPTER 1

Introduction

In this monograph we develop the spectral theory of an nth order two-point differential operator L in the Hilbert space $L^2[0,1]$. This initial chapter introduces the basic definitions, notations, and properties of the differential operators, and it summarizes the main results that are established. There remain many unsolved problems for this spectral theory.

1.1. Definitions and Notations

Throughout this monograph we work in the complex Hilbert space $L^2[0,1]$ with its standard inner product $(\ ,\ )$ and norm $\|\ \|$. Let n be a positive integer with $n \geq 2$, and let $H^n[0,1]$ denote the subspace of $L^2[0,1]$ consisting of all functions $u \in C^{n-1}[0,1]$ with $u^{(n-1)}$ absolutely continuous on $[0,1]$ and with $u^{(n)} \in L^2[0,1]$. The subspace $H^n[0,1]$ becomes a Banach space under the norm

$$|u|_{H^n} := \sum_{p=0}^{n-1} \|u^{(p)}\|_\infty + \|u^{(n)}\|,$$

with this norm structure referred to as the H^n*-structure* or the H^n*-Sobolev structure* for $H^n[0,1]$. Let

$$\ell := \sum_{p=0}^{n} a_p(t)\left(\frac{d}{dt}\right)^p$$

be an nth order formal differential operator on $[0,1]$, where the leading coefficients are assumed to be $a_n(t) = 1/i^n$ and $a_{n-1}(t) = 0$; let

$$B_i(u) := \sum_{p=0}^{n-1} \alpha_{ip} u^{(p)}(0) + \sum_{p=0}^{n-1} \beta_{ip} u^{(p)}(1), \qquad i = 1, \ldots, n,$$

be a set of n linearly independent boundary values on $H^n[0,1]$; and let L be the nth order two-point differential operator in $L^2[0,1]$ defined by

$$\mathcal{D}(L) := \{u \in H^n[0,1] \mid B_i(u) = 0,\ i = 1, \ldots, n\}, \qquad Lu := \ell u.$$

We assume that the coefficient functions a_p are infinitely differentiable on $[0,1]$. The differential operator L is a prime example of a Fredholm operator of index 0.

For each $\lambda \in \mathbb{C}$ we form the subspace $\mathcal{M}_\lambda := \bigcup_{k=1}^\infty \mathcal{N}((\lambda I - L)^k)$. Relevant to this subspace is its dimension

$$\nu(\lambda) := \dim \mathcal{M}_\lambda = \lim_{k\to\infty} \dim \mathcal{N}((\lambda I - L)^k),$$

which is the algebraic multiplicity of λ; and relevant to the differential operator $\lambda I - L$ is its ascent $m(\lambda) := \alpha(\lambda I - L)$, which is the smallest integer $k \geq 0$ such that $\mathcal{N}((\lambda I - L)^k) = \mathcal{N}((\lambda I - L)^{k+1})$, with $m(\lambda) = \infty$ when no such integer k

exists. In case $\mathcal{M}_\lambda \neq \{0\}$, then λ is an eigenvalue of L and $\mathcal{M}_\lambda$ is the generalized eigenspace of L corresponding to λ. The spectrum $\sigma(L)$ is precisely the set of all eigenvalues of L. Because L is a Fredholm operator of index 0, $\sigma(L)$ is either a countable set having no limit points in $\mathbb{C}$ or it is equal to all of $\mathbb{C}$. The complement of the spectrum is the resolvent set $\rho(L)$. For each λ belonging to $\rho(L)$, the resolvent $R_\lambda(L) = (\lambda I - L)^{-1}$ is an L^2-integral operator on $L^2[0,1]$, and the Green's function $G(t,s;\lambda)$ for $\lambda I - L$ is the L^2-kernel of $R_\lambda(L)$:

$$R_\lambda(L)u(t) = \int_0^1 G(t,s;\lambda)u(s)\,ds, \qquad 0 \le t \le 1, \tag{1.1}$$

for $u \in L^2[0,1]$. Thus, for each $\lambda \in \rho(L)$ the resolvent $R_\lambda(L)$ is a Hilbert-Schmidt operator on $L^2[0,1]$, and the differential operator L is a Hilbert-Schmidt discrete operator in case $\rho(L) \neq \emptyset$. See [**34**] for a detailed discussion of these concepts.

Introducing the formal differential operators

$$\tau := a_n(t)\left(\frac{d}{dt}\right)^n = \frac{1}{i^n}\left(\frac{d}{dt}\right)^n \quad \text{and} \quad \sigma := \sum_{p=0}^{n-2} a_p(t)\left(\frac{d}{dt}\right)^p,$$

we define differential operators T and S in $L^2[0,1]$ by

$$\mathcal{D}(T) := \mathcal{D}(L) = \{u \in H^n[0,1] \mid B_i(u) = 0,\ i = 1,\dots,n\},$$
$$Tu := \tau u = i^{-n}u^{(n)},$$

and

$$\mathcal{D}(S) := H^{n-2}[0,1], \quad Su := \sigma u = \sum_{p=0}^{n-2} a_p(t)u^{(p)}.$$

Clearly $\ell = \tau + \sigma$ and $L = T + S$. The differential operator T is the *principal part* of the differential operator L, and we can consider the operator L as a perturbation of the operator T by the operator S. The assumption that $a_n(t) = 1/i^n$ produces a relatively simple principal part T, with τ being formally self-adjoint. The net effect of this will be to locate the spectrum of L near the real axis in the complex plane.

Let us introduce the $n \times 2n$ *boundary coefficient matrix* associated with the boundary values $B_1, \dots, B_n$:

$$A := \begin{pmatrix} \alpha_{1\,n-1} & \beta_{1\,n-1} & \alpha_{1\,n-2} & \beta_{1\,n-2} & \cdots & \alpha_{1\,0} & \beta_{1\,0} \\ \alpha_{2\,n-1} & \beta_{2\,n-1} & \alpha_{2\,n-2} & \beta_{2\,n-2} & \cdots & \alpha_{2\,0} & \beta_{2\,0} \\ \vdots & \vdots & \vdots & \vdots & & \vdots & \vdots \\ \alpha_{n\,n-1} & \beta_{n\,n-1} & \alpha_{n\,n-2} & \beta_{n\,n-2} & \cdots & \alpha_{n\,0} & \beta_{n\,0} \end{pmatrix}.$$

Without loss of generality we can assume that the matrix A is in reduced row echelon form with rank n. This corresponds to a normalization of the boundary values $B_1, \dots, B_n$. For $i = 1, \dots, n$ let m_i denote the order of the boundary value B_i. Then $0 \le m_i \le n-1$ and $m_1 \ge m_2 \ge \dots \ge m_{n-1} \ge m_n$. Set

$$p_0 := \sum_{i=1}^n m_i.$$

The integer p_0 plays an important role in the spectral theory of the differential operator L.

1.2. Summary of Results

The goal of this monograph is to develop a classification scheme for the differential operator L consisting of the regular class and the irregular class, with the irregular class subdivided into the simply irregular class and the degenerate irregular class, and then to develop the spectral theory for L belonging to these various classes. The regular and irregular classes are determined completely by the boundary values $B_1, \ldots, B_n$, while the subdivision into the simply irregular and degenerate irregular classes depends on both the boundary values $B_1, \ldots, B_n$ and the formal differential operator ℓ. The regular class of differential operators has been studied extensively, and these operators have a more or less complete spectral theory. In contrast, for the irregular class there are some results for the special case $n = 2$ and a few results for the general case $n > 2$. Our major contribution is the identification of the simply irregular class for arbitrary $n \geq 2$, and for L belonging to this very large class, the establishment of a large portion of the spectral theory for such a differential operator: asymptotic expansion of solutions of the differential equation $(\rho^n I - \ell)u = 0$, construction of the characteristic determinant and Green's function, a calculation of the eigenvalues of L and the corresponding algebraic multiplicities and ascents, and a demonstration that the generalized eigenfunctions of L are complete in $L^2[0, 1]$. As we derive these new results for the simply irregular class, we will simultaneously derive the corresponding well-known results for the regular class. At the end of this monograph we give some examples of degenerate irregular differential operators in the special case $L = T$; these examples illustrate some of the unusual features of the spectrum that occur in the degenerate irregular case.

The results presented here are a natural extension of our previous work in the four-part series [**30, 31, 32, 33**] and the monograph [**34**], and in our joint work with Patrick Lang [**21, 22, 23, 24, 25, 26, 35**]. From the literature on the spectral theory of differential operators, this new work has been most directly influenced by the papers of Birkhoff [**3, 4**], Stone [**45, 46**], and Benzinger [**2**] and by the books of Naimark [**37**] and Dunford and Schwartz [**8**]. The literature on the spectral theory of non-self-adjoint differential operators is enormous—we refer the reader to the bibliography in the new book by Mennicken and Möller [**36**]. Interesting applications of this spectral theory are given in [**36**] and [**51**].

Let us describe the main features of our classification scheme and state our principal results. Throughout we express the order n of the differential operator L in the form $n = 2\nu$ for n even and the form $n = 2\nu - 1$ for n odd, and let $\omega_k = e^{i2\pi k/n}$, $k = 0, \pm 1, \pm 2, \ldots$, denote the nth roots of unity.

First, let $c_{0\ell p}(t)$, $\ell = 0, 1, \ldots, n$, $p = 0, 1, \ldots, n - \ell$, be the triangular array of functions formed from the coefficients of the formal differential operator ℓ as follows:

$$c_{000}(t) := 0, \qquad c_{00p}(t) := a_{n-p}(t)\, i^{2n-p}, \quad p = 1, 2, \ldots, n,$$

$$c_{0\ell p}(t) := a_{n-p}(t) \binom{n-p}{\ell} i^{2n-\ell-p}, \quad \ell = 1, 2, \ldots, n,\ p = 0, 1, \ldots, n - \ell,$$

and for $q = 2, \ldots, n$ let ℓ_q be the qth order formal differential operator defined by

$$\ell_q u(t) := \sum_{p=0}^{q} c_{0p\,q-p}(t) u^{(p)}(t).$$

The leading coefficient of ℓ_q is $c_{0q0}(t) = \binom{n}{q} i^{n-q}$, and for $q = n$ we have $\ell_n = i^n \ell$. Let $z_{0j}(t)$, $j = 0, 1, 2, \ldots$, be the sequence of functions defined by the equations $z_{00}(t) := 1$ and

$$z_{0\,s-1}(t) := \int_0^t z'_{0\,s-1}(\xi)\,d\xi \quad \text{for } s = 2, 3, \ldots, \tag{1.2}$$

where the derivatives satisfy the recursion relations

$$z'_{0\,s-1}(t) = -\frac{1}{n i^{n-1}} \sum_{j=0}^{s-2} \ell_{s-j}\, z_{0j}(t), \qquad 2 \le s \le n, \tag{1.3}$$

$$z'_{0\,s-1}(t) = -\frac{1}{n i^{n-1}} \sum_{j=s-n}^{s-2} \ell_{s-j}\, z_{0j}(t), \qquad n+1 \le s < \infty. \tag{1.4}$$

The derivative $z'_{0\,s-1}(t)$ is determined by the $n-1$ preceding functions $z_{0\,s-n}(t)$, $z_{0\,s-n+1}(t)$, $\ldots$, $z_{0\,s-2}(t)$ and their derivatives up to order n. For the term $z_{01}(t)$ we have $z'_{0\,1}(t) = -1/(n i^{n-1})\, \ell_2\, z_{00}(t) = -(i^{n-1}/n)\, a_{n-2}(t)$ and

$$z_{01}(t) = -\frac{i^{n-1}}{n} \int_0^t a_{n-2}(\xi)\,d\xi, \qquad 0 \le t \le 1. \tag{1.5}$$

Finally, for $k = 0, 1, \ldots, n-1$ let $z_{kj}(t)$, $j = 0, 1, 2, \ldots$, be the sequence of functions defined by

$$z_{kj}(t) := z_{0j}(t)(\omega_k)^{-j}, \qquad j = 0, 1, 2, \ldots.$$

We will see that the functions $z_{kj}(t)$ appear naturally as the coefficients in the Birkhoff approximate solutions of the differential equation $(\rho^n I - \ell)u = 0$. They depend strongly on the coefficients of the formal differential operator ℓ.

Second, for $i = 1, \ldots, n$ and $k = 0, 1, \ldots, \nu - 1$ we form the two sequences of constants

$$p_{iks} := \sum_{p=s}^{m_i} \sum_{\ell=0}^{p-s} \alpha_{ip} \binom{p}{\ell} (i\omega_k)^{p-\ell} z^{(\ell)}_{k\,p-\ell-s}(0), \qquad s = 0, 1, \ldots, m_i,$$

$$p_{iks} := \sum_{p=0}^{m_i} \sum_{\ell=0}^{p} \alpha_{ip} \binom{p}{\ell} (i\omega_k)^{p-\ell} z^{(\ell)}_{k\,p-\ell-s}(0), \qquad s = -1, -2, \ldots,$$

and

$$q_{iks} := \sum_{p=s}^{m_i} \sum_{\ell=0}^{p-s} \beta_{ip} \binom{p}{\ell} (i\omega_k)^{p-\ell} z^{(\ell)}_{k\,p-\ell-s}(1), \qquad s = 0, 1, \ldots, m_i,$$

$$q_{iks} := \sum_{p=0}^{m_i} \sum_{\ell=0}^{p} \beta_{ip} \binom{p}{\ell} (i\omega_k)^{p-\ell} z^{(\ell)}_{k\,p-\ell-s}(1), \qquad s = -1, -2, \ldots.$$

Similarly, for $i = 1, \ldots, n$ and $k = \nu, \ldots, n-1$ we form two analogous sequences:

$$p_{iks} := \sum_{p=s}^{m_i} \sum_{\ell=0}^{p-s} \beta_{ip} \binom{p}{\ell} (i\omega_k)^{p-\ell} z^{(\ell)}_{k\,p-\ell-s}(1), \qquad s = 0, 1, \ldots, m_i,$$

$$p_{iks} := \sum_{p=0}^{m_i} \sum_{\ell=0}^{p} \beta_{ip} \binom{p}{\ell} (i\omega_k)^{p-\ell} z^{(\ell)}_{k\,p-\ell-s}(1), \qquad s = -1, -2, \ldots,$$

and

$$q_{iks} := \sum_{p=s}^{m_i} \sum_{\ell=0}^{p-s} \alpha_{ip} \binom{p}{\ell} (i\omega_k)^{p-\ell} z_{k\,p-\ell-s}^{(\ell)}(0), \qquad s = 0, 1, \ldots, m_i,$$

$$q_{iks} := \sum_{p=0}^{m_i} \sum_{\ell=0}^{p} \alpha_{ip} \binom{p}{\ell} (i\omega_k)^{p-\ell} z_{k\,p-\ell-s}^{(\ell)}(0), \qquad s = -1, -2, \ldots.$$

These constants depend on both the boundary values $B_1, \ldots, B_n$ and the formal differential operator ℓ. We will see that they appear naturally when the boundary values are applied to the Birkhoff approximate solutions.

The leading terms in these sequences are given as follows: for $i = 1, \ldots, n$ and $k = 0, 1, \ldots, \nu - 1$,

$$p_{ikm_i} = \alpha_{im_i}(i\omega_k)^{m_i}, \qquad q_{ikm_i} = \beta_{im_i}(i\omega_k)^{m_i}, \tag{1.6}$$

while for $i = 1, \ldots, n$ and $k = \nu, \ldots, n - 1$,

$$p_{ikm_i} = \beta_{im_i}(i\omega_k)^{m_i}, \qquad q_{ikm_i} = \alpha_{im_i}(i\omega_k)^{m_i}. \tag{1.7}$$

These leading coefficients are determined exclusively by the normalized boundary values $B_1, \ldots, B_n$. They are independent of the formal differential operator ℓ that determines L, and they are identical for both L and for its principal part T.

Third, the classification scheme depends on two sequences of constants a_κ, $\kappa = p_0, p_0 - 1, \ldots, 1, 0, -1, \ldots$, and b_κ, $\kappa = p_0, p_0 - 1, \ldots, 1, 0, -1, \ldots$. The formulation of these constants is slightly different for the cases n even and n odd.

Assume that n is even: $n = 2\nu$. For integers $s_1, \ldots, s_n$ with $-\infty < s_i \le m_i$ for $i = 1, \ldots, n$, we introduce the matrices

$$\Pi_2(s_1, \ldots, s_n) := \begin{pmatrix} q_{1\,0\,s_1} & p_{1\,1\,s_1} & \cdots & p_{1\,\nu-1\,s_1} & q_{1\,\nu\,s_1} & p_{1\,\nu+1\,s_1} & \cdots & p_{1\,n-1\,s_1} \\ \vdots & \vdots & & \vdots & \vdots & \vdots & & \vdots \\ q_{n\,0\,s_n} & p_{n\,1\,s_n} & \cdots & p_{n\,\nu-1\,s_n} & q_{n\,\nu\,s_n} & p_{n\,\nu+1\,s_n} & \cdots & p_{n\,n-1\,s_n} \end{pmatrix},$$

$$\Pi_1(s_1, \ldots, s_n) := \begin{pmatrix} p_{1\,0\,s_1} & p_{1\,1\,s_1} & \cdots & p_{1\,\nu-1\,s_1} & q_{1\,\nu\,s_1} & p_{1\,\nu+1\,s_1} & \cdots & p_{1\,n-1\,s_1} \\ \vdots & \vdots & & \vdots & \vdots & \vdots & & \vdots \\ p_{n\,0\,s_n} & p_{n\,1\,s_n} & \cdots & p_{n\,\nu-1\,s_n} & q_{n\,\nu\,s_n} & p_{n\,\nu+1\,s_n} & \cdots & p_{n\,n-1\,s_n} \end{pmatrix},$$

$$\Pi^1(s_1, \ldots, s_n) := \begin{pmatrix} q_{1\,0\,s_1} & p_{1\,1\,s_1} & \cdots & p_{1\,\nu-1\,s_1} & p_{1\,\nu\,s_1} & p_{1\,\nu+1\,s_1} & \cdots & p_{1\,n-1\,s_1} \\ \vdots & \vdots & & \vdots & \vdots & \vdots & & \vdots \\ q_{n\,0\,s_n} & p_{n\,1\,s_n} & \cdots & p_{n\,\nu-1\,s_n} & p_{n\,\nu\,s_n} & p_{n\,\nu+1\,s_n} & \cdots & p_{n\,n-1\,s_n} \end{pmatrix},$$

$$\Pi_0(s_1, \ldots, s_n) := \begin{pmatrix} p_{1\,0\,s_1} & p_{1\,1\,s_1} & \cdots & p_{1\,\nu-1\,s_1} & p_{1\,\nu\,s_1} & p_{1\,\nu+1\,s_1} & \cdots & p_{1\,n-1\,s_1} \\ \vdots & \vdots & & \vdots & \vdots & \vdots & & \vdots \\ p_{n\,0\,s_n} & p_{n\,1\,s_n} & \cdots & p_{n\,\nu-1\,s_n} & p_{n\,\nu\,s_n} & p_{n\,\nu+1\,s_n} & \cdots & p_{n\,n-1\,s_n} \end{pmatrix}.$$

Then form the infinite sequences of constants

$$a_\kappa := \sum_{s_1+\cdots+s_n=\kappa} \det \Pi_2(s_1, \ldots, s_n),$$

$$b_\kappa := \sum_{s_1+\cdots+s_n=\kappa} [\det \Pi_1(s_1, \ldots, s_n) + \det \Pi^1(s_1, \ldots, s_n)],$$

$$c_\kappa := \sum_{s_1+\cdots+s_n=\kappa} \det \Pi_0(s_1, \ldots, s_n),$$

$\kappa = p_0, p_0 - 1, \ldots, 1, 0, -1, \ldots$, where the indices $s_1, \ldots, s_n$ are restricted to satisfy the conditions $m_i - (p_0 - \kappa) \leq s_i \leq m_i$ for $i = 1, \ldots, n$. The third sequence is actually determined from the first sequence by the relation

$$c_\kappa = -(\omega_\kappa)^{-1} a_\kappa \quad \text{for } \kappa = p_0, p_0 - 1, \ldots, 1, 0, -1, \ldots. \tag{1.8}$$

The constants a_κ, b_κ, c_κ appear naturally when the approximate characteristic determinant is formed using the Birkhoff approximate solutions.

In terms of these sequences we now formulate our classification scheme.

DEFINITION 1.1. In the case $n = 2\nu$ even with the boundary values $B_1, \ldots, B_n$ in normalized form and $p_0 = m_1 + \cdots + m_n$, the differential operator L is said to be:

(i) *regular* if $a_{p_0} \neq 0$.
(ii) *simply irregular* if $a_{p_0} = 0$ and $a_\kappa \neq 0$ for some integer κ with $-\infty < \kappa < p_0$.
(iii) *degenerate irregular* if $a_\kappa = 0$ for $\kappa = p_0, p_0 - 1, \ldots, 1, 0, -1, \ldots$.

From the definition $a_{p_0} = \det \Pi_2(m_1, \ldots, m_n)$, and hence, by (1.6), (1.7)

$$a_{p_0} = i^{p_0} \det \begin{pmatrix} \beta_{1m_1} & \overbrace{\alpha_{1m_1}\omega_k^{m_1}}^{1 \leq k \leq \nu - 1} & \alpha_{1m_1}(-1)^{m_1} & \overbrace{\beta_{1m_1}\omega_k^{m_1}}^{\nu+1 \leq k \leq n-1} \\ \vdots & \vdots & \vdots & \vdots \\ \beta_{nm_n} & \alpha_{nm_n}\omega_k^{m_n} & \alpha_{nm_n}(-1)^{m_n} & \beta_{nm_n}\omega_k^{m_n} \end{pmatrix}. \tag{1.9}$$

The condition for the differential operator L to be regular is that the above determinant be nonzero. Cf. Naimark [**37**, p. 57].

Assume that n is odd: $n = 2\nu - 1$. For integers $s_1, \ldots, s_n$ with $-\infty < s_i \leq m_i$ for $i = 1, \ldots, n$, introduce the matrices

$$\Pi_1(s_1, \ldots, s_n) := \begin{pmatrix} q_{1\,0\,s_1} & p_{1\,1\,s_1} & \cdots & p_{1\,\nu-1\,s_1} & p_{1\,\nu\,s_1} & \cdots & p_{1\,n-1\,s_1} \\ \vdots & \vdots & & \vdots & \vdots & & \vdots \\ q_{n\,0\,s_n} & p_{n\,1\,s_n} & \cdots & p_{n\,\nu-1\,s_n} & p_{n\,\nu\,s_n} & \cdots & p_{n\,n-1\,s_n} \end{pmatrix},$$

$$\Pi_0(s_1, \ldots, s_n) := \begin{pmatrix} p_{1\,0\,s_1} & p_{1\,1\,s_1} & \cdots & p_{1\,\nu-1\,s_1} & p_{1\,\nu\,s_1} & \cdots & p_{1\,n-1\,s_1} \\ \vdots & \vdots & & \vdots & \vdots & & \vdots \\ p_{n\,0\,s_n} & p_{n\,1\,s_n} & \cdots & p_{n\,\nu-1\,s_n} & p_{n\,\nu\,s_n} & \cdots & p_{n\,n-1\,s_n} \end{pmatrix},$$

and then form the infinite sequences of constants

$$a_\kappa := \sum_{s_1 + \cdots + s_n = \kappa} \det \Pi_1(s_1, \ldots, s_n),$$

$$b_\kappa := \sum_{s_1 + \cdots + s_n = \kappa} \det \Pi_0(s_1, \ldots, s_n),$$

$\kappa = p_0, p_0 - 1, \ldots, 1, 0, -1, \ldots$, where the indices $s_1, \ldots, s_n$ are restricted to satisfy the conditions $m_i - (p_0 - \kappa) \leq s_i \leq m_i$ for $i = 1, \ldots, n$. The constants a_κ, b_κ appear naturally when forming the approximate characteristic determinant.

Our classification scheme is formulated in terms of these constants.

DEFINITION 1.2. In the case $n = 2\nu - 1$ odd with the boundary values $B_1, \ldots, B_n$ in normalized form and $p_0 = m_1 + \cdots + m_n$, the differential operator L is said to be:

(i) *regular* if $a_{p_0} \neq 0$ and $b_{p_0} \neq 0$.

(ii) *simply irregular* if either $a_{p_0} = 0$ or $b_{p_0} = 0$, and $a_\kappa \neq 0$ and $b_\ell \neq 0$ for some integers κ, ℓ with $-\infty < \kappa, \ell \leq p_0$.

(iii) *degenerate irregular* if either $a_\kappa = 0$ for $\kappa = p_0, p_0 - 1, \ldots, 1, 0, -1, \ldots$ or $b_\kappa = 0$ for $\kappa = p_0, p_0 - 1, \ldots, 1, 0, -1, \ldots$.

From the definitions $a_{p_0} = \det \Pi_1(m_1, \ldots, m_n)$ and $b_{p_0} = \det \Pi_0(m_1, \ldots, m_n)$, and hence by (1.6), (1.7)

$$a_{p_0} = i^{p_0} \det \begin{pmatrix} & 1 \leq k \leq \nu - 1 & \nu \leq k \leq n - 1 \\ \beta_{1m_1} & \alpha_{1m_1}\omega_k^{m_1} & \beta_{1m_1}\omega_k^{m_1} \\ \vdots & \vdots & \vdots \\ \beta_{nm_n} & \alpha_{nm_n}\omega_k^{m_n} & \beta_{nm_n}\omega_k^{m_n} \end{pmatrix}, \tag{1.10}$$

$$b_{p_0} = i^{p_0} \det \begin{pmatrix} & 1 \leq k \leq \nu - 1 & \nu \leq k \leq n - 1 \\ \alpha_{1m_1} & \alpha_{1m_1}\omega_k^{m_1} & \beta_{1m_1}\omega_k^{m_1} \\ \vdots & \vdots & \vdots \\ \alpha_{nm_n} & \alpha_{nm_n}\omega_k^{m_n} & \beta_{nm_n}\omega_k^{m_n} \end{pmatrix}. \tag{1.11}$$

The conditions for the differential operator L to be regular are that the above two determinants be nonzero. Cf. Naimark [**37**, p. 56].

For both n even and n odd we draw the following conclusions: (i) the definitions for the differential operator L being regular given here are consistent with the definitions of regular boundary values given in the literature (see [**34**]); (ii) the differential operator L is regular if and only if its principal part T is regular; and (iii) in checking the conditions for regularity, only the boundary values $B_1, \ldots, B_n$ are relevant — the coefficients of the formal differential operator ℓ play no role.

The initial step, and most important step, in our development of the spectral theory is the classification of the differential operator L according to Definition 1.1 or Definition 1.2. This first step is completed before calculating any solutions of the differential equation $(\rho^n I - \ell)u = 0$, and prior to the construction of the characteristic determinant and the Green's function. In fact, in classifying L we obtain the parameters needed for the asymptotic expansion of solutions and the construction of the characteristic determinant and Green's function. To classify L and obtain the key parameters, we proceed to recursively calculate the constants a_κ, b_κ for $\kappa = p_0, p_0 - 1, \ldots, 1, 0, -1, \ldots$. From the definitions we see that to calculate a_κ, b_κ, we must first compute the constants p_{iks_i}, q_{iks_i} for $i = 1, \ldots, n$, $k = 0, 1, \ldots, n-1$, $m_i - (p_0 - \kappa) \leq s_i \leq m_i$; to determine these latter constants it is sufficient to calculate the functions $z_{kj}(t) = z_{0j}(t)(\omega_k)^{-j}$ for $k = 0, 1, \ldots, n-1$ and $j = 0, 1, \ldots, p_0 - \kappa$, or equivalently, it is sufficient to calculate the functions $z_{0j}(t)$, $j = 0, 1, \ldots, p_0 - \kappa$, using equations (1.2)–(1.4). For example, to compute a_{p_0} and b_{p_0}, we need to calculate the constants p_{ikm_i}, q_{ikm_i} and the function $z_{00}(t) = 1$. Cf. equations (1.9) and (1.10), (1.11). To compute a_{p_0-1} and b_{p_0-1}, we need the constants p_{ikm_i}, q_{ikm_i}, p_{ikm_i-1}, q_{ikm_i-1} and the functions

$$z_{00}(t) = 1, \qquad z_{01}(t) = -\frac{i^{n-1}}{n}\int_0^t a_{n-2}(\xi)\,d\xi.$$

If the differential operator L is either regular or simply irregular, then after a finite number of steps we will find a constant $a_p \neq 0$ for the case n even and a pair of constants $a_p \neq 0$, $b_q \neq 0$ for the case n odd. No additional constants a_κ, b_κ are needed in order to apply the results of this monograph. In many applications it is only necessary to calculate the first few a_κ, b_κ. On the other hand, if it turns out that $a_\kappa = 0$ for all κ for the case n even, or either $a_\kappa = 0$ for all κ or $b_\kappa = 0$ for all κ for the case n odd, then the differential operator L is degenerate irregular and the results herein do not apply.

Our principal results are summarized below.

Assume that n is even, $n = 2\nu \geq 2$, and that the differential operator L is either regular or simply irregular. Let p be the largest integer with $a_p \neq 0$, so $-\infty < p \leq p_0$. From (1.8) we have $c_p = -(\omega_p)^{-1}a_p \neq 0$ and $a_\kappa = c_\kappa = 0$ for $\kappa = p+1, \ldots, p_0$. At this point we define a second integer q as follows: First, if $b_\kappa = 0$ for $\kappa = p+1, \ldots, p_0$, then set $q := p$, so $b_q = b_p$ in this case and the constant b_q may be either zero or nonzero. Second, if $b_\kappa \neq 0$ for some integer κ with $p+1 \leq \kappa \leq p_0$, then let q be defined to be the largest such integer, so $p < q \leq p_0$ in this case and the constant b_q is nonzero. In terms of the integers p, q and the constants a_p, b_q, c_p, in the sequel we will establish the following results for the spectrum of the differential operator L.

THEOREM 1.3. *Let the differential operator L be either regular or simply irregular with $n = 2\nu \geq 2$, let the integers p and q satisfy the additional condition $p = q$, and let the quadratic polynomial $Q(z) = a_p z^2 + b_p z + c_p$ have roots ξ_0 and η_0 with $\xi_0 \neq \eta_0$ (so $|\eta_0| = 1/|\xi_0|$ and $\arg\eta_0 = -\arg\xi_0 - 2\pi p/n + \pi$). Then the elements of the spectrum $\sigma(L)$ can be listed as two distinct sequences*

$$\lambda'_k = (\rho'_k)^n, \quad k = k_0, k_0+1, \ldots, \qquad \lambda''_k = (\rho''_k)^n, \quad k = k_0, k_0+1, \ldots,$$

plus a finite number of additional points, where k_0 is a positive integer and

$$\rho'_k = (2\pi k + \operatorname{Arg}\xi_0) - i\ln|\xi_0| + \epsilon'_k, \quad k = k_0, k_0+1, \ldots,$$
$$\rho''_k = (2\pi k + \operatorname{Arg}\eta_0) + i\ln|\xi_0| + \epsilon''_k, \quad k = k_0, k_0+1, \ldots,$$

with $|\epsilon'_k| \leq \gamma/k$ and $|\epsilon''_k| \leq \gamma/k$ for $k = k_0, k_0+1, \ldots$. Moreover, the corresponding algebraic multiplicities and ascents are

$$\nu(\lambda'_k) = m(\lambda'_k) = 1, \quad k = k_0, k_0+1, \ldots,$$

$$\nu(\lambda''_k) = m(\lambda''_k) = 1, \quad k = k_0, k_0+1, \ldots.$$

THEOREM 1.4. *Let the differential operator L be either regular or simply irregular with $n = 2\nu \geq 2$, let the integers p and q satisfy the additional condition $p = q$, and let the quadratic polynomial $Q(z) = a_p z^2 + b_p z + c_p$ have roots ξ_0 and η_0 with $\xi_0 = \eta_0$ (so $\xi_0 = \eta_0 = \pm i/\sqrt{\omega_p}$). Then the elements of the spectrum $\sigma(L)$ can be listed as two sequences*

$$\lambda'_k = (\rho'_k)^n, \quad k = k_0, k_0+1, \ldots, \qquad \lambda''_k = (\rho''_k)^n, \quad k = k_0, k_0+1, \ldots,$$

plus a finite number of additional points, where k_0 *is a positive integer and*

$$\rho_k' = 2\pi k + \operatorname{Arg}\xi_0 + \epsilon_k', \quad k = k_0, k_0+1, \ldots,$$
$$\rho_k'' = 2\pi k + \operatorname{Arg}\xi_0 + \epsilon_k'', \quad k = k_0, k_0+1, \ldots,$$

with $|\epsilon_k'| \le \gamma/\sqrt{k}$ *and* $|\epsilon_k''| \le \gamma/\sqrt{k}$ *for* $k = k_0, k_0+1, \ldots$. *For each* $k \ge k_0$ *if* $\rho_k' \ne \rho_k''$, *then* $\lambda_k' \ne \lambda_k''$ *and the algebraic multiplicities and ascents are*

$$\nu(\lambda_k') = m(\lambda_k') = 1 \quad \textit{and} \quad \nu(\lambda_k'') = m(\lambda_k'') = 1,$$

while if $\rho_k' = \rho_k''$, *then* $\lambda_k' = \lambda_k''$ *and the algebraic multiplicities and ascents are*

$$\nu(\lambda_k') = 2 \quad \textit{and} \quad m(\lambda_k') = 1 \ \textit{or}\ m(\lambda_k') = 2.$$

In Theorem 1.4 we encounter the possibility of multiple eigenvalues.

THEOREM 1.5. *Let the differential operator* L *be either regular or simply irregular with* $n = 2\nu \ge 2$, *let the integers* p *and* q *satisfy the additional condition* $p < q$, *and let* $n_0 = q - p$, $\mu_0 = -b_q/c_p \ne 0$, *and* $\mu_1 = -b_q/a_p \ne 0$ (*so* $|\mu_1| = |\mu_0|$ *and* $\arg\mu_1 = \arg\mu_0 - 2\pi p/n + \pi$). *Then the elements of the spectrum* $\sigma(L)$ *can be listed as two distinct sequences*

$$\lambda_k' = (\rho_k')^n, \quad k = k_0, k_0+1, \ldots, \qquad \lambda_k'' = (\rho_k'')^n, \quad k = k_0, k_0+1, \ldots,$$

plus a finite number of additional points, where k_0 *is a positive integer and*

$$\rho_k' = (2\pi k - \operatorname{Arg}\mu_0) + in_0 \ln[|\mu_0|^{1/n_0}(2\pi k - \operatorname{Arg}\mu_0)] + \epsilon_k', \quad k = k_0, k_0+1, \ldots,$$
$$\rho_k'' = (2\pi k + \operatorname{Arg}\mu_1) - in_0 \ln[|\mu_0|^{1/n_0}(2\pi k + \operatorname{Arg}\mu_1)] + \epsilon_k'', \quad k = k_0, k_0+1, \ldots,$$

with $|\epsilon_k'| \le \gamma \ln k/k$ *and* $|\epsilon_k''| \le \gamma \ln k/k$ *for* $k = k_0, k_0+1, \ldots$. *Moreover, the corresponding algebraic multiplicities and ascents are*

$$\nu(\lambda_k') = m(\lambda_k') = 1, \quad k = k_0, k_0+1, \ldots,$$
$$\nu(\lambda_k'') = m(\lambda_k'') = 1, \quad k = k_0, k_0+1, \ldots.$$

Assume that n is odd, $n = 2\nu - 1 \ge 3$, and that the differential operator L is either regular or simply irregular. Let p be the largest integer with $a_p \ne 0$, so $-\infty < p \le p_0$, and let q be the largest integer with $b_q \ne 0$, so $-\infty < q \le p_0$. Then $a_\kappa = 0$ for $\kappa = p+1, \ldots, p_0$, and $b_\kappa = 0$ for $\kappa = q+1, \ldots, p_0$. In terms of the integers p, q and the nonzero constants a_p, b_q, in the sequel we will establish the following results for the spectrum of the differential operator L.

THEOREM 1.6. *Let the differential operator* L *be either regular or simply irregular with* $n = 2\nu - 1 \ge 3$, *let the integers* p *and* q *satisfy the additional condition* $p = q$, *and let* $\xi_0 = -b_p/a_p \ne 0$ *and* $\eta_0 = 1/(\omega_p\xi_0) \ne 0$ (*so* $|\eta_0| = 1/|\xi_0|$, $\arg\eta_0 = -\arg\xi_0 - 2\pi p/n$). *Then the elements of the spectrum* $\sigma(L)$ *can be listed as two sequences*

$$\lambda_k' = (\rho_k')^n, \quad k = k_0, k_0+1, \ldots, \qquad \lambda_k'' = (\rho_k'')^n, \quad k = k_0, k_0+1, \ldots,$$

plus a finite number of additional points, where k_0 *is a positive integer and*

$$\rho_k' = (2\pi k + \operatorname{Arg}\xi_0) - i\ln|\xi_0| + \epsilon_k', \qquad k = k_0, k_0+1, \ldots,$$
$$\rho_k'' = -(2\pi k + \operatorname{Arg}\eta_0) - i\ln|\xi_0| + \epsilon_k'', \quad k = k_0, k_0+1, \ldots,$$

with $|\epsilon'_k| \le \gamma/k$ *and* $|\epsilon''_k| \le \gamma/k$ *for* $k = k_0, k_0+1, \ldots$. *Moreover, the corresponding algebraic multiplicities and ascents are*

$$\nu(\lambda'_k) = m(\lambda'_k) = 1, \quad k = k_0, k_0+1, \ldots,$$
$$\nu(\lambda''_k) = m(\lambda''_k) = 1, \quad k = k_0, k_0+1, \ldots.$$

THEOREM 1.7. *Let the differential operator L be either regular or simply irregular with $n = 2\nu - 1 \ge 3$, let the integers p and q satisfy the additional condition $p < q$, and let $n_0 = q - p$, $\mu_0 = -b_q/a_p \ne 0$, and $\mu_1 = \omega_{n_0\nu+p}\,\mu_0 \ne 0$ (so $|\mu_1| = |\mu_0|$ and $\arg\mu_1 = \arg\mu_0 + 2\pi(n_0\nu + p)/n$). Then the elements of the spectrum $\sigma(L)$ can be listed as two sequences*

$$\lambda'_k = (\rho'_k)^n, \quad k = k_0, k_0+1, \ldots, \qquad \lambda''_k = (\rho''_k)^n, \quad k = k_0, k_0+1, \ldots,$$

plus a finite number of additional points, where k_0 is a positive integer and

$$\rho'_k = (2\pi k + \operatorname{Arg}\mu_0) - i n_0 \ln[|\mu_0|^{1/n_0}(2\pi k + \operatorname{Arg}\mu_0)] + \epsilon'_k,$$
$$k = k_0, k_0+1, \ldots,$$
$$\rho''_k = -(2\pi k - \operatorname{Arg}\mu_1 + \pi n_0) - i n_0 \ln[|\mu_0|^{1/n_0}(2\pi k - \operatorname{Arg}\mu_1 + \pi n_0)] + \epsilon''_k,$$
$$k = k_0, k_0+1, \ldots,$$

with $|\epsilon'_k| \le \gamma \ln k/k$ and $|\epsilon''_k| \le \gamma \ln k/k$ for $k = k_0, k_0+1, \ldots$. In addition, the corresponding algebraic multiplicities and ascents are

$$\nu(\lambda'_k) = m(\lambda'_k) = 1, \quad k = k_0, k_0+1, \ldots,$$
$$\nu(\lambda''_k) = m(\lambda''_k) = 1, \quad k = k_0, k_0+1, \ldots.$$

THEOREM 1.8. *Let the differential operator L be either regular or simply irregular with $n = 2\nu - 1 \ge 3$, let the integers p and q satisfy the additional condition $q < p$, and let $n_0 = p - q$, $\mu_0 = -a_p/b_q \ne 0$, and $\mu_1 = \omega_{n_0\nu-p}\,\mu_0 \ne 0$ (so $|\mu_1| = |\mu_0|$ and $\arg\mu_1 = \arg\mu_0 + 2\pi(n_0\nu - p)/n$). Then the elements of the spectrum $\sigma(L)$ can be listed as two sequences*

$$\lambda'_k = (\rho'_k)^n, \quad k = k_0, k_0+1, \ldots, \qquad \lambda''_k = (\rho''_k)^n, \quad k = k_0, k_0+1, \ldots,$$

plus a finite number of additional points, where k_0 is a positive integer and

$$\rho'_k = (2\pi k - \operatorname{Arg}\mu_0) + i n_0 \ln[|\mu_0|^{1/n_0}(2\pi k - \operatorname{Arg}\mu_0)] + \epsilon'_k,$$
$$k = k_0, k_0+1, \ldots,$$
$$\rho''_k = -(2\pi k + \operatorname{Arg}\mu_1 + \pi n_0) + i n_0 \ln[|\mu_0|^{1/n_0}(2\pi k + \operatorname{Arg}\mu_1 + \pi n_0)] + \epsilon''_k,$$
$$k = k_0, k_0+1, \ldots,$$

with $|\epsilon'_k| \le \gamma \ln k/k$ and $|\epsilon''_k| \le \gamma \ln k/k$ for $k = k_0, k_0+1, \ldots$. In addition, the corresponding algebraic multiplicities and ascents are

$$\nu(\lambda'_k) = m(\lambda'_k) = 1, \quad k = k_0, k_0+1, \ldots,$$
$$\nu(\lambda''_k) = m(\lambda''_k) = 1, \quad k = k_0, k_0+1, \ldots.$$

Finally, assume that the differential operator L is either regular or simply irregular, where the order n may be either even or odd. Then the above theorems show that the resolvent set $\rho(L)$ is nonempty and give a detailed description of the spectrum $\sigma(L)$. Let

$$\sigma(L) = \{\lambda_i\}_{i=1}^{\infty}$$

be any enumeration of $\sigma(L)$, let m_i $(0 < m_i < \infty)$ denote the ascent of the operator $\lambda_i I - L$ for $i = 1, 2, \ldots$, and let P_i, $i = 1, 2, \ldots$, denote the projection of $L^2[0,1]$ onto the generalized eigenspace $\mathcal{N}((\lambda_i I - L)^{m_i})$ along the range $\mathcal{R}((\lambda_i I - L)^{m_i})$. Let $\mathrm{sp}(L)$ denote the subspace of $L^2[0,1]$ spanned by the generalized eigenfunctions of L, and let us introduce the subspaces

$$S_\infty(L) = \Big\{u \in L^2[0,1] \;\Big|\; u = \sum_{i=1}^{\infty} P_i u\Big\}$$

and

$$M_\infty(L) = \{u \in L^2[0,1] \mid P_i u = 0 \text{ for } i = 1, 2, \ldots\}.$$

Then $M_\infty(L)$ is closed, $\mathrm{sp}(L)$ is a subset of $S_\infty(L)$, and $\overline{\mathrm{sp}(L)} = \overline{S_\infty(L)}$. In the sequel we will derive the following completeness theorem, generalizing Theorem 6.2.3 of [**34**].

THEOREM 1.9. *Let the differential operator L be either regular or simply irregular with $n \geq 2$. Then the spectrum $\sigma(L)$ is an infinite countable subset of $\mathbb{C}$ having no limit points in $\mathbb{C}$, and if $\sigma(L) = \{\lambda_i\}_{i=1}^{\infty}$ is any enumeration of $\sigma(L)$ and if $S_\infty(L)$ and $M_\infty(L)$ are the corresponding subspaces, then*

$$\overline{\mathrm{sp}(L)} = \overline{S_\infty(L)} = L^2[0,1] \quad \textit{and} \quad M_\infty(L) = \{0\}.$$

REMARK 1.10. If the differential operator L is regular, then Theorems 6.4.1, 6.5.1, and 6.6.1 of [**34**] are applicable. These three theorems give stronger versions of Theorem 1.9 for the special case when L is regular:

$$\mathbb{S}_\infty(L) = L^2[0,1] \quad \text{and} \quad \mathbb{M}_\infty(L) = \{0\}, \tag{1.12}$$

giving a complete solution to the L^2-expansion problem.

Let us illustrate these results by applying them to a simple example.

EXAMPLE 1.11. Consider the 2nd order differential operator L determined by the formal differential operator $\ell = -(d/dt)^2 + a_0(t)$ and the boundary values

$$B_1(u) = u'(0) + u'(1), \qquad B_2(u) = u(0) - u(1),$$

where we assume that $a_0(0) \neq a_0(1)$. Then the boundary coefficient matrix is

$$A = \begin{pmatrix} 1 & 1 & 0 & 0 \\ 0 & 0 & 1 & -1 \end{pmatrix},$$

the integer p_0 is equal to 1, and from the definitions of the constants a_κ, b_κ, c_κ we calculate $a_1 = c_1 = 0$, $a_0 = -c_0 = 0$,

$$a_{-1} = c_{-1} = -\frac{i}{2}\left[a_0(1) - a_0(0)\right] \neq 0,$$

and $b_1 = b_0 = b_{-1} = 0$. Thus, the differential operator L is simply irregular with $p = q = -1$. From Theorem 1.3 it follows that $Q(z) = a_{-1}z^2 + a_{-1}$, $\xi_0 = i$ and $\eta_0 = -i$, and the spectrum $\sigma(L)$ consists of two sequences of points $\lambda'_k = (\rho'_k)^2$, $k = k_0, k_0 + 1, \ldots$, and $\lambda''_k = (\rho''_k)^2$, $k = k_0, k_0 + 1, \ldots$, plus a finite number of additional points, where

$$\rho'_k = (2\pi k + \pi/2) + \epsilon'_k, \quad k = k_0, k_0 + 1, \ldots,$$
$$\rho''_k = (2\pi k - \pi/2) + \epsilon''_k, \quad k = k_0, k_0 + 1, \ldots,$$

with $|\epsilon'_k| \le \gamma/k$ and $|\epsilon''_k| \le \gamma/k$ for $k = k_0, k_0+1, \ldots$. Also, by Theorem 1.9 the generalized eigenfunctions of L are complete in $L^2[0,1]$. On the other hand, for the principal part of L, which is the differential operator T determined by $\tau = -(d/dt)^2$ and by the same boundary values B_1, B_2, it is well-known that the spectrum is

$$\sigma(T) = \mathbb{C}.$$

See [**30**, p. 300] or [**34**, p. 28]. For the principal part T, equations (1.2)–(1.4) yield $z_{0j}(t) = 0$ for $j = 1, 2, \ldots$, and this produces the constants $a_\kappa = b_\kappa = c_\kappa = 0$ for $\kappa = 1, 0, -1, \ldots$. Thus, the principal part T is a degenerate irregular differential operator. See also Example 10.2.

This example illustrates an important feature of irregular differential operators: unlike regular differential operators, the spectral theory of an irregular differential operator L is not necessarily a perturbation of the spectral theory of its principal part T.

In the following chapters we will develop the mathematics leading up to the classification scheme and to the above theorems. The program is as follows: We begin by introducing the Birkhoff approximate solutions of the differential equation $(\rho^n I - \ell)u = 0$ and forming the corresponding approximate characteristic determinant. As a part of this process we obtain the two infinite sequences of constants a_κ, $\kappa = p_0, p_0-1, \ldots, 1, 0, -1, \ldots$, and b_κ, $\kappa = p_0, p_0-1, \ldots, 1, 0, -1, \ldots$, in terms of which the differential operator L is classified as being either regular, simply irregular, or degenerate irregular. Assuming that L is either regular or simply irregular, we proceed to develop asymptotic expansions of solutions of the differential equation $(\rho^n I - \ell)u = 0$ on carefully chosen sectors of the ρ plane. The methods used in deriving these asymptotic expansions are quite different from those commonly found in the literature; they are much simpler, and they immediately yield the necessary regularity properties of the solutions. Next, using these solutions, we then form the characteristic determinant and Green's function for L, calculate the eigenvalues of L and the corresponding algebraic multiplicities and ascents, and show that the generalized eigenfunctions of L are complete in $L^2[0,1]$. For the degenerate irregular class we give an example where $\sigma(L) = \emptyset$ and another where $\sigma(L) = \mathbb{C}$, and then a very unusual 4th order example where $\sigma(L)$ consists of a single sequence of eigenvalues that approach in the limit the *negative* real axis.

The implementation of this program involves working through many cases, with the methods being quite similar in each case. In an attempt to keep the presentation at a reasonable length, we will work through one or two cases in detail and simply sketch the treatment of the remaining cases. A complete version of the manuscript containing all the technical details for all cases is available for downloading at the author's website: www.math.colostate.edu/~locker/. A hard copy of the complete manuscript has also been posted in the Library of Colorado State University.

CHAPTER 2

Birkhoff Approximate Solutions

In this chapter we construct approximate solutions to the differential equation

$$(2.1)\qquad (\rho^n I - \ell)u(t) = \rho^n u(t) - \sum_{p=0}^{n} a_p(t)u^{(p)}(t) = 0, \qquad 0 \le t \le 1,$$

for $\rho \neq 0$ in $\mathbb{C}$, where it assumed that $a_n(t) = 1/i^n$ and $a_{n-1}(t) = 0$. We will follow the pioneering work of Birkhoff [**4**]. Set

$$b_p(t,\rho) := \frac{i^n a_p(t)}{\rho^{n-p}}, \quad p = 1,\ldots,n, \qquad b_0(t,\rho) := \frac{i^n a_0(t)}{\rho^n} - i^n,$$

so $b_n(t,\rho) = 1$, $b_{n-1}(t,\rho) = 0$, and

$$-\frac{i^n}{\rho^n}(\rho^n I - \ell)u(t) = \sum_{p=0}^{n} \frac{i^n a_p(t)}{\rho^{n-p}} \frac{u^{(p)}(t)}{\rho^p} - i^n u(t) = \sum_{p=0}^{n} b_p(t,\rho)\frac{u^{(p)}(t)}{\rho^p}.$$

Thus, for $\rho \neq 0$ in $\mathbb{C}$ we can replace (2.1) by the equivalent differential equation

$$(2.2)\qquad \sum_{p=0}^{n} b_p(t,\rho)\frac{u^{(p)}(t)}{\rho^p} = 0, \qquad 0 \le t \le 1.$$

2.1. Birkhoff Approximate Solutions

Fix $\rho \neq 0$ in $\mathbb{C}$, and fix an integer k with $0 \le k \le n-1$ and an integer m with $m > n$. We look for an approximate solution to the differential equation (2.2) of the form

$$z_k(t,\rho) = z_k(t,\rho,m) = e^{i\rho\omega_k t} \sum_{j=0}^{m-1} z_{kj}(t)\rho^{-j},$$

where the coefficient functions $z_{kj}(t)$, $j = 0,1,\ldots,m-1$, are to be determined. In our notation for the approximate solution $z_k(t,\rho) = z_k(t,\rho,m)$, we always display the ρ dependence, but generally surpress the m dependence, using the simpler notation $z_k(t,\rho)$. The coefficient functions $z_{kj}(t)$, $j = 0,1,\ldots,m-1$, are to be selected (if possible) independent of both ρ and m.

Substituting $z_k(t,\rho)$ into the left side of (2.2), we see that the quantities $b_p(t,\rho)$ and $z_k^{(p)}(t,\rho)/\rho^p$ appear as finite sums of the powers $\rho^0, \rho^{-1}, \rho^{-2}, \ldots$, and upon collecting like powers of ρ, the left side of (2.2) becomes a finite sum of the powers $\rho^0, \rho^{-1}, \rho^{-2}, \ldots, \rho^{-(n+m-1)}$. The algorithm for calculating the coefficients $z_{kj}(t)$, $j = 0,1,\ldots,m-1$, is to force the coefficients of the powers $\rho^0, \rho^{-1}, \rho^{-2}, \ldots, \rho^{-m}$ in the collected sum to be identically zero on the interval $[0,1]$; the terms involving the powers $\rho^{-(m+1)}, \rho^{-(m+2)}, \ldots, \rho^{-(n+m-1)}$ are not eliminated and form residual quantities.

Here are the details of the algorithm. For $p = 0, 1, \ldots, n$ and for $\ell = 0, 1, \ldots, p$, set $\alpha_{kp\ell} := \binom{p}{\ell}(i\omega_k)^{p-\ell}$, and then form the functions

$$C_{k0}(t,\rho) := \sum_{p=0}^{n} b_p(t,\rho)\alpha_{kp0} = \frac{i^n a_{n-2}(t)\alpha_{k\,n-2\,0}}{\rho^2} + \cdots + \frac{i^n a_0(t)\alpha_{k00}}{\rho^n}$$

and for $\ell = 1, \ldots, n$

$$C_{k\ell}(t,\rho) := \sum_{p=\ell}^{n} b_p(t,\rho)\alpha_{kp\ell} = \alpha_{kn\ell} + \frac{i^n a_{n-2}(t)\alpha_{k\,n-2\,\ell}}{\rho^2} + \cdots + \frac{i^n a_\ell(t)\alpha_{k\ell\ell}}{\rho^{n-\ell}}.$$

By Leibniz's rule $z_k^{(p)}(t,\rho) = e^{i\rho\omega_k t}\sum_{\ell=0}^{p}\sum_{j=0}^{m-1}\alpha_{kp\ell}\, z_{kj}^{(\ell)}(t)\rho^{p-\ell-j}$, and hence,

$$\begin{aligned}
\sum_{p=0}^{n} b_p(t,\rho)\frac{z_k^{(p)}(t,\rho)}{\rho^p} &= e^{i\rho\omega_k t}\sum_{p=0}^{n}\sum_{\ell=0}^{p}\sum_{j=0}^{m-1} b_p(t,\rho)\alpha_{kp\ell} z_{kj}^{(\ell)}(t)\rho^{-\ell-j} \\
&= e^{i\rho\omega_k t}\sum_{\ell=0}^{n}\sum_{p=\ell}^{n} b_p(t,\rho)\alpha_{kp\ell}\cdot\sum_{j=0}^{m-1} z_{kj}^{(\ell)}(t)\rho^{-\ell-j} \\
&= e^{i\rho\omega_k t}\sum_{\ell=0}^{n}\sum_{j=0}^{m-1} C_{k\ell}(t,\rho) z_{kj}^{(\ell)}(t)\rho^{-\ell-j}.
\end{aligned}$$

For $\ell = 0, 1, \ldots, n$ and $p = 0, 1, \ldots, n-\ell$, let $c_{k\ell p}(t)$ be the functions defined by

$$\begin{aligned}
&c_{k00}(t) := 0, \\
&c_{k0p}(t) := i^n\, a_{n-p}(t)\,\alpha_{k\,n-p\,0} = a_{n-p}(t)\, i^{2n-p}\,(\omega_k)^{n-p}, \quad p = 1, 2, \ldots, n, \\
&c_{k\ell p}(t) := i^n\, a_{n-p}(t)\,\alpha_{k\,n-p\,\ell} = a_{n-p}(t)\binom{n-p}{\ell} i^{2n-\ell-p}\,(\omega_k)^{n-\ell-p}, \\
&\qquad\qquad \ell = 1, 2, \ldots, n, \quad p = 0, 1, \ldots, n-\ell.
\end{aligned}$$

Note that for $k = 0$ these functions are precisely the functions $c_{0\ell p}(t)$ defined in Chapter 1, the functions $C_{k\ell}(t,\rho)$ can be expressed in the simpler form

$$C_{k\ell}(t,\rho) = \sum_{p=0}^{n-\ell} c_{k\ell p}(t)\rho^{-p}$$

for $\ell = 0, 1, \ldots, n$, and the above equation becomes

$$\begin{aligned}
\sum_{p=0}^{n} b_p(t,\rho)\frac{z_k^{(p)}(t,\rho)}{\rho^p} &= e^{i\rho\omega_k t}\sum_{\ell=0}^{n}\sum_{j=0}^{m-1}\sum_{p=0}^{n-\ell} c_{k\ell p}(t) z_{kj}^{(\ell)}(t)\rho^{-\ell-p-j} \\
&= e^{i\rho\omega_k t}\sum_{s=0}^{n+m-1}\left[\sum_{\ell+p+j=s} c_{k\ell p}(t) z_{kj}^{(\ell)}(t)\right]\rho^{-s}.
\end{aligned} \tag{2.3}$$

Thus, the coefficients in the approximate solution $z_k(t,\rho)$ are determined by the equations

$$\sum_{\ell+p+j=s} c_{k\ell p}(t) z_{kj}^{(\ell)}(t) = 0, \qquad s = 0, 1, \ldots, m. \tag{2.4}$$

As we proceed to solve the system (2.4), it will be convenient to set

$$\begin{aligned} c_{k\ell p}(t) &:= 0, \qquad \ell = 0, 1, \ldots, n, \qquad && p = n - \ell + 1, n - \ell + 2, \ldots, \\ c_{k\ell p}(t) &:= 0, \qquad \ell = n+1, n+2, \ldots, && p = 0, 1, 2, \ldots. \end{aligned}$$

Then in (2.4) we allow the ranges $\ell \geq 0$, $p \geq 0$, and $0 \leq j \leq m - 1$.

Several observations are in order which will assist us in solving the system (2.4). First, for $s = 0$ the corresponding equation in (2.4) is $c_{k00}(t) z_{k0}(t) = 0$, which is automatically valid because $c_{k00}(t) = 0$. Second, for $s = 1$ the corresponding equation in (2.4) reads

$$\underbrace{c_{k10}(t)}_{\alpha_{kn1}} z'_{k0}(t) + \underbrace{c_{k01}(t)}_{0} z_{k0}(t) + \underbrace{c_{k00}(t)}_{0} z_{k1}(t) = 0$$

or $n(i\omega_k)^{n-1} z'_{k0}(t) = 0$. This implies that $z_{k0}(t)$ must be a constant: we will choose $z_{k0}(t) := 1$. Third, for the integer s with $2 \leq s \leq m-1$, the corresponding equation in (2.4) can be written as

$$\sum_{j=0}^{s} \sum_{\ell=0}^{s-j} c_{k\,\ell\, s-\ell-j}(t) z_{kj}^{(\ell)}(t) = 0, \tag{2.5}$$

with $z_{k0}(t) = 1$. When $j = s$ in (2.5), we obtain the term $c_{k00}(t) z_{ks}(t)$, which is identically zero because $c_{k00}(t) = 0$. On the other hand, when $j = s - 1$ in (2.5), we get the terms

$$\underbrace{c_{k01}(t)}_{0} z_{k\,s-1}(t) + \underbrace{c_{k10}(t)}_{\alpha_{kn1}} z'_{k\,s-1}(t) = \alpha_{kn1}\, z'_{k\,s-1}(t),$$

so (2.5) simplifies to

$$\alpha_{kn1}\, z'_{k\,s-1}(t) + \sum_{j=0}^{s-2} \sum_{\ell=0}^{s-j} c_{k\,\ell\, s-\ell-j}(t) z_{kj}^{(\ell)}(t) = 0 \tag{2.6}$$

for $s = 2, \ldots, m-1$ with $z_{k0}(t) = 1$. Fourth, it is simple to check that (2.6) is also valid for $s = m$, and hence, the system (2.4) can be expressed in the simpler form

$$z'_{k\,s-1}(t) = -\frac{1}{\alpha_{kn1}} \sum_{j=0}^{s-2} \sum_{\ell=0}^{s-j} c_{k\,\ell\, s-\ell-j}(t)\, z_{kj}^{(\ell)}(t), \qquad s = 2, \ldots, m, \tag{2.7}$$

where $z_{k0}(t) = 1$ and $\alpha_{kn1} = n(i\omega_k)^{n-1} \neq 0$. Equation (2.7) expresses the derivative $z'_{k\,s-1}(t)$ in terms of the functions $z_{k0}(t), z_{k1}(t), \ldots, z_{k\,s-2}(t)$ and their derivatives, and hence, we have a recursive scheme going here. If we set

$$\begin{aligned} z_{k\,s-1}(t) &:= \int_0^t z'_{k\,s-1}(\xi)\, d\xi \\ &= -\frac{1}{\alpha_{kn1}} \sum_{j=0}^{s-2} \sum_{\ell=0}^{s-j} \int_0^t c_{k\,\ell\, s-\ell-j}(\xi)\, z_{kj}^{(\ell)}(\xi)\, d\xi \end{aligned} \tag{2.8}$$

for $s = 2, \ldots, m$, with $z_{k0}(t) = 1$, then the functions $z_{k1}(t), \ldots, z_{k\,m-1}(t)$ are uniquely determined by equation (2.8). Since the functions $c_{k\ell p}(t)$ are infinitely differentiable on the interval $[0, 1]$, the functions $z_{kj}(t)$, $j = 0, 1, \ldots, m-1$, are also infinitely differentiable on $[0, 1]$.

Note that the recursive scheme (2.7)–(2.8) is not restricted in any way to the range $s = 2, \dots, m$, but can be used for all values $s = 2, 3, \dots$. Thus, we can use (2.7) and (2.8) to construct an infinite sequence of functions $z_{k0}(t), z_{k1}(t), z_{k2}(t), \dots$. These functions are independent of both ρ and m. In forming the approximate solution $z_k(t,\rho) = z_k(t,\rho,m)$, only the terms $z_{k0}(t), z_{k1}(t), \dots, z_{k\,m-1}(t)$ are used from the infinite sequence.

For $s = 2$ in (2.7), we have

$$z'_{k1}(t) = -\frac{1}{\alpha_{kn1}}\left[c_{k02}(t)z_{k0}(t) + c_{k11}(t)z'_{k0}(t) + c_{k20}(t)z''_{k0}(t)\right] = -\frac{i^{n-1}a_{n-2}(t)}{n\omega_k},$$

and hence,

$$z_{k1}(t) = -\frac{i^{n-1}}{n\omega_k}\int_0^t a_{n-2}(\xi)\,d\xi, \qquad 0 \le t \le 1. \tag{2.9}$$

Let us summarize the above results as a theorem; this result is a refinement of Lemma I appearing in [**4**].

THEOREM 2.1. *For each integer k with $0 \le k \le n-1$, there exists an infinite sequence of functions $z_{k0}(t), z_{k1}(t), z_{k2}(t), \dots$, which are infinitely differentiable on $[0,1]$, such that $z_{k0}(t) = 1$ and such that for any $\rho \ne 0$ in $\mathbb{C}$ and for any integer m with $m > n$, if the function*

$$z_k(t,\rho) = z_k(t,\rho,m) = e^{i\rho\omega_k t}\sum_{j=0}^{m-1} z_{kj}(t)\rho^{-j}$$

is substituted into the differential expression $-(i^n/\rho^n)(\rho^n I - \ell)u(t)$, then the coefficients of the terms $e^{i\rho\omega_k t}\rho^{-s}$, $s = 0, 1, \dots, m$, all vanish identically on the interval $[0,1]$ (the terms involving $e^{i\rho\omega_k t}\rho^{-s}$, $s = m+1, m+2, \dots, n+m-1$, still remain). Moreover, the functions $z_{kj}(t)$, $j = 1, 2, \dots$, can be calculated recursively using equations (2.7) and (2.8).

Fix an integer m with $m > n$, and let us consider the n functions

$$z_k(t,\rho) = z_k(t,\rho,m) = e^{i\rho\omega_k t}\sum_{j=0}^{m-1} z_{kj}(t)\rho^{-j}, \qquad k = 0, 1, \dots, n-1,$$

determined by Theorem 2.1 and defined for $0 \le t \le 1$ and for $\rho \ne 0$ in $\mathbb{C}$. We will refer to these functions as the *mth order Birkhoff approximate solutions* of the differential equation (2.1). From (2.3) and (2.4) we have

$$(\rho^n I - \ell)z_k(t,\rho) = -\frac{\rho^n}{i^n}\sum_{p=0}^{n} b_p(t,\rho)\frac{z_k^{(p)}(t,\rho)}{\rho^p} = -\frac{\rho^n}{i^n}e^{i\rho\omega_k t}\sum_{s=m+1}^{n+m-1}\left[\sum_{\ell+p+j=s} c_{k\ell p}(t)z_{kj}^{(\ell)}(t)\right]\rho^{-s}$$

for $k = 0, 1, \dots, n-1$. For $\rho \ne 0$ in $\mathbb{C}$ and for $k = 0, 1, \dots, n-1$, introduce the functions

$$\eta_k(t,\rho) = \eta_k(t,\rho,m) := -\frac{\rho^n}{i^n}\sum_{s=m+1}^{n+m-1}\left[\sum_{\ell+p+j=s} c_{k\ell p}(t)z_{kj}^{(\ell)}(t)\right]\rho^{-s}.$$

In terms of these functions the last equation can be rewritten as

$$(\rho^n I - \ell)z_k(t,\rho) = e^{i\rho\omega_k t}\eta_k(t,\rho), \qquad 0 \le t \le 1, \tag{2.10}$$

for $\rho \neq 0$ in $\mathbb{C}$ and for $k = 0, 1, \ldots, n-1$. The functions $\eta_k(t,\rho)$, $k = 0, 1, \ldots, n-1$, will be referred to as the *mth order residual functions*. They are linear combinations of the powers $\rho^{-(m+1-n)}, \rho^{-(m+2-n)}, \ldots, \rho^{-(m+n-1-n)}$. In Chapter 4 we will construct actual solutions of the differential equation (2.1), and in this construction both the Birkhoff approximate solutions and the residual functions will play important roles.

Several remarks are in order that can greatly simplify the calculation of the Birkhoff approximate solutions.

REMARK 2.2. Fix the integer m with $m > n$, and consider the Birkhoff approximate solutions $z_k(t,\rho) = e^{i\rho\omega_k t}\sum_{j=0}^{m-1} z_{kj}(t)\rho^{-j}$, $k = 0, 1, \ldots, n-1$. We assert that

$$z_k(t,\rho) = z_0(t,\rho\omega_k) = e^{i\rho\omega_k t}\sum_{j=0}^{m-1} z_{0j}(t)(\rho\omega_k)^{-j}, \qquad k = 1, \ldots, n-1. \tag{2.11}$$

Indeed, using induction on j, we can easily show that

$$z_{kj}(t) = z_{0j}(t)(\omega_k)^{-j} \quad \text{for } j = 0, 1, \ldots, m-1, \tag{2.12}$$

from which (2.11) is an immediate consequence. Thus, to calculate the Birkhoff approximate solutions, we need only calculate $z_0(t,\rho)$. This result has been pointed out by Stone [**45**, p. 707].

REMARK 2.3. Let us look more carefully at the system (2.7) for the case $k = 0$, which in combination with (2.8) determines the functions $z_{0j}(t)$, $j = 0, 1, \ldots, m-1$. We start with $z_{00}(t) = 1$ and $\alpha_{0n1} = n i^{n-1}$, and from the definitions of the functions $c_{0\ell p}(t)$, we have $c_{0\ell p}(t) = 0$ when $\ell + p > n$. For the integer s with $2 \le s \le n$, the corresponding equation in (2.7) simplifies to

$$z'_{0\,s-1}(t) = -\frac{1}{n i^{n-1}}\sum_{j=0}^{s-2}\sum_{\ell=0}^{s-j} c_{0\,\ell\,s-\ell-j}(t)\, z_{0j}^{(\ell)}(t). \tag{2.13}$$

On the other hand, for the integer s satisfying $n+1 \le s \le m$, for indices j and ℓ with $0 \le j \le s-n-1$ and $0 \le \ell \le s-j$, we have $\ell + (s-\ell-j) = s-j \ge n+1$ and $c_{0\,\ell\,s-\ell-j}(t) = 0$, and hence, the corresponding equation in (2.7) simplifies to

$$z'_{0\,s-1}(t) = -\frac{1}{n i^{n-1}}\sum_{j=s-n}^{s-2}\sum_{\ell=0}^{s-j} c_{0\,\ell\,s-\ell-j}(t)\, z_{0j}^{(\ell)}(t). \tag{2.14}$$

Now for the case $2 \le s \le n$, for any j with $0 \le j \le s-2$ we have $2 \le s-j \le n$, while for the case $n+1 \le s \le m$, for any j with $s-n \le j \le s-2$ we also have $2 \le s-j \le n$. Thus, in terms of the formal differential operators ℓ_q, $q = 2, \ldots, n$, introduced in Chapter 1, we can rewrite equations (2.13) and (2.14) as

$$z'_{0\,s-1}(t) = -\frac{1}{n i^{n-1}}\sum_{j=0}^{s-2}\ell_{s-j}\, z_{0j}(t), \qquad 2 \le s \le n, \tag{2.15}$$

$$z'_{0\,s-1}(t) = -\frac{1}{n i^{n-1}}\sum_{j=s-n}^{s-2}\ell_{s-j}\, z_{0j}(t), \qquad n+1 \le s \le m. \tag{2.16}$$

In equations (2.15) and (2.16) we have obtained our most simplified form of the system (2.7) for the index $k = 0$. These equations are precisely equations (1.3) and (1.4) of Chapter 1, and equation (2.8) with $k = 0$ reduces to equation (1.2). This final system has a banded structure, with the derivative $z'_{0\,s-1}(t)$ determined by the $n-1$ preceding functions $z_{0\,s-n}(t)$, $z_{0\,s-n+1}(t)$, ..., $z_{0\,s-2}(t)$ and their derivatives up to order n. Once again we see that the functions $z_{0j}(t)$, $j = 0, 1, 2, \ldots$, are independent of ρ and m. The integer m specifies how many terms to include in forming the Birkhoff approximate solutions; it does not affect the values of the coefficients in these approximate solutions. In the next chapter we will see how to choose m; its selection is determined by both the formal differential operator ℓ and by the boundary values $B_1, \ldots, B_n$. The selection of the integer m is a very subtle feature in our development of the spectral theory.

2.2. Special Case: $n = 2$

Let us consider the case $n = 2$ and $\ell = -(d/dt)^2 + q(t)$, where the coefficient $q(t) := a_0(t)$ is an infinitely differentiable function on $[0,1]$. We proceed to calculate the two Birkhoff approximate solutions $z_0(t,\rho) = e^{i\rho t}\sum_{j=0}^{m-1} z_{0j}(t)\rho^{-j}$ and $z_1(t,\rho) = z_0(t,-\rho) = e^{-i\rho t}\sum_{j=0}^{m-1} z_{0j}(t)(-\rho)^{-j}$. For $n = 2$ we have only the formal differential operator $\ell_2 = -\ell = (d/dt)^2 - q(t)$. Thus, starting with $z_{00}(t) = 1$, equation (2.15) gives $z'_{01}(t) = -(1/2i)\ell_2 z_{00}(t) = (1/2i)q(t)$, and hence,

$$z_{01}(t) = \frac{1}{2i}\int_0^t q(\xi)\,d\xi := \frac{1}{2i}\,Q(t). \tag{2.17}$$

Cf. (1.5) and (2.9). For $3 \le s \le m$ equation (2.16) gives the recursion equation

$$z'_{0\,s-1}(t) = -\frac{1}{2i}\,\ell_2 z_{0\,s-2}(t) = -\frac{1}{2i}\left[z''_{0\,s-2}(t) - q(t)z_{0\,s-2}(t)\right]. \tag{2.18}$$

Thus, for $s = 3$ we get

$$\begin{aligned} z'_{02}(t) &= -\frac{1}{2i}\left[z''_{01}(t) - q(t)z_{01}(t)\right] = \frac{1}{4}\left[q'(t) - q(t)Q(t)\right], \\ z_{02}(t) &= \frac{1}{4}\int_0^t \left[q'(\xi) - q(\xi)Q(\xi)\right]d\xi = \frac{1}{4}\left[q(t) - q(0)\right] - \frac{1}{8}\,Q(t)^2, \end{aligned} \tag{2.19}$$

and for $s = 4$

$$\begin{aligned} z'_{03}(t) &= -\frac{1}{2i}\left[z''_{02}(t) - q(t)z_{02}(t)\right] \\ &= -\frac{1}{8i}\left[q''(t) - q'(t)Q(t) - 2q(t)^2 + q(t)q(0) + \frac{1}{2}\,q(t)Q(t)^2\right], \\ z_{03}(t) &= -\frac{1}{8i}\int_0^t \left[q''(\xi) - q'(\xi)Q(\xi) - 2q(\xi)^2 + q(\xi)q(0) + \frac{1}{2}\,q(\xi)Q(\xi)^2\right]d\xi. \end{aligned} \tag{2.20}$$

Therefore, for $m = 3$ the Birkhoff approximate solutions $z_0(t,\rho)$, $z_1(t,\rho)$ are

$$\begin{aligned} z_0(t,\rho) &= e^{i\rho t}\left\{1 + \frac{1}{2i}\,Q(t)\rho^{-1} + \left[\frac{1}{4}\,q(t) - \frac{1}{4}\,q(0) - \frac{1}{8}\,Q(t)^2\right]\rho^{-2}\right\}, \\ z_1(t,\rho) &= z_0(t,-\rho) \\ &= e^{-i\rho t}\left\{1 - \frac{1}{2i}\,Q(t)\rho^{-1} + \left[\frac{1}{4}\,q(t) - \frac{1}{4}\,q(0) - \frac{1}{8}\,Q(t)^2\right]\rho^{-2}\right\} \end{aligned} \tag{2.21}$$

for $0 \le t \le 1$ and for $\rho \ne 0$ in $\mathbb{C}$.

CHAPTER 3

The Approximate Characteristic Determinant: Classification

For the nth order differential operator L, we are assuming that n is expressed in the form $n = 2\nu$ for n even and in the form $n = 2\nu - 1$ for n odd. Fix an integer m with $m > n$ and $m > p_0$, and form the mth order Birkhoff approximate solutions $z_k(t,\rho) = z_k(t,\rho,m) = e^{i\rho\omega_k t}\sum_{j=0}^{m-1} z_{kj}(t)\rho^{-j}$, $k = 0, 1, \ldots, n-1$, as in Theorem 2.1.

3.1. The Approximate Characteristic Determinant

To simplify the discussion, we modify the Birkhoff approximate solutions by introducing the functions

$$y_k(t,\rho) := z_k(t,\rho) = e^{i\rho\omega_k t}\sum_{j=0}^{m-1} z_{kj}(t)\rho^{-j}, \qquad k = 0, 1, \ldots, \nu - 1,$$

$$y_k(t,\rho) := e^{-i\rho\omega_k} z_k(t,\rho) = e^{i\rho\omega_k (t-1)}\sum_{j=0}^{m-1} z_{kj}(t)\rho^{-j}, \quad k = \nu, \ldots, n-1,$$

for $0 \le t \le 1$ and for $\rho \ne 0$ in $\mathbb{C}$. These functions are again approximate solutions of the differential equation (2.1) in the sense of Theorem 2.1. In terms of the boundary values and these modified approximate solutions, we then form the functions $B_i(y_k(\,\cdot\,,\rho))$. Indeed, for $i = 1, \ldots, n$ and $k = 0, 1, \ldots, \nu - 1$, we have

$$\begin{aligned} B_i(y_k(\,\cdot\,,\rho)) &= \sum_{p=0}^{m_i}\sum_{\ell=0}^{p}\sum_{j=0}^{m-1} \alpha_{ip}\binom{p}{\ell}(i\omega_k)^{p-\ell} z_{kj}^{(\ell)}(0)\rho^{p-\ell-j} \\ &\quad + e^{i\rho\omega_k}\sum_{p=0}^{m_i}\sum_{\ell=0}^{p}\sum_{j=0}^{m-1} \beta_{ip}\binom{p}{\ell}(i\omega_k)^{p-\ell} z_{kj}^{(\ell)}(1)\rho^{p-\ell-j} \\ &:= \widehat{P}_{ik}(\rho) + \widehat{Q}_{ik}(\rho)e^{i\rho\omega_k} \end{aligned} \tag{3.1}$$

for $\rho \ne 0$ in $\mathbb{C}$, while for $i = 1, \ldots, n$ and $k = \nu, \ldots, n-1$

$$\begin{aligned} B_i(y_k(\,\cdot\,,\rho)) &= e^{-i\rho\omega_k}\sum_{p=0}^{m_i}\sum_{\ell=0}^{p}\sum_{j=0}^{m-1} \alpha_{ip}\binom{p}{\ell}(i\omega_k)^{p-\ell} z_{kj}^{(\ell)}(0)\rho^{p-\ell-j} \\ &\quad + \sum_{p=0}^{m_i}\sum_{\ell=0}^{p}\sum_{j=0}^{m-1} \beta_{ip}\binom{p}{\ell}(i\omega_k)^{p-\ell} z_{kj}^{(\ell)}(1)\rho^{p-\ell-j} \\ &:= \widehat{P}_{ik}(\rho) + \widehat{Q}_{ik}(\rho)e^{-i\rho\omega_k} \end{aligned} \tag{3.2}$$

for $\rho \neq 0$ in $\mathbb{C}$. The functions $\widehat{P}_{ik}(\rho)$, $\widehat{Q}_{ik}(\rho)$ are defined and analytic for $\rho \neq 0$ in $\mathbb{C}$, and they can be calculated explicitly once the integer m has been selected.

Fix indices i and k with $1 \leq i \leq n$ and $0 \leq k \leq \nu - 1$, and let us consider the functions $\widehat{P}_{ik}(\rho)$, $\widehat{Q}_{ik}(\rho)$ defined by equation (3.1), where we express them in a format that incorporates their dependence on the integer m:

(3.3)

$$\widehat{P}_{ik}(\rho, m) = \sum_{p=0}^{m_i} \sum_{\ell=0}^{p} \sum_{j=0}^{m-1} \alpha_{ip} \binom{p}{\ell} (i\omega_k)^{p-\ell} z_{kj}^{(\ell)}(0) \rho^{p-\ell-j} := \sum_{s=-(m-1)}^{m_i} \hat{p}_{iks}(m) \rho^s,$$

$$\widehat{Q}_{ik}(\rho, m) = \sum_{p=0}^{m_i} \sum_{\ell=0}^{p} \sum_{j=0}^{m-1} \beta_{ip} \binom{p}{\ell} (i\omega_k)^{p-\ell} z_{kj}^{(\ell)}(1) \rho^{p-\ell-j} := \sum_{s=-(m-1)}^{m_i} \hat{q}_{iks}(m) \rho^s$$

for $\rho \neq 0$ in $\mathbb{C}$. We are going to derive precise formulas for the constants $\hat{p}_{iks}(m)$, $\hat{q}_{iks}(m)$, determining their exact dependence on the integer m. Many of them turn out to be independent of m. This analysis involves looking at three cases.

Case 1. Fix the integer s with $0 \leq s \leq m_i$. Suppose p, ℓ, j are integers satisfying $0 \leq p \leq m_i$, $0 \leq \ell \leq p$, $0 \leq j \leq m-1$, and $p - \ell - j = s$. Then $p = s + \ell + j \geq s$, so $s \leq p \leq m_i$; $\ell = p - s - j \leq p - s$, so $0 \leq \ell \leq p - s$; and $p - \ell - j = s$. Conversely, suppose p, ℓ, j are integers satisfying $s \leq p \leq m_i$, $0 \leq \ell \leq p - s$, and $p - \ell - j = s$. Then $0 \leq s \leq p$, so $0 \leq p \leq m_i$; $\ell \leq p - s \leq p$, so $0 \leq \ell \leq p$; $j = p - s - \ell \geq 0$ and $j = p - s - \ell \leq p \leq m_i \leq p_0 \leq m - 1$, so $0 \leq j \leq m - 1$; and $p - \ell - j = s$. Thus, the conditions $0 \leq p \leq m_i$, $0 \leq \ell \leq p$, $0 \leq j \leq m - 1$, $p - \ell - j = s$ are equivalent to the conditions $s \leq p \leq m_i$, $0 \leq \ell \leq p - s$, $p - \ell - j = s$. It follows that the coefficients $\hat{p}_{iks}(m)$, $\hat{q}_{iks}(m)$ are given by

(3.4)

$$\hat{p}_{iks}(m) = \sum_{p=s}^{m_i} \sum_{\ell=0}^{p-s} \alpha_{ip} \binom{p}{\ell} (i\omega_k)^{p-\ell} z_{k\,p-\ell-s}^{(\ell)}(0),$$

$$\hat{q}_{iks}(m) = \sum_{p=s}^{m_i} \sum_{\ell=0}^{p-s} \beta_{ip} \binom{p}{\ell} (i\omega_k)^{p-\ell} z_{k\,p-\ell-s}^{(\ell)}(1)$$

for the case $s = 0, 1, \ldots, m_i$. These coefficients are *independent* of the integer m. In terms of these results we introduce the constants

$$p_{iks} := \sum_{p=s}^{m_i} \sum_{\ell=0}^{p-s} \alpha_{ip} \binom{p}{\ell} (i\omega_k)^{p-\ell} z_{k\,p-\ell-s}^{(\ell)}(0),$$

$$q_{iks} := \sum_{p=s}^{m_i} \sum_{\ell=0}^{p-s} \beta_{ip} \binom{p}{\ell} (i\omega_k)^{p-\ell} z_{k\,p-\ell-s}^{(\ell)}(1)$$

for $s = 0, 1, \ldots, m_i$. The p_{iks}, q_{iks} are constants that are independent of the integer m, and from the above $\hat{p}_{iks}(m) = p_{iks}$, $\hat{q}_{iks}(m) = q_{iks}$ for $s = 0, 1, \ldots, m_i$.

Case 2. Fix the integer s with $-(m - m_i - 1) \leq s \leq -1$. Then a similar argument shows that the conditions $0 \leq p \leq m_i$, $0 \leq \ell \leq p$, $0 \leq j \leq m - 1$, $p - \ell - j = s$ are equivalent to the smaller set of conditions $0 \leq p \leq m_i$, $0 \leq \ell \leq p$,

$p-\ell-j=s$. It follows that the coefficients $\hat{p}_{iks}(m)$, $\hat{q}_{iks}(m)$ are given by

$$\begin{aligned} \hat{p}_{iks}(m) &= \sum_{p=0}^{m_i}\sum_{\ell=0}^{p} \alpha_{ip}\binom{p}{\ell}(i\omega_k)^{p-\ell} z_{k\,p-\ell-s}^{(\ell)}(0), \\ \hat{q}_{iks}(m) &= \sum_{p=0}^{m_i}\sum_{\ell=0}^{p} \beta_{ip}\binom{p}{\ell}(i\omega_k)^{p-\ell} z_{k\,p-\ell-s}^{(\ell)}(1) \end{aligned} \tag{3.5}$$

for the case $s=-(m-m_i-1),\dots,-2,-1$. These coefficients are also *independent* of the integer m. Based on this analysis we introduce the constants

$$\begin{aligned} p_{iks} &:= \sum_{p=0}^{m_i}\sum_{\ell=0}^{p} \alpha_{ip}\binom{p}{\ell}(i\omega_k)^{p-\ell} z_{k\,p-\ell-s}^{(\ell)}(0), \\ q_{iks} &:= \sum_{p=0}^{m_i}\sum_{\ell=0}^{p} \beta_{ip}\binom{p}{\ell}(i\omega_k)^{p-\ell} z_{k\,p-\ell-s}^{(\ell)}(1) \end{aligned}$$

for $s=-1,-2,\dots$. The p_{iks}, $s=-1,-2,\dots$, and q_{iks}, $s=-1,-2,\dots$, are infinite sequences of constants that are independent of the integer m, and $\hat{p}_{iks}(m)=p_{iks}$, $\hat{q}_{iks}(m)=q_{iks}$ for $s=-(m-m_i-1),\dots,-2,-1$. A finite number of terms from these sequences appear in (3.3).

Case 3. Fix the integer s with $-(m-1)\le s\le -(m-m_i)$. Then the conditions $0\le p\le m_i$, $0\le \ell\le p$, $0\le j\le m-1$, $p-\ell-j=s$ are equivalent to the pair of conditions $0\le p\le s+(m-1)$, $0\le \ell\le p$, $p-\ell-j=s$ or $s+m\le p\le m_i$, $p-s-(m-1)\le \ell\le p$, $p-\ell-j=s$. It follows that the coefficients $\hat{p}_{iks}(m)$, $\hat{q}_{iks}(m)$ are given by

$$\begin{aligned} \hat{p}_{iks}(m) &= \sum_{p=0}^{s+(m-1)}\sum_{\ell=0}^{p} \alpha_{ip}\binom{p}{\ell}(i\omega_k)^{p-\ell} z_{k\,p-\ell-s}^{(\ell)}(0) \\ &\quad + \sum_{p=s+m}^{m_i}\sum_{\ell=p-s-(m-1)}^{p} \alpha_{ip}\binom{p}{\ell}(i\omega_k)^{p-\ell} z_{k\,p-\ell-s}^{(\ell)}(0), \\ \hat{q}_{iks}(m) &= \sum_{p=0}^{s+(m-1)}\sum_{\ell=0}^{p} \beta_{ip}\binom{p}{\ell}(i\omega_k)^{p-\ell} z_{k\,p-\ell-s}^{(\ell)}(1) \\ &\quad + \sum_{p=s+m}^{m_i}\sum_{\ell=p-s-(m-1)}^{p} \beta_{ip}\binom{p}{\ell}(i\omega_k)^{p-\ell} z_{k\,p-\ell-s}^{(\ell)}(1) \end{aligned} \tag{3.6}$$

for the case $s=-(m-1),-(m-2),\dots,-(m-m_i)$. These coefficients are *dependent* on the integer m.

Applying the results from the three cases, we see that (3.3) can be rewritten in terms of the constants p_{iks}, q_{iks} that are independent of the integer m as follows:

$$\begin{aligned} \widehat{P}_{ik}(\rho,m) &= \sum_{s=-(m-m_i-1)}^{m_i} p_{iks}\rho^s + \sum_{s=-(m-1)}^{-(m-m_i)} \hat{p}_{iks}(m)\rho^s, \\ \widehat{Q}_{ik}(\rho,m) &= \sum_{s=-(m-m_i-1)}^{m_i} q_{iks}\rho^s + \sum_{s=-(m-1)}^{-(m-m_i)} \hat{q}_{iks}(m)\rho^s \end{aligned} \tag{3.7}$$

for $\rho \neq 0$ in $\mathbb{C}$. These are our results for the structure of the functions $\widehat{P}_{ik}(\rho, m)$, $\widehat{Q}_{ik}(\rho, m)$ for the indices $i = 1, \ldots, n$ and $k = 0, 1, \ldots, \nu - 1$. Note that the constants p_{iks}, q_{iks} that appear in (3.7) are precisely the constants p_{iks}, q_{iks} that were introduced in Chapter 1.

Next, fix indices i and k with $1 \leq i \leq n$ and $\nu \leq k \leq n-1$. Then the corresponding functions $\widehat{P}_{ik}(\rho)$, $\widehat{Q}_{ik}(\rho)$ defined by (3.2) can be expressed in the form

$$(3.8)\quad \begin{aligned} \widehat{P}_{ik}(\rho, m) &= \sum_{p=0}^{m_i} \sum_{\ell=0}^{p} \sum_{j=0}^{m-1} \beta_{ip} \binom{p}{\ell} (i\omega_k)^{p-\ell} z_{kj}^{(\ell)}(1) \rho^{p-\ell-j} := \sum_{s=-(m-1)}^{m_i} \hat{p}_{iks}(m) \rho^s, \\ \widehat{Q}_{ik}(\rho, m) &= \sum_{p=0}^{m_i} \sum_{\ell=0}^{p} \sum_{j=0}^{m-1} \alpha_{ip} \binom{p}{\ell} (i\omega_k)^{p-\ell} z_{kj}^{(\ell)}(0) \rho^{p-\ell-j} := \sum_{s=-(m-1)}^{m_i} \hat{q}_{iks}(m) \rho^s \end{aligned}$$

for $\rho \neq 0$ in $\mathbb{C}$, where the m-dependence has been incorporated into these equations. Proceeding as above, we set

$$p_{iks} := \sum_{p=s}^{m_i} \sum_{\ell=0}^{p-s} \beta_{ip} \binom{p}{\ell} (i\omega_k)^{p-\ell} z_{k\,p-\ell-s}^{(\ell)}(1),$$

$$q_{iks} := \sum_{p=s}^{m_i} \sum_{\ell=0}^{p-s} \alpha_{ip} \binom{p}{\ell} (i\omega_k)^{p-\ell} z_{k\,p-\ell-s}^{(\ell)}(0)$$

for $s = 0, 1, \ldots, m_i$, and

$$p_{iks} := \sum_{p=0}^{m_i} \sum_{\ell=0}^{p} \beta_{ip} \binom{p}{\ell} (i\omega_k)^{p-\ell} z_{k\,p-\ell-s}^{(\ell)}(1),$$

$$q_{iks} := \sum_{p=0}^{m_i} \sum_{\ell=0}^{p} \alpha_{ip} \binom{p}{\ell} (i\omega_k)^{p-\ell} z_{k\,p-\ell-s}^{(\ell)}(0)$$

for $s = -1, -2, \ldots$. These are infinite sequences of constants whose terms are independent of m, and $\hat{p}_{iks}(m) = p_{iks}$, $\hat{q}_{iks}(m) = q_{iks}$ for $s = -(m-m_i-1), \ldots, m_i$. Thus, the functions appearing in (3.8) can be expressed in the form

$$(3.9)\quad \begin{aligned} \widehat{P}_{ik}(\rho, m) &= \sum_{s=-(m-m_i-1)}^{m_i} p_{iks} \rho^s + \sum_{s=-(m-1)}^{-(m-m_i)} \hat{p}_{iks}(m) \rho^s, \\ \widehat{Q}_{ik}(\rho, m) &= \sum_{s=-(m-m_i-1)}^{m_i} q_{iks} \rho^s + \sum_{s=-(m-1)}^{-(m-m_i)} \hat{q}_{iks}(m) \rho^s \end{aligned}$$

for $\rho \neq 0$ in $\mathbb{C}$. These are our results for $\widehat{P}_{ik}(\rho, m)$, $\widehat{Q}_{ik}(\rho, m)$ for $i = 1, \ldots, n$, $k = \nu, \ldots, n-1$. Again we note that the constants p_{iks}, q_{iks} that appear in (3.9) are precisely the constants p_{iks}, q_{iks} that were introduced in Chapter 1.

REMARK 3.1. For any integer m with $m > n$ and $m > p_0$, we can form the corresponding Birkhoff approximate solutions $z_k(t, \rho) = z_k(t, \rho, m)$, $k = 0, 1, \ldots, n-1$, and at the same time form the coefficient functions $z_{kj}(t)$, $k = 0, 1, \ldots, n-1$, $j = 0, 1, \ldots, m-1$. The constants p_{iks}, q_{iks} can then be computed from their definitions for any values of i, k, and s with $-(m-m_i-1) \leq s \leq m_i$. By taking the integer m sufficiently large, any of the constants p_{iks}, q_{iks} can be calculated explicitly.

In terms of the functions $B_i(y_k(\,\cdot\,,\rho))$ and the functions $\widehat{P}_{ik}(\rho)$, $\widehat{Q}_{ik}(\rho)$, the *approximate characteristic determinant* is defined by

$$\widehat{\Delta}(\rho) = \widehat{\Delta}(\rho, m) := \det(B_i(y_k(\,\cdot\,,\rho)))$$

for $\rho \neq 0$ in $\mathbb{C}$. It is also defined and analytic for $\rho \neq 0$ in $\mathbb{C}$, and can be calculated explicitly once m has been chosen. For the case $n = 2\nu$ even, we express the approximate characteristic determinant in the form

(3.10)

$$\begin{aligned}\widehat{\Delta}(\rho) &= \det(B_i(y_k(\,\cdot\,,\rho)))\\ &= \det\begin{pmatrix} \widehat{P}_{10}(\rho)+\widehat{Q}_{10}(\rho)e^{i\rho} & \overbrace{\widehat{P}_{1k}(\rho)+\widehat{Q}_{1k}(\rho)e^{i\rho\omega_k}}^{1\le k\le\nu-1} & \widehat{P}_{1\nu}(\rho)+\widehat{Q}_{1\nu}(\rho)e^{i\rho} & \overbrace{\widehat{P}_{1k}(\rho)+\widehat{Q}_{1k}(\rho)e^{-i\rho\omega_k}}^{\nu+1\le k\le n-1} \\ \vdots & \vdots & \vdots & \vdots \\ \widehat{P}_{n0}(\rho)+\widehat{Q}_{n0}(\rho)e^{i\rho} & \widehat{P}_{nk}(\rho)+\widehat{Q}_{nk}(\rho)e^{i\rho\omega_k} & \widehat{P}_{n\nu}(\rho)+\widehat{Q}_{n\nu}(\rho)e^{i\rho} & \widehat{P}_{nk}(\rho)+\widehat{Q}_{nk}(\rho)e^{-i\rho\omega_k} \end{pmatrix}\end{aligned}$$

for $\rho \neq 0$ in $\mathbb{C}$, where the future emphasis is on the 0th and the νth columns of this matrix. Similarly, for the case $n = 2\nu - 1$ odd, we express it in the form

$$\begin{aligned}\widehat{\Delta}(\rho) &= \det(B_i(y_k(\,\cdot\,,\rho)))\\ &= \det\begin{pmatrix} \widehat{P}_{10}(\rho)+\widehat{Q}_{10}(\rho)e^{i\rho} & \overbrace{\widehat{P}_{1k}(\rho)+\widehat{Q}_{1k}(\rho)e^{i\rho\omega_k}}^{1\le k\le\nu-1} & \overbrace{\widehat{P}_{1k}(\rho)+\widehat{Q}_{1k}(\rho)e^{-i\rho\omega_k}}^{\nu\le k\le n-1} \\ \vdots & \vdots & \vdots \\ \widehat{P}_{n0}(\rho)+\widehat{Q}_{n0}(\rho)e^{i\rho} & \widehat{P}_{nk}(\rho)+\widehat{Q}_{nk}(\rho)e^{i\rho\omega_k} & \widehat{P}_{nk}(\rho)+\widehat{Q}_{nk}(\rho)e^{-i\rho\omega_k} \end{pmatrix}\end{aligned} \tag{3.11}$$

for $\rho \neq 0$ in $\mathbb{C}$, with the future emphasis focusing on the 0th column of this matrix.

While the analysis of the structure of $\widehat{\Delta}(\rho)$ is slightly different for the two cases $n = 2\nu$ and $n = 2\nu - 1$, the general approach is the same. Consequently, we examine the case $n = 2\nu$ in detail, and then simply outline the case $n = 2\nu - 1$.

3.2. Classification for n Even

Assume that n is even: $n = 2\nu$. Let us begin by expanding the determinant in equation (3.10) using the linearity of the determinant function in the 0th and νth columns:

$$\widehat{\Delta}(\rho) = \widehat{D}_2(\rho)e^{2i\rho} + \widehat{D}_1(\rho)e^{i\rho} + \widehat{D}_0(\rho) \tag{3.12}$$

for $\rho \neq 0$ in $\mathbb{C}$, where

$$\widehat{D}_2(\rho) := \det\begin{pmatrix} \widehat{Q}_{10}(\rho) & \overbrace{\widehat{P}_{1k}(\rho)+\widehat{Q}_{1k}(\rho)e^{i\rho\omega_k}}^{1\le k\le\nu-1} & \widehat{Q}_{1\nu}(\rho) & \overbrace{\widehat{P}_{1k}(\rho)+\widehat{Q}_{1k}(\rho)e^{-i\rho\omega_k}}^{\nu+1\le k\le n-1} \\ \vdots & \vdots & \vdots & \vdots \\ \widehat{Q}_{n0}(\rho) & \widehat{P}_{nk}(\rho)+\widehat{Q}_{nk}(\rho)e^{i\rho\omega_k} & \widehat{Q}_{n\nu}(\rho) & \widehat{P}_{nk}(\rho)+\widehat{Q}_{nk}(\rho)e^{-i\rho\omega_k} \end{pmatrix},$$

$$\widehat{D}_1(\rho) := \det\begin{pmatrix} \widehat{P}_{10}(\rho) & \overbrace{\widehat{P}_{1k}(\rho)+\widehat{Q}_{1k}(\rho)e^{i\rho\omega_k}}^{1\le k\le \nu-1} & \widehat{Q}_{1\nu}(\rho) & \overbrace{\widehat{P}_{1k}(\rho)+\widehat{Q}_{1k}(\rho)e^{-i\rho\omega_k}}^{\nu+1\le k\le n-1} \\ \vdots & \vdots & \vdots & \vdots \\ \widehat{P}_{n0}(\rho) & \widehat{P}_{nk}(\rho)+\widehat{Q}_{nk}(\rho)e^{i\rho\omega_k} & \widehat{Q}_{n\nu}(\rho) & \widehat{P}_{nk}(\rho)+\widehat{Q}_{nk}(\rho)e^{-i\rho\omega_k} \end{pmatrix}$$

$$+ \det\begin{pmatrix} \widehat{Q}_{10}(\rho) & \overbrace{\widehat{P}_{1k}(\rho)+\widehat{Q}_{1k}(\rho)e^{i\rho\omega_k}}^{1\le k\le \nu-1} & \widehat{P}_{1\nu}(\rho) & \overbrace{\widehat{P}_{1k}(\rho)+\widehat{Q}_{1k}(\rho)e^{-i\rho\omega_k}}^{\nu+1\le k\le n-1} \\ \vdots & \vdots & \vdots & \vdots \\ \widehat{Q}_{n0}(\rho) & \widehat{P}_{nk}(\rho)+\widehat{Q}_{nk}(\rho)e^{i\rho\omega_k} & \widehat{P}_{n\nu}(\rho) & \widehat{P}_{nk}(\rho)+\widehat{Q}_{nk}(\rho)e^{-i\rho\omega_k} \end{pmatrix},$$

and

$$\widehat{D}_0(\rho) := \det\begin{pmatrix} \widehat{P}_{10}(\rho) & \overbrace{\widehat{P}_{1k}(\rho)+\widehat{Q}_{1k}(\rho)e^{i\rho\omega_k}}^{1\le k\le \nu-1} & \widehat{P}_{1\nu}(\rho) & \overbrace{\widehat{P}_{1k}(\rho)+\widehat{Q}_{1k}(\rho)e^{-i\rho\omega_k}}^{\nu+1\le k\le n-1} \\ \vdots & \vdots & \vdots & \vdots \\ \widehat{P}_{n0}(\rho) & \widehat{P}_{nk}(\rho)+\widehat{Q}_{nk}(\rho)e^{i\rho\omega_k} & \widehat{P}_{n\nu}(\rho) & \widehat{P}_{nk}(\rho)+\widehat{Q}_{nk}(\rho)e^{-i\rho\omega_k} \end{pmatrix}$$

for $\rho \neq 0$ in $\mathbb{C}$.

Now consider the analytic functions $\widehat{D}_i(\rho)$, $i = 0, 1, 2$. Suppose we expand $\widehat{D}_2(\rho)$ using the linearity of the determinant in the columns with indices 1 through $\nu-1$ and $\nu+1$ through $n-1$. Then $\widehat{D}_2(\rho)$ becomes the sum of 2^{n-2} determinants, starting with the determinant

$$\widehat{\pi}_2(\rho) := \det\begin{pmatrix} \widehat{Q}_{10}(\rho) & \widehat{P}_{11}(\rho) & \cdots & \widehat{P}_{1\nu-1}(\rho) & \widehat{Q}_{1\nu}(\rho) & \widehat{P}_{1\nu+1}(\rho) & \cdots & \widehat{P}_{1n-1}(\rho) \\ \vdots & \vdots & & \vdots & \vdots & \vdots & & \vdots \\ \widehat{Q}_{n0}(\rho) & \widehat{P}_{n1}(\rho) & \cdots & \widehat{P}_{n\nu-1}(\rho) & \widehat{Q}_{n\nu}(\rho) & \widehat{P}_{n\nu+1}(\rho) & \cdots & \widehat{P}_{nn-1}(\rho) \end{pmatrix},$$

which is defined and analytic for $\rho \neq 0$ in $\mathbb{C}$. Thus, $\widehat{D}_2(\rho)$ can be expressed in the form

$$\widehat{D}_2(\rho) = \widehat{\pi}_2(\rho) + \widehat{\Phi}_2(\rho), \qquad \rho \neq 0 \text{ in } \mathbb{C},$$

where the function $\widehat{\Phi}_2(\rho)$ is defined and analytic for $\rho \neq 0$ in $\mathbb{C}$, and where $\widehat{\Phi}_2(\rho)$ is the sum of $2^{n-2}-1$ determinants with each determinant expressible in the form of a product of some of the exponentials $e^{i\rho\omega_k}$, $k = 1, \ldots, \nu-1$, or $e^{-i\rho\omega_k}$, $k = \nu+1, \ldots, n-1$, (at least one of these exponentials appears in each product) times the determinant of an $n \times n$ matrix whose entries are selected from among the functions $\widehat{P}_{ik}$, $\widehat{Q}_{ik}$. Similarly, the functions $\widehat{D}_1(\rho)$, $\widehat{D}_0(\rho)$ can be expressed in the form

$$\widehat{D}_1(\rho) = \widehat{\pi}_1(\rho) + \widehat{\Phi}_1(\rho), \qquad \widehat{D}_0(\rho) = \widehat{\pi}_0(\rho) + \widehat{\Phi}_0(\rho)$$

for $\rho \neq 0$ in $\mathbb{C}$, where

$$\widehat{\pi}_1(\rho) := \det\begin{pmatrix} \widehat{P}_{10}(\rho) & \widehat{P}_{11}(\rho) & \cdots & \widehat{P}_{1\nu-1}(\rho) & \widehat{Q}_{1\nu}(\rho) & \widehat{P}_{1\nu+1}(\rho) & \cdots & \widehat{P}_{1n-1}(\rho) \\ \vdots & \vdots & & \vdots & \vdots & \vdots & & \vdots \\ \widehat{P}_{n0}(\rho) & \widehat{P}_{n1}(\rho) & \cdots & \widehat{P}_{n\nu-1}(\rho) & \widehat{Q}_{n\nu}(\rho) & \widehat{P}_{n\nu+1}(\rho) & \cdots & \widehat{P}_{nn-1}(\rho) \end{pmatrix}$$

$$+ \det\begin{pmatrix} \widehat{Q}_{10}(\rho) & \widehat{P}_{11}(\rho) & \cdots & \widehat{P}_{1\nu-1}(\rho) & \widehat{P}_{1\nu}(\rho) & \widehat{P}_{1\nu+1}(\rho) & \cdots & \widehat{P}_{1n-1}(\rho) \\ \vdots & \vdots & & \vdots & \vdots & \vdots & & \vdots \\ \widehat{Q}_{n0}(\rho) & \widehat{P}_{n1}(\rho) & \cdots & \widehat{P}_{n\nu-1}(\rho) & \widehat{P}_{n\nu}(\rho) & \widehat{P}_{n\nu+1}(\rho) & \cdots & \widehat{P}_{nn-1}(\rho) \end{pmatrix},$$

$$\widehat{\pi}_0(\rho) := \det \begin{pmatrix} \widehat{P}_{10}(\rho) & \widehat{P}_{11}(\rho) & \cdots & \widehat{P}_{1\nu-1}(\rho) & \widehat{P}_{1\nu}(\rho) & \widehat{P}_{1\nu+1}(\rho) & \cdots & \widehat{P}_{1n-1}(\rho) \\ \vdots & \vdots & & \vdots & \vdots & \vdots & & \vdots \\ \widehat{P}_{n0}(\rho) & \widehat{P}_{n1}(\rho) & \cdots & \widehat{P}_{n\nu-1}(\rho) & \widehat{P}_{n\nu}(\rho) & \widehat{P}_{n\nu+1}(\rho) & \cdots & \widehat{P}_{nn-1}(\rho) \end{pmatrix}$$

for $\rho \neq 0$ in $\mathbb{C}$, and where the functions $\widehat{\Phi}_1(\rho)$, $\widehat{\Phi}_0(\rho)$ have the same structure as the function $\widehat{\Phi}_2(\rho)$. We can now rewrite the approximate characteristic determinant in the form

$$\widehat{\Delta}(\rho) = \widehat{\pi}_2(\rho)e^{2i\rho} + \widehat{\pi}_1(\rho)e^{i\rho} + \widehat{\pi}_0(\rho) + \widehat{\Phi}_2(\rho)e^{2i\rho} + \widehat{\Phi}_1(\rho)e^{i\rho} + \widehat{\Phi}_0(\rho) \tag{3.13}$$

for $\rho \neq 0$ in $\mathbb{C}$.

The functions $\widehat{\pi}_2(\rho)$ and $\widehat{\pi}_0(\rho)$ appearing in (3.13) are intimately related to each other. Indeed, consider the matrix that appears in the definition of $\widehat{\pi}_0(\rho)$. For $k = 0, 1, \dots, n-2$ equation (2.12) shows that

$$z^{(\ell)}_{k+1\,j}(t) = z^{(\ell)}_{0j}(t)(\omega_1\omega_k)^{-j} = z^{(\ell)}_{kj}(t)(\omega_1)^{-j} \quad \text{for } j, \ell = 0, 1, 2, \dots,$$

while

$$z^{(\ell)}_{0j}(t) = z^{(\ell)}_{0j}(t)(\omega_1\omega_{n-1})^{-j} = z^{(\ell)}_{n-1\,j}(t)(\omega_1)^{-j} \quad \text{for } j, \ell = 0, 1, 2, \dots.$$

Consequently, from the definitions of the functions $\widehat{P}_{ik}(\rho)$, $\widehat{Q}_{ik}(\rho)$, we obtain the following results: for $i = 1, \dots, n$ and $k = 0, 1, \dots, \nu-2$

$$\widehat{P}_{ik}(\rho\omega_1) = \sum_{p=0}^{m_i}\sum_{\ell=0}^{p}\sum_{j=0}^{m-1} \alpha_{ip}\binom{p}{\ell}(i\omega_k)^{p-\ell} z^{(\ell)}_{kj}(0)(\rho\omega_1)^{p-\ell-j} = \widehat{P}_{i\,k+1}(\rho) \tag{3.14}$$

for $\rho \neq 0$ in $\mathbb{C}$; for $i = 1, \dots, n$ and $k = \nu - 1$

$$\widehat{P}_{i\,\nu-1}(\rho\omega_1) = \sum_{p=0}^{m_i}\sum_{\ell=0}^{p}\sum_{j=0}^{m-1} \alpha_{ip}\binom{p}{\ell}(i\omega_{\nu-1})^{p-\ell} z^{(\ell)}_{\nu-1\,j}(0)(\rho\omega_1)^{p-\ell-j} = \widehat{Q}_{i\nu}(\rho) \tag{3.15}$$

for $\rho \neq 0$ in $\mathbb{C}$; for $i = 1, \dots, n$ and $k = \nu, \dots, n-2$

$$\widehat{P}_{ik}(\rho\omega_1) = \sum_{p=0}^{m_i}\sum_{\ell=0}^{p}\sum_{j=0}^{m-1} \beta_{ip}\binom{p}{\ell}(i\omega_k)^{p-\ell} z^{(\ell)}_{kj}(1)(\rho\omega_1)^{p-\ell-j} = \widehat{P}_{i\,k+1}(\rho) \tag{3.16}$$

for $\rho \neq 0$ in $\mathbb{C}$; and for $i = 1, \dots, n$ and $k = n-1$

$$\widehat{P}_{i\,n-1}(\rho\omega_1) = \sum_{p=0}^{m_i}\sum_{\ell=0}^{p}\sum_{j=0}^{m-1} \beta_{ip}\binom{p}{\ell}(i\omega_{n-1})^{p-\ell} z^{(\ell)}_{n-1\,j}(1)(\rho\omega_1)^{p-\ell-j} = \widehat{Q}_{i0}(\rho) \tag{3.17}$$

for $\rho \neq 0$ in $\mathbb{C}$. If we substitute (3.14)–(3.17) into the definition of $\widehat{\pi}_0(\rho)$, then we get

$$\begin{aligned} \widehat{\pi}_0(\rho\omega_1) &= \det \begin{pmatrix} \widehat{P}_{11}(\rho) & \widehat{P}_{12}(\rho) & \cdots & \widehat{P}_{1\,\nu-1}(\rho) & \widehat{Q}_{1\nu}(\rho) & \widehat{P}_{1\,\nu+1}(\rho) & \cdots & \widehat{P}_{1n-1}(\rho) & \widehat{Q}_{10}(\rho) \\ \vdots & \vdots & & \vdots & \vdots & \vdots & & \vdots & \vdots \\ \widehat{P}_{n1}(\rho) & \widehat{P}_{n2}(\rho) & \cdots & \widehat{P}_{n\,\nu-1}(\rho) & \widehat{Q}_{n\nu}(\rho) & \widehat{P}_{n\,\nu+1}(\rho) & \cdots & \widehat{P}_{n\,n-1}(\rho) & \widehat{Q}_{n0}(\rho) \end{pmatrix} \\ &= (-1)^{n-1}\widehat{\pi}_2(\rho) \end{aligned}$$

for $\rho \neq 0$ in $\mathbb{C}$. Thus,

$$\widehat{\pi}_2(\rho) = -\,\widehat{\pi}_0(\rho\omega_1) \quad \text{for } \rho \neq 0 \text{ in } \mathbb{C}. \tag{3.18}$$

Next, we examine in detail the structure of the functions $\widehat{\pi}_i(\rho)$, $i = 0, 1, 2$, determining their dependence on the integer m. We will show that associated with each of these functions is an infinite sequence of constants that are independent of the integer m. These sequences can be calculated explicitly using the functions $z_{k0}(t)$, $z_{k1}(t)$, $z_{k2}(t)$, ..., $k = 0, 1, \ldots, n-1$, and the boundary values $B_1, \ldots, B_n$; they form important invariants for the differential operator L. In terms of these sequences of constants we will (a) classify the differential operator L as being regular, simply irregular, or degenerate irregular (this chapter); (b) determine the exact sectors T_0, T_1 to be used in the sequel to develop asymptotic expansions for actual solutions of the differential equation (2.1) (Chapter 4); (c) develop the characteristic determinant (Chapter 5); and (d) derive the basic theory for the eigenvalues of the differential operator L (Chapter 7).

Consider the function $\widehat{\pi}_2(\rho) = \widehat{\pi}_2(\rho, m)$ that appears in the representation (3.13) of the approximate characteristic determinant. Upon substituting the representations (3.3) and (3.8) into the matrix appearing in the definition of $\widehat{\pi}_2(\rho, m)$, we observe that each row of this matrix is a linear combination of row vectors. Hence, appealing to the linearity of the determinant function in its rows, we see that

$$\widehat{\pi}_2(\rho, m) = \sum_{s_1=-(m-1)}^{m_1} \cdots \sum_{s_n=-(m-1)}^{m_n} \rho^{s_1+\cdots+s_n} \det \widehat{\Pi}_2(s_1, \ldots, s_n, m), \tag{3.19}$$

where $\widehat{\Pi}_2(s_1, \ldots, s_n, m)$ is the $n \times n$ matrix of constants defined by

$$\widehat{\Pi}_2(s_1, \ldots, s_n, m) :=$$

$$\begin{pmatrix} \hat{q}_{1\,0\,s_1}(m) & \hat{p}_{1\,1\,s_1}(m) & \cdots & \hat{p}_{1\,\nu-1\,s_1}(m) & \hat{q}_{1\,\nu\,s_1}(m) & \hat{p}_{1\,\nu+1\,s_1}(m) & \cdots & \hat{p}_{1\,n-1\,s_1}(m) \\ \vdots & \vdots & & \vdots & \vdots & \vdots & & \vdots \\ \hat{q}_{n\,0\,s_n}(m) & \hat{p}_{n\,1\,s_n}(m) & \cdots & \hat{p}_{n\,\nu-1\,s_n}(m) & \hat{q}_{n\,\nu\,s_n}(m) & \hat{p}_{n\,\nu+1\,s_n}(m) & \cdots & \hat{p}_{n\,n-1\,s_n}(m) \end{pmatrix}.$$

It is immediate that $\widehat{\pi}_2(\rho, m)$ can be represented in the form

$$\widehat{\pi}_2(\rho, m) = \sum_{\kappa=-n(m-1)}^{p_0} a_\kappa(m) \rho^\kappa \tag{3.20}$$

for $\rho \neq 0$ in $\mathbb{C}$, where the $a_\kappa(m)$ are constants that depend on the integer m. Specifically, for $\kappa = -n(m-1), \ldots, p_0$ we have

$$a_\kappa(m) = \sum_{s_1+\cdots+s_n=\kappa} \det \widehat{\Pi}_2(s_1, \ldots, s_n, m), \tag{3.21}$$

where the indices $s_1, \ldots, s_n$ satisfy the conditions $-(m-1) \leq s_i \leq m_i$ for $i = 1, \ldots, n$. We will show that many of the $a_\kappa(m)$ are actually independent of m.

Indeed, in equation (3.21) the constant $a_\kappa(m)$ depends upon the integer m in two ways: first, through the entries of the matrices $\widehat{\Pi}_2(s_1, \ldots, s_n, m)$, and second, through the conditions $-(m-1) \leq s_i \leq m_i$ on the indices $s_1, \ldots, s_n$. Fix an index κ that satisfies the condition $-(m-p_0-1) \leq \kappa \leq p_0$. Let us examine how the constant $a_\kappa(m)$ is formed. How small can the index s_1 be and still make a contribution to forming $a_\kappa(m)$? Initially we have the condition $-(m-1) \leq s_1 \leq m_1$. Note that

$$m_1 - (p_0 - \kappa) \geq (m_1 - p_0) - (m - p_0 - 1) = m_1 - (m-1) \geq -(m-1).$$

Consider an index s_1 with $-(m-1) \leq s_1 \leq m_1 - (p_0 - \kappa) - 1$. Then the largest possible power of $\rho^{s_1+\cdots+s_n}$ that can be produced is

$$s_1 + m_2 + \cdots + m_n \leq m_1 - (p_0 - \kappa) - 1 + m_2 + \cdots + m_n = \kappa - 1,$$

and hence, these powers make no contribution to the computation of $a_\kappa(m)$. Thus, in forming $a_\kappa(m)$ it can be assumed that $m_1 - (p_0 - \kappa) \le s_1 \le m_1$. We obtain similar conditions for the indices $s_2, \dots, s_n$. Summarizing, for any index κ with $-(m-p_0-1) \le \kappa \le p_0$, in forming the constant $a_\kappa(m)$ by means of equation (3.21), the indices $s_1, \dots, s_n$ can be assumed to satisfy the more restrictive conditions $m_i - (p_0 - \kappa) \le s_i \le m_i$ for $i = 1, \dots, n$; these conditions no longer depend on the integer m.

For integers $s_1, \dots, s_n$ with $-\infty < s_i \le m_i$ for $i = 1, \dots, n$, introduce the matrices

$$\Pi_2(s_1, \dots, s_n) := \begin{pmatrix} q_{1\,0\,s_1} & p_{1\,1\,s_1} & \cdots & p_{1\,\nu-1\,s_1} & q_{1\,\nu\,s_1} & p_{1\,\nu+1\,s_1} & \cdots & p_{1\,n-1\,s_1} \\ \vdots & \vdots & & \vdots & \vdots & \vdots & & \vdots \\ q_{n\,0\,s_n} & p_{n\,1\,s_n} & \cdots & p_{n\,\nu-1\,s_n} & q_{n\,\nu\,s_n} & p_{n\,\nu+1\,s_n} & \cdots & p_{n\,n-1\,s_n} \end{pmatrix},$$

which are matrices with constant entries independent of the integer m. Now continuing the discussion for κ with $-(m - p_0 - 1) \le \kappa \le p_0$, for integers i, k, s_i with $1 \le i \le n$, $0 \le k \le n-1$, and $m_i - (p_0 - \kappa) \le s_i \le m_i$, we have

$$m_i - (p_0 - \kappa) \ge -(m - p_0 - 1) - (p_0 - m_i) = -(m - m_i - 1),$$

and hence, by the representations (3.7) and (3.9),

$$\hat{p}_{i\,k\,s_i}(m) = p_{i\,k\,s_i} \qquad \hat{q}_{i\,k\,s_i}(m) = q_{i\,k\,s_i}, \tag{3.22}$$

which are constants independent of m. Thus, the matrix $\widehat{\Pi}_2(s_1, \dots, s_n, m)$ simplifies to

$$\widehat{\Pi}_2(s_1, \dots, s_n, m) = \Pi_2(s_1, \dots, s_n),$$

and (3.21) in turn simplifies to

$$a_\kappa(m) = \sum_{s_1+\cdots+s_n=\kappa} \det \Pi_2(s_1, \dots, s_n), \tag{3.23}$$

where the indices $s_1, \dots, s_n$ satisfy the restricted conditions $m_i - (p_0 - \kappa) \le s_i \le m_i$ for $i = 1, \dots, n$. Thus, the constants $a_\kappa(m)$, $\kappa = -(m - p_0 - 1), \dots, p_0$, are *independent* of the integer m, and at the same time we have obtained an explicit formula for calculating these constants.

In terms of the above analysis, set

$$a_\kappa := \sum_{s_1+\cdots+s_n=\kappa} \det \Pi_2(s_1, \dots, s_n)$$

for $\kappa = p_0, p_0 - 1, \dots, 1, 0, -1, \dots$, where the indices $s_1, \dots, s_n$ are restricted to satisfy the conditions $m_i - (p_0 - \kappa) \le s_i \le m_i$ for $i = 1, \dots, n$. Clearly this yields an infinite sequence of constants that are independent of the integer m, and from the above $a_\kappa(m) = a_\kappa$ for $\kappa = -(m - p_0 - 1), \dots, p_0$. Therefore, the function $\widehat{\pi}_2(\rho, m)$ has the representation

$$\widehat{\pi}_2(\rho, m) = \sum_{\kappa=-(m-p_0-1)}^{p_0} a_\kappa \rho^\kappa + \sum_{\kappa=-n(m-1)}^{-(m-p_0)} a_\kappa(m) \rho^\kappa \tag{3.24}$$

for $\rho \ne 0$ in $\mathbb{C}$. We emphasize again that the constants a_κ, $\kappa = p_0, p_0 - 1, \dots, 1, 0, -1, \dots$, can be calculated explicitly using the functions $z_{k0}(t)$, $z_{k1}(t)$, $z_{k2}(t)$, $\dots$, $k = 0, 1, \dots, n-1$, and the boundary values $B_1, \dots, B_n$. The infinite sequence a_κ, $\kappa = p_0, p_0 - 1, \dots, 1, 0, -1, \dots$, forms an important invariant for the differential operator L.

A similar discussion can be carried out for the functions $\widehat{\pi}_1(\rho, m)$ and $\widehat{\pi}_0(\rho, m)$, and we will simply state the results. For integers $s_1, \ldots, s_n$ with $-\infty < s_i \le m_i$ for $i = 1, \ldots, n$, introduce the matrices

$$\Pi_1(s_1, \ldots, s_n) := \begin{pmatrix} p_{1\,0\,s_1} & p_{1\,1\,s_1} & \cdots & p_{1\,\nu-1\,s_1} & q_{1\,\nu\,s_1} & p_{1\,\nu+1\,s_1} & \cdots & p_{1\,n-1\,s_1} \\ \vdots & \vdots & & \vdots & \vdots & \vdots & & \vdots \\ p_{n\,0\,s_n} & p_{n\,1\,s_n} & \cdots & p_{n\,\nu-1\,s_n} & q_{n\,\nu\,s_n} & p_{n\,\nu+1\,s_n} & \cdots & p_{n\,n-1\,s_n} \end{pmatrix},$$

$$\Pi^1(s_1, \ldots, s_n) := \begin{pmatrix} q_{1\,0\,s_1} & p_{1\,1\,s_1} & \cdots & p_{1\,\nu-1\,s_1} & p_{1\,\nu\,s_1} & p_{1\,\nu+1\,s_1} & \cdots & p_{1\,n-1\,s_1} \\ \vdots & \vdots & & \vdots & \vdots & \vdots & & \vdots \\ q_{n\,0\,s_n} & p_{n\,1\,s_n} & \cdots & p_{n\,\nu-1\,s_n} & p_{n\,\nu\,s_n} & p_{n\,\nu+1\,s_n} & \cdots & p_{n\,n-1\,s_n} \end{pmatrix},$$

and

$$\Pi_0(s_1, \ldots, s_n) := \begin{pmatrix} p_{1\,0\,s_1} & p_{1\,1\,s_1} & \cdots & p_{1\,\nu-1\,s_1} & p_{1\,\nu\,s_1} & p_{1\,\nu+1\,s_1} & \cdots & p_{1\,n-1\,s_1} \\ \vdots & \vdots & & \vdots & \vdots & \vdots & & \vdots \\ p_{n\,0\,s_n} & p_{n\,1\,s_n} & \cdots & p_{n\,\nu-1\,s_n} & p_{n\,\nu\,s_n} & p_{n\,\nu+1\,s_n} & \cdots & p_{n\,n-1\,s_n} \end{pmatrix},$$

which are matrices with constant entries independent of the integer m. Then form the infinite sequences of constants

$$b_\kappa := \sum_{s_1+\cdots+s_n=\kappa} [\det \Pi_1(s_1, \ldots, s_n) + \det \Pi^1(s_1, \ldots, s_n)],$$

$$c_\kappa := \sum_{s_1+\cdots+s_n=\kappa} \det \Pi_0(s_1, \ldots, s_n),$$

$\kappa = p_0, p_0 - 1, \ldots, 1, 0, -1, \ldots$, where the indices $s_1, \ldots, s_n$ are restricted to satisfy the conditions $m_i - (p_0 - \kappa) \le s_i \le m_i$ for $i = 1, \ldots, n$. The functions $\widehat{\pi}_1(\rho, m)$, $\widehat{\pi}_0(\rho, m)$ can be represented in the form

$$\widehat{\pi}_1(\rho, m) = \sum_{\kappa=-n(m-1)}^{p_0} b_\kappa(m)\rho^\kappa, \tag{3.25}$$

$$\widehat{\pi}_0(\rho, m) = \sum_{\kappa=-n(m-1)}^{p_0} c_\kappa(m)\rho^\kappa \tag{3.26}$$

for $\rho \ne 0$ in $\mathbb{C}$, where the $b_\kappa(m)$, $c_\kappa(m)$ are constants that depend on the integer m. Since $b_\kappa(m) = b_\kappa$ and $c_\kappa(m) = c_\kappa$ for $\kappa = -(m - p_0 - 1), \ldots, p_0$, this produces the representations

$$\widehat{\pi}_1(\rho, m) = \sum_{\kappa=-(m-p_0-1)}^{p_0} b_\kappa\rho^\kappa + \sum_{\kappa=-n(m-1)}^{-(m-p_0)} b_\kappa(m)\rho^\kappa, \tag{3.27}$$

$$\widehat{\pi}_0(\rho, m) = \sum_{\kappa=-(m-p_0-1)}^{p_0} c_\kappa\rho^\kappa + \sum_{\kappa=-n(m-1)}^{-(m-p_0)} c_\kappa(m)\rho^\kappa \tag{3.28}$$

for $\rho \ne 0$ in $\mathbb{C}$. The infinite sequences b_κ, $\kappa = p_0, p_0 - 1, \ldots, 1, 0, -1, \ldots$, and c_κ, $\kappa = p_0, p_0 - 1, \ldots, 1, 0, -1, \ldots$, are also important invariants for the differential operator L.

The constants a_κ, b_κ, c_κ defined above are precisely the constants a_κ, b_κ, c_κ introduced in Chapter 1 for the case n even.

From equation (3.18) we see that

$$a_\kappa(m) = -\,\omega_1^\kappa c_\kappa(m) = -\,\omega_\kappa c_\kappa(m) \quad \text{for } \kappa = -n(m-1), \dots, p_0, \tag{3.29}$$

and hence, $a_\kappa = -\,\omega_\kappa c_\kappa$ for $\kappa = -(m - p_0 - 1), \dots, p_0$. Since m can be chosen arbitrarily large, it follows that

$$a_\kappa = -\,\omega_\kappa c_\kappa \quad \text{for } \kappa = p_0, p_0 - 1, \dots, 1, 0, -1, \dots. \tag{3.30}$$

We conclude this discussion of the m-dependence by defining the functions

$$\pi_2(\rho, m) := \sum_{\kappa=-(m-p_0-1)}^{p_0} a_\kappa \rho^\kappa,$$

$$\pi_1(\rho, m) := \sum_{\kappa=-(m-p_0-1)}^{p_0} b_\kappa \rho^\kappa,$$

$$\pi_0(\rho, m) := \sum_{\kappa=-(m-p_0-1)}^{p_0} c_\kappa \rho^\kappa$$

for $\rho \neq 0$ in $\mathbb{C}$. The only dependence of these functions on the integer m is in the lower limit $-(m - p_0 - 1)$ of the summations, i.e., m determines how many terms to use in forming the functions $\pi_i(\rho, m)$, $i = 0, 1, 2$.

Finally, we use the three sequences $a_\kappa, b_\kappa, c_\kappa$, $\kappa = p_0, p_0 - 1, \dots, 1, 0, -1, \dots$, to formulate our classification scheme for the differential operator L. This is the formulation given previously in Definition 1.1, and we repeat it here for the sake of continuity.

DEFINITION 3.2. In the case $n = 2\nu$ even with the boundary values $B_1, \dots, B_n$ in normalized form and $p_0 = m_1 + \cdots + m_n$, the differential operator L is said to be:

(i) *regular* if $a_{p_0} \neq 0$.
(ii) *simply irregular* if $a_{p_0} = 0$ and $a_\kappa \neq 0$ for some integer κ with $-\infty < \kappa < p_0$.
(iii) *degenerate irregular* if $a_\kappa = 0$ for $\kappa = p_0, p_0 - 1, \dots, 1, 0, -1, \dots$.

Henceforth, we will assume that for the case n even, $n = 2\nu \geq 2$, the differential operator L is either regular or simply irregular. Let p be the largest integer with $a_p \neq 0$, so $-\infty < p \leq p_0$. From (3.30) we have $c_p = -\,(\omega_p)^{-1} a_p \neq 0$ and $a_\kappa = c_\kappa = 0$ for $\kappa = p+1, \dots, p_0$. At this point we define a second integer q as follows: First, if $b_\kappa = 0$ for $\kappa = p + 1, \dots, p_0$, then set $q := p$, so $b_q = b_p$ in this case and the constant b_q may be either zero or nonzero. Second, if $b_\kappa \neq 0$ for some integer κ with $p + 1 \leq \kappa \leq p_0$, then let q be defined to be the largest such integer, so $p < q \leq p_0$ in this case and the constant b_q is nonzero. In either case we clearly have $b_\kappa = 0$ for $\kappa = q + 1, \dots, p_0$. The integers p, q and the constants a_p, b_q, c_p will play crucial roles in the sequel.

In the ρ plane let us introduce the two sectors

$$S_0\colon \text{ all } \rho = |\rho| e^{i\theta} \in \mathbb{C} \text{ with } 0 \leq \theta \leq \frac{\pi}{n},$$

$$S_1\colon \text{ all } \rho = |\rho| e^{i\theta} \in \mathbb{C} \text{ with } -\frac{\pi}{n} \leq \theta \leq 0\,.$$

Each of these sectors has angular opening π/n. In Chapter 5 we will show that the exponentials $e^{i\rho\omega_1}, \dots, e^{i\rho\omega_{\nu-1}}, e^{-i\rho\omega_{\nu+1}}, \dots, e^{-i\rho\omega_{n-1}}$ go to zero very rapidly on the sectors S_0 and S_1. For the case $p = q$, choose a constant $d > 0$ such that

$$|a_p|e^{-2d} + |b_p|e^{-d} + |c_p|e^{-2d} \leq \frac{1}{4}|a_p| = \frac{1}{4}|c_p|, \tag{3.31}$$

and in terms of the constant d introduce the horizontal strip

$$\Gamma := \{\rho = a + ib \in \mathbb{C} \mid a \geq -\pi \text{ and } |b| \leq d\}.$$

Then select complex constants τ_0 and τ_1 and form the translated sectors

$$T_0 := \{\rho - \tau_0 \mid \rho \in S_0\} \quad \text{and} \quad T_1 := \{\rho - \tau_1 \mid \rho \in S_1\}$$

with the following properties: for the case $p = q$ we require that the sectors S_0, S_1 lie in the interiors of T_0, T_1, respectively, and that the horizontal strip Γ lies in the interiors of both T_0 and T_1; for the case $p < q$ we require only that the sectors S_0, S_1 lie in the interiors of T_0, T_1, respectively. The translated sectors T_0 and T_1 also have angular opening π/n. They have been constructed by utilizing the constants a_p, b_q, c_p determined by the differential operator L, and their construction is *independent* of the integer m. The asymptotic expansions developed in the next chapter will take place in the sectors T_0 and T_1.

Once the integers p, q and the constants a_p, b_q, c_p have been determined and the sectors T_0, T_1 selected, then the integer m can be fixed once and for all: choose any integer m with $m > n$, $m > p_0$, and $-(m - p_0 - 1) \leq p \leq p_0$. For this choice of m we proceed to form the Birkhoff approximate solutions $z_k(t,\rho) = z_k(t,\rho,m)$, $k = 0, 1, \ldots, n-1$, the modified functions $y_k(t,\rho) = y_k(t,\rho,m)$, $k = 0, 1, \ldots, n-1$, and the approximate characteristic determinant $\widehat{\Delta}(\rho) = \widehat{\Delta}(\rho, m)$. Lastly, we introduce the functions

$$\pi_2(\rho) := \pi_2(\rho, m) = \sum_{\kappa=-(m-p_0-1)}^{p} a_\kappa \rho^\kappa,$$

$$\pi_1(\rho) := \pi_1(\rho, m) = \sum_{\kappa=-(m-p_0-1)}^{q} b_\kappa \rho^\kappa,$$

$$\pi_0(\rho) := \pi_0(\rho, m) = \sum_{\kappa=-(m-p_0-1)}^{p} c_\kappa \rho^\kappa$$

for $\rho \neq 0$ in $\mathbb{C}$. In anticipation of our future work, we rewrite the representation (3.13) of the approximate characteristic determinant in its final form:

$$\begin{aligned}\widehat{\Delta}(\rho) &= \pi_2(\rho)e^{2i\rho} + \pi_1(\rho)e^{i\rho} + \pi_0(\rho) \\ &\quad + \Big[\sum_{\kappa=-n(m-1)}^{-(m-p_0)} a_\kappa(m)\rho^\kappa + \widehat{\Phi}_2(\rho,m)\Big]e^{2i\rho} \\ &\quad + \Big[\sum_{\kappa=-n(m-1)}^{-(m-p_0)} b_\kappa(m)\rho^\kappa + \widehat{\Phi}_1(\rho,m)\Big]e^{i\rho} \\ &\quad + \Big[\sum_{\kappa=-n(m-1)}^{-(m-p_0)} c_\kappa(m)\rho^\kappa + \widehat{\Phi}_0(\rho,m)\Big]\end{aligned} \tag{3.32}$$

for $\rho \neq 0$ in $\mathbb{C}$.

For the case $n = 2\nu$ even, all of the quantities introduced in the last three paragraphs will remain fixed in the sequel. In particular, the integer m will be held fixed with $m > n$, $m > p_0$, and $-(m-p_0-1) \le p \le p_0$.

3.3. Classification for n Odd

Assume that n is odd: $n = 2\nu - 1$. Let us proceed to outline the analogous theory for this case. Recall that we are assuming $n \ge 3$, so $\nu \ge 2$. The starting point for the odd order case is the representation (3.11) of the approximate characteristic determinant $\widehat{\Delta}(\rho)$. Expanding the determinant for $\widehat{\Delta}(\rho)$ using linearity in the 0th column, $\widehat{\Delta}(\rho)$ can be expressed in the form

$$\widehat{\Delta}(\rho) = \widehat{D}_1(\rho)e^{i\rho} + \widehat{D}_0(\rho) \tag{3.33}$$

for $\rho \ne 0$ in $\mathbb{C}$, where the functions $\widehat{D}_1(\rho)$, $\widehat{D}_0(\rho)$ are given by

$$\widehat{D}_1(\rho) = \widehat{\pi}_1(\rho) + \widehat{\Phi}_1(\rho), \qquad \widehat{D}_0(\rho) = \widehat{\pi}_0(\rho) + \widehat{\Phi}_0(\rho)$$

for $\rho \ne 0$ in $\mathbb{C}$. In this last equation the functions $\widehat{\pi}_1(\rho)$, $\widehat{\pi}_0(\rho)$ are defined by

$$\widehat{\pi}_1(\rho) := \det\begin{pmatrix} \widehat{Q}_{10}(\rho) & \widehat{P}_{11}(\rho) & \cdots & \widehat{P}_{1\nu-1}(\rho) & \widehat{P}_{1\nu}(\rho) & \cdots & \widehat{P}_{1n-1}(\rho) \\ \vdots & \vdots & & \vdots & \vdots & & \vdots \\ \widehat{Q}_{n0}(\rho) & \widehat{P}_{n1}(\rho) & \cdots & \widehat{P}_{n\nu-1}(\rho) & \widehat{P}_{n\nu}(\rho) & \cdots & \widehat{P}_{nn-1}(\rho) \end{pmatrix},$$

$$\widehat{\pi}_0(\rho) := \det\begin{pmatrix} \widehat{P}_{10}(\rho) & \widehat{P}_{11}(\rho) & \cdots & \widehat{P}_{1\nu-1}(\rho) & \widehat{P}_{1\nu}(\rho) & \cdots & \widehat{P}_{1n-1}(\rho) \\ \vdots & \vdots & & \vdots & \vdots & & \vdots \\ \widehat{P}_{n0}(\rho) & \widehat{P}_{n1}(\rho) & \cdots & \widehat{P}_{n\nu-1}(\rho) & \widehat{P}_{n\nu}(\rho) & \cdots & \widehat{P}_{nn-1}(\rho) \end{pmatrix}$$

for $\rho \ne 0$ in $\mathbb{C}$, and the functions $\widehat{\Phi}_1(\rho)$, $\widehat{\Phi}_0(\rho)$ are the sums of $2^{n-1} - 1$ determinants with each determinant expressible in the form of a product of some of the exponentials $e^{i\rho\omega_k}$, $k = 1, \ldots, \nu - 1$, or $e^{-i\rho\omega_k}$, $k = \nu, \ldots, n - 1$, (at least one of these exponentials appears in each product) times the determinant of an $n \times n$ matrix whose entries are selected from among the functions $\widehat{P}_{ik}(\rho)$, $\widehat{Q}_{ik}(\rho)$. Thus, the approximate characteristic determinant can be rewritten in the form

$$\widehat{\Delta}(\rho) = \widehat{\pi}_1(\rho)e^{i\rho} + \widehat{\pi}_0(\rho) + \widehat{\Phi}_1(\rho)e^{i\rho} + \widehat{\Phi}_0(\rho) \tag{3.34}$$

for $\rho \ne 0$ in $\mathbb{C}$.

The approximate characteristic determinant $\widehat{\Delta}(\rho)$ and the functions $\widehat{\pi}_i(\rho)$, $\widehat{\Phi}_i(\rho)$ are all defined and analytic for $\rho \ne 0$ in $\mathbb{C}$, and these functions can be calculated explicitly once the integer m has been fixed.

Next, we examine the structure of the functions $\widehat{\pi}_i(\rho)$, $i = 0, 1$, determining how each of these functions depends on the integer m. We will show that associated with each of these functions is an infinite sequence of constants that are independent of the integer m. In terms of these sequences of constants we will (a) classify the differential operator L as being regular, simply irregular, or degenerate irregular (this chapter); (b) determine the exact sectors T_0, T_1 to be used in the sequel to develop asymptotic expansions for actual solutions of the differential equation (2.1) (Chapter 4); (c) develop the characteristic determinant (Chapter 5); and (d) derive the basic theory for the eigenvalues of the differential operator L (Chapter 8).

Consider the function $\widehat{\pi}_1(\rho) = \widehat{\pi}_1(\rho, m)$. Appealing to the linearity of the determinant function in its rows, we see that

$$\begin{aligned}\widehat{\pi}_1(\rho, m) &= \sum_{s_1=-(m-1)}^{m_1} \cdots \sum_{s_n=-(m-1)}^{m_n} \rho^{s_1+\cdots+s_n} \det \widehat{\Pi}_1(s_1, \ldots, s_n, m) \\ &= \sum_{\kappa=-n(m-1)}^{p_0} a_\kappa(m)\rho^\kappa\end{aligned} \tag{3.35}$$

for $\rho \neq 0$ in $\mathbb{C}$, where $\widehat{\Pi}_1(s_1, \ldots, s_n, m)$ is the $n \times n$ matrix

$$\widehat{\Pi}_1(s_1, \ldots, s_n, m) := \begin{pmatrix} \hat{q}_{1\,0\,s_1}(m) & \hat{p}_{1\,1\,s_1}(m) & \cdots & \hat{p}_{1\,\nu-1\,s_1}(m) & \hat{p}_{1\,\nu\,s_1}(m) & \cdots & \hat{p}_{1\,n-1\,s_1}(m) \\ \vdots & \vdots & & \vdots & \vdots & & \vdots \\ \hat{q}_{n\,0\,s_n}(m) & \hat{p}_{n\,1\,s_n}(m) & \cdots & \hat{p}_{n\,\nu-1\,s_n}(m) & \hat{p}_{n\,\nu\,s_n}(m) & \cdots & \hat{p}_{n\,n-1\,s_n}(m) \end{pmatrix}$$

and the $a_\kappa(m)$ are the constants given by

$$a_\kappa(m) := \sum_{s_1+\cdots+s_n=\kappa} \det \widehat{\Pi}_1(s_1, \ldots, s_n, m) \tag{3.36}$$

for $\kappa = -n(m-1), \ldots, p_0$. Many of the constants $a_\kappa(m)$ turn out to be independent of m.

Indeed, let us introduce the matrices

$$\Pi_1(s_1, \ldots, s_n) := \begin{pmatrix} q_{1\,0\,s_1} & p_{1\,1\,s_1} & \cdots & p_{1\,\nu-1\,s_1} & p_{1\,\nu\,s_1} & \cdots & p_{1\,n-1\,s_1} \\ \vdots & \vdots & & \vdots & \vdots & & \vdots \\ q_{n\,0\,s_n} & p_{n\,1\,s_n} & \cdots & p_{n\,\nu-1\,s_n} & p_{n\,\nu\,s_n} & \cdots & p_{n\,n-1\,s_n} \end{pmatrix}$$

for integers $s_1, \ldots, s_n$, where $-\infty < s_i \leq m_i$ for $i = 1, \ldots, n$, and then set

$$a_\kappa := \sum_{s_1+\cdots+s_n=\kappa} \det \Pi_1(s_1, \ldots, s_n)$$

for $\kappa = p_0, p_0 - 1, \ldots, 1, 0, -1, \ldots$, where the indices $s_1, \ldots, s_n$ are restricted to satisfy the conditions $m_i - (p_0 - \kappa) \leq s_i \leq m_i$ for $i = 1, \ldots, n$. The a_κ form an infinite sequence of constants that are independent of m, and $a_\kappa(m) = a_\kappa$ for $\kappa = -(m - p_0 - 1), \ldots, p_0$. Therefore, the function $\widehat{\pi}_1(\rho, m)$ has the representation

$$\widehat{\pi}_1(\rho, m) = \sum_{\kappa=-(m-p_0-1)}^{p_0} a_\kappa \rho^\kappa + \sum_{\kappa=-n(m-1)}^{-(m-p_0)} a_\kappa(m)\rho^\kappa \tag{3.37}$$

for $\rho \neq 0$ in $\mathbb{C}$. The constants a_κ, $\kappa = p_0, p_0 - 1, \ldots, 1, 0, -1, \ldots$, can be calculated explicitly using the functions $z_{k0}(t)$, $z_{k1}(t)$, $z_{k2}(t)$, $\ldots$, $k = 0, 1, \ldots, n-1$, and the boundary values $B_1, \ldots, B_n$. This infinite sequence is an important invariant for the differential operator L.

A similar discussion can be carried out for the function $\widehat{\pi}_0(\rho, m)$. For integers $s_1, \ldots, s_n$ with $-\infty < s_i \leq m_i$ for $i = 1, \ldots, n$, introduce the $n \times n$ matrices

$$\Pi_0(s_1, \ldots, s_n) := \begin{pmatrix} p_{1\,0\,s_1} & p_{1\,1\,s_1} & \cdots & p_{1\,\nu-1\,s_1} & p_{1\,\nu\,s_1} & \cdots & p_{1\,n-1\,s_1} \\ \vdots & \vdots & & \vdots & \vdots & & \vdots \\ p_{n\,0\,s_n} & p_{n\,1\,s_n} & \cdots & p_{n\,\nu-1\,s_n} & p_{n\,\nu\,s_n} & \cdots & p_{n\,n-1\,s_n} \end{pmatrix},$$

and then form the infinite sequence of constants

$$b_\kappa := \sum_{s_1+\cdots+s_n=\kappa} \det \Pi_0(s_1,\ldots,s_n),$$

$\kappa = p_0, p_0-1,\ldots,1,0,-1,\ldots$, where the indices $s_1,\ldots,s_n$ are restricted to satisfy the conditions $m_i-(p_0-\kappa)\le s_i\le m_i$ for $i=1,\ldots,n$. The b_κ are independent of the integer m. In terms of these constants the function $\widehat{\pi}_0(\rho,m)$ can be represented in the form

$$\widehat{\pi}_0(\rho,m) = \sum_{\kappa=-(m-p_0-1)}^{p_0} b_\kappa\rho^\kappa + \sum_{\kappa=-n(m-1)}^{-(m-p_0)} b_\kappa(m)\rho^\kappa \tag{3.38}$$

for $\rho\ne 0$ in $\mathbb{C}$, where the $b_\kappa(m)$ are constants that do depend on m. The infinite sequence b_κ, $\kappa=p_0,p_0-1,\ldots,1,0,-1,\ldots$, forms another important invariant for the differential operator L.

The constants a_κ, b_κ defined above are precisely the constants a_κ, b_κ introduced in Chapter 1 for the case n odd.

To conclude this discussion of the m-dependence, we form the two functions

$$\pi_1(\rho,m) := \sum_{\kappa=-(m-p_0-1)}^{p_0} a_\kappa\rho^\kappa, \qquad \pi_0(\rho,m) := \sum_{\kappa=-(m-p_0-1)}^{p_0} b_\kappa\rho^\kappa$$

for $\rho\ne 0$ in $\mathbb{C}$. Their only dependence on the integer m is in the lower limit $-(m-p_0-1)$ of the summations.

The sequences a_κ, $\kappa=p_0,p_0-1,\ldots,1,0,-1,\ldots$, and b_κ, $\kappa=p_0,p_0-1,\ldots,1,0,-1,\ldots$, are used to formulate our classification scheme for the differential operator L. This is the formulation given previously in Definition 1.2.

DEFINITION 3.3. In the case $n=2\nu-1$ odd with the boundary values $B_1,\ldots,B_n$ in normalized form and $p_0=m_1+\cdots+m_n$, the differential operator L is said to be:

(i) *regular* if $a_{p_0}\ne 0$ and $b_{p_0}\ne 0$.

(ii) *simply irregular* if either $a_{p_0}=0$ or $b_{p_0}=0$, and $a_\kappa\ne 0$ and $b_\ell\ne 0$ for some integers κ,ℓ with $-\infty<\kappa,\ell\le p_0$.

(iii) *degenerate irregular* if either $a_\kappa=0$ for $\kappa=p_0,p_0-1,\ldots,1,0,-1,\ldots$ or $b_\kappa=0$ for $\kappa=p_0,p_0-1,\ldots,1,0,-1,\ldots$.

Next, in the ρ plane we introduce the three sectors

$$\begin{aligned}
&S_0\colon\ \text{all } \rho=|\rho|e^{i\theta}\in\mathbb{C} \text{ with } -\frac{\pi}{2n}\le\theta\le\frac{\pi}{2n},\\
&S_\diamond\colon\ \text{all } \rho=|\rho|e^{i\theta}\in\mathbb{C} \text{ with } \frac{\pi}{2n}\le\theta\le\frac{3\pi}{2n},\\
&S_1\colon\ \text{all } \rho=|\rho|e^{i\theta}\in\mathbb{C} \text{ with } \pi-\frac{\pi}{2n}\le\theta\le\pi+\frac{\pi}{2n}.
\end{aligned}$$

Each of these sectors has angular opening π/n, with the sectors S_0 and S_1 symmetric in the real axis. Observe that the sector S_0 has been altered from its earlier form for the case n even. The reason for this change is to establish a sector where the exponentials $e^{i\rho\omega_1},\ldots,e^{i\rho\omega_{\nu-1}},e^{-i\rho\omega_\nu},\ldots,e^{-i\rho\omega_{n-1}}$ go to zero very rapidly. Indeed, on the original sector S_0 the exponential $e^{i\rho\omega_{\nu-1}}$ does not go to zero when n is odd: for any point $\rho=|\rho|e^{i\pi/n}$ we have

$$\left|e^{i\rho\omega_{\nu-1}}\right| = e^{|\rho|\cos[\arg(i\rho\omega_{\nu-1})]} = e^{|\rho|\cos\left[\frac{\pi}{2}+\frac{\pi}{n}+\frac{2\pi(\nu-1)}{n}\right]} = e^{|\rho|\cos[3\pi/2]} = 1.$$

The new sector S_0 corrects this problem (see Chapter 5).

Now consider the sectors $S_\diamond$ and S_1. For $\rho = |\rho|e^{i\theta} \neq 0$ we have

$$\begin{aligned}\rho \in S_\diamond &\iff \frac{\pi}{2n} \le \theta \le \frac{3\pi}{2n}\\ &\iff \frac{\pi}{2n} + \frac{2\pi}{n}(\nu-1) \le \theta + \frac{2\pi}{n}(\nu-1) \le \frac{3\pi}{2n} + \frac{2\pi}{n}(\nu-1)\\ &\iff \frac{\pi}{2n} + \frac{\pi}{n}(n-1) \le \arg(\rho\omega_{\nu-1}) \le \frac{3\pi}{2n} + \frac{\pi}{n}(n-1)\\ &\iff \pi - \frac{\pi}{2n} \le \arg(\rho\omega_{\nu-1}) \le \pi + \frac{\pi}{2n} \iff \rho\omega_{\nu-1} \in S_1.\end{aligned}$$

Equivalently, $\rho \in S_1 \iff \rho\omega_{\nu-1}^{-1} \in S_\diamond$. Thus, $\rho \in S_\diamond$ and ρ^n is an eigenvalue of L if and only if $\rho\omega_{\nu-1} \in S_1$ and $(\rho\omega_{\nu-1})^n = \rho^n$ is an eigenvalue of L. It follows that finding the eigenvalues $\lambda = \rho^n$ of L with $\rho \in S_\diamond$ is equivalent to finding the eigenvalues $\lambda = \rho^n$ of L with $\rho \in S_1$. In the sequel we choose to work on the sector S_1 because of the simpler geometry.

The modified Birkhoff approximate solutions $y_k(t,\rho)$, $k = 0, 1, \ldots, n-1$, and the approximate characteristic determinant $\widehat{\Delta}(\rho)$ are designed to develop the spectral theory of the differential operator L relative to the sector S_0. In working on the sector S_1, we must replace these quantities with alternate forms. We begin by introducing the functions

$$x_k(t,\rho) := e^{-i\rho\omega_k} z_k(t,\rho) = e^{i\rho\omega_k(t-1)} \sum_{j=0}^{m-1} z_{kj}(t)\rho^{-j}, \quad k = 0, 1, \ldots, \nu-1,$$

$$x_k(t,\rho) := z_k(t,\rho) = e^{i\rho\omega_k t} \sum_{j=0}^{m-1} z_{kj}(t)\rho^{-j}, \qquad k = \nu, \ldots, n-1,$$

for $0 \le t \le 1$ and for $\rho \neq 0$ in $\mathbb{C}$. These functions are again approximate solutions of the differential equation (2.1) in the sense of Theorem 2.1. They are related to the earlier modified Birkhoff approximate solutions by the equations

$$\begin{aligned} x_k(t,\rho) &= e^{-i\rho\omega_k} y_k(t,\rho), \quad && k = 0, 1, \ldots, \nu-1,\\ x_k(t,\rho) &= e^{i\rho\omega_k} y_k(t,\rho), && k = \nu, \ldots, n-1,\end{aligned}$$

for $0 \le t \le 1$ and for $\rho \neq 0$ in $\mathbb{C}$.

Applying the boundary values, for $i = 1, \ldots, n$ and $k = 0, 1, \ldots, \nu-1$ we have

$$B_i(x_k(\cdot,\rho)) = e^{-i\rho\omega_k} B_i(y_k(\cdot,\rho)) = \widehat{P}_{ik}(\rho)e^{-i\rho\omega_k} + \widehat{Q}_{ik}(\rho) \tag{3.39}$$

for $\rho \neq 0$ in $\mathbb{C}$, while for $i = 1, \ldots, n$ and $k = \nu, \ldots, n-1$

$$B_i(x_k(\cdot,\rho)) = e^{i\rho\omega_k} B_i(y_k(\cdot,\rho)) = \widehat{P}_{ik}(\rho)e^{i\rho\omega_k} + \widehat{Q}_{ik}(\rho) \tag{3.40}$$

for $\rho \neq 0$ in $\mathbb{C}$. In terms of these functions we then form a new *approximate characteristic determinant* by defining

$$\widetilde{\Delta}(\rho) = \widetilde{\Delta}(\rho, m) := \det(B_i(x_k(\cdot,\rho)))$$

for $\rho \neq 0$ in $\mathbb{C}$. It is also defined and analytic for $\rho \neq 0$ in $\mathbb{C}$; it can be calculated explicitly once the integer m has been chosen; and it is particularly well-suited for work on the sector S_1.

The new approximate characteristic determinant has the representation

(3.41) $$\widetilde{\Delta}(\rho) = \det \begin{pmatrix} \widehat{P}_{10}(\rho)e^{-i\rho}+\widehat{Q}_{10}(\rho) & \overset{1 \leq k \leq \nu-1}{\widehat{P}_{1k}(\rho)e^{-i\rho\omega_k}+\widehat{Q}_{1k}(\rho)} & \overset{\nu \leq k \leq n-1}{\widehat{P}_{1k}(\rho)e^{i\rho\omega_k}+\widehat{Q}_{1k}(\rho)} \\ \vdots & \vdots & \vdots \\ \widehat{P}_{n0}(\rho)e^{-i\rho}+\widehat{Q}_{n0}(\rho) & \widehat{P}_{nk}(\rho)e^{-i\rho\omega_k}+\widehat{Q}_{nk}(\rho) & \widehat{P}_{nk}(\rho)e^{i\rho\omega_k}+\widehat{Q}_{nk}(\rho) \end{pmatrix}$$

for $\rho \neq 0$ in $\mathbb{C}$. Setting $\eta = 1 + \omega_1 + \cdots + \omega_{\nu-1} - \omega_\nu - \cdots - \omega_{n-1}$, we can factor out the exponentials appearing in the columns of the matrix in equation (3.41):

(3.42) $$\widetilde{\Delta}(\rho) = e^{-i\rho}e^{-i\rho\omega_1}\cdots e^{-i\rho\omega_{\nu-1}}e^{i\rho\omega_\nu}\cdots e^{i\rho\omega_{n-1}}\widehat{\Delta}(\rho) = e^{-i\rho\eta}\widehat{\Delta}(\rho)$$

for $\rho \neq 0$ in $\mathbb{C}$. Using equation (3.41), we proceed to expand the determinant for $\widetilde{\Delta}(\rho)$ using the linearity in the 0th column. This produces the representation

(3.43) $$\widetilde{\Delta}(\rho) = \widetilde{\pi}_1(\rho)e^{-i\rho} + \widetilde{\pi}_0(\rho) + \widetilde{\Phi}_1(\rho)e^{-i\rho} + \widetilde{\Phi}_0(\rho)$$

for $\rho \neq 0$ in $\mathbb{C}$, where the functions $\widetilde{\pi}_1(\rho)$, $\widetilde{\pi}_0(\rho)$ are defined by

$$\widetilde{\pi}_1(\rho) := \det \begin{pmatrix} \widehat{P}_{10}(\rho) & \widehat{Q}_{11}(\rho) & \cdots & \widehat{Q}_{1\,\nu-1}(\rho) & \widehat{Q}_{1\nu}(\rho) & \cdots & \widehat{Q}_{1\,n-1}(\rho) \\ \vdots & \vdots & & \vdots & \vdots & & \vdots \\ \widehat{P}_{n0}(\rho) & \widehat{Q}_{n1}(\rho) & \cdots & \widehat{Q}_{n\,\nu-1}(\rho) & \widehat{Q}_{n\nu}(\rho) & \cdots & \widehat{Q}_{n\,n-1}(\rho) \end{pmatrix},$$

$$\widetilde{\pi}_0(\rho) := \det \begin{pmatrix} \widehat{Q}_{10}(\rho) & \widehat{Q}_{11}(\rho) & \cdots & \widehat{Q}_{1\,\nu-1}(\rho) & \widehat{Q}_{1\nu}(\rho) & \cdots & \widehat{Q}_{1\,n-1}(\rho) \\ \vdots & \vdots & & \vdots & \vdots & & \vdots \\ \widehat{Q}_{n0}(\rho) & \widehat{Q}_{n1}(\rho) & \cdots & \widehat{Q}_{n\,\nu-1}(\rho) & \widehat{Q}_{n\nu}(\rho) & \cdots & \widehat{Q}_{n\,n-1}(\rho) \end{pmatrix}$$

for $\rho \neq 0$ in $\mathbb{C}$, and the functions $\widetilde{\Phi}_1(\rho)$, $\widetilde{\Phi}_0(\rho)$ are the sums of $2^{n-1}-1$ determinants with each determinant expressible in the form of a product of some of the exponentials $e^{-i\rho\omega_k}$, $k = 1, \ldots, \nu-1$, or some of the exponentials $e^{i\rho\omega_k}$, $k = \nu, \ldots, n-1$, (at least one of these exponentials appears in each product), times the determinant of an $n \times n$ matrix whose entries are selected from among the functions $\widehat{P}_{ik}(\rho)$, $\widehat{Q}_{ik}(\rho)$.

For integers $s_1, \ldots, s_n$ with $-\infty < s_i \leq m_i$ for $i = 1, \ldots, n$, introduce the $n \times n$ matrices

$$\Pi'_1(s_1,\ldots,s_n) := \begin{pmatrix} p_{1\,0\,s_1} & q_{1\,1\,s_1} & \cdots & q_{1\,\nu-1\,s_1} & q_{1\,\nu\,s_1} & \cdots & q_{1\,n-1\,s_1} \\ \vdots & \vdots & & \vdots & \vdots & & \vdots \\ p_{n\,0\,s_n} & q_{n\,1\,s_n} & \cdots & q_{n\,\nu-1\,s_n} & q_{n\,\nu\,s_n} & \cdots & q_{n\,n-1\,s_n} \end{pmatrix},$$

$$\Pi'_0(s_1,\ldots,s_n) := \begin{pmatrix} q_{1\,0\,s_1} & q_{1\,1\,s_1} & \cdots & q_{1\,\nu-1\,s_1} & q_{1\,\nu\,s_1} & \cdots & q_{1\,n-1\,s_1} \\ \vdots & \vdots & & \vdots & \vdots & & \vdots \\ q_{n\,0\,s_n} & q_{n\,1\,s_n} & \cdots & q_{n\,\nu-1\,s_n} & q_{n\,\nu\,s_n} & \cdots & q_{n\,n-1\,s_n} \end{pmatrix},$$

and then set

$$a'_\kappa := \sum_{s_1+\cdots+s_n=\kappa} \det \Pi'_1(s_1,\ldots,s_n),$$

$$b'_\kappa := \sum_{s_1+\cdots+s_n=\kappa} \det \Pi'_0(s_1,\ldots,s_n)$$

for $\kappa = p_0, p_0 - 1, \ldots, 1, 0, -1, \ldots$, where the indices $s_1, \ldots, s_n$ are restricted to satisfy the conditions $m_i - (p_0 - \kappa) \leq s_i \leq m_i$ for $i = 1, \ldots, n$. The a'_κ, b'_κ form infinite sequences of constants that are independent of the integer m, and in terms of them the functions $\widetilde{\pi}_1(\rho) = \widetilde{\pi}_1(\rho, m)$, $\widetilde{\pi}_0(\rho) = \widetilde{\pi}_0(\rho, m)$ have the representations

$$\widetilde{\pi}_1(\rho, m) = \sum_{\kappa=-(m-p_0-1)}^{p_0} a'_\kappa \rho^\kappa + \sum_{\kappa=-n(m-1)}^{-(m-p_0)} a'_\kappa(m)\rho^\kappa, \tag{3.44}$$

$$\widetilde{\pi}_0(\rho, m) = \sum_{\kappa=-(m-p_0-1)}^{p_0} b'_\kappa \rho^\kappa + \sum_{\kappa=-n(m-1)}^{-(m-p_0)} b'_\kappa(m)\rho^\kappa \tag{3.45}$$

for $\rho \neq 0$ in $\mathbb{C}$. Here the constants $a'_\kappa(m)$, $b'_\kappa(m)$ do depend on the integer m. The constants a'_κ, b'_κ can be calculated explicitly, and they form invariants for the differential operator L. The functions $\widehat{\pi}_1(\rho)$, $\widehat{\pi}_0(\rho)$ appearing in (3.34) and the functions $\widetilde{\pi}_1(\rho)$, $\widetilde{\pi}_0(\rho)$ appearing in (3.43) are related by the equations

$$\widetilde{\pi}_1(\rho) = \widehat{\pi}_0(\rho\omega_\nu), \qquad \widetilde{\pi}_0(\rho) = \widehat{\pi}_1(\rho\omega_{\nu-1}) \tag{3.46}$$

for $\rho \neq 0$ in $\mathbb{C}$. We conclude the discussion of m-dependence by introducing the functions

$$\pi'_1(\rho, m) := \sum_{\kappa=-(m-p_0-1)}^{p_0} a'_\kappa \rho^\kappa, \qquad \pi'_0(\rho, m) := \sum_{\kappa=-(m-p_0-1)}^{p_0} b'_\kappa \rho^\kappa$$

for $\rho \neq 0$ in $\mathbb{C}$.

Henceforth, we will assume that for the case n odd, $n = 2\nu - 1 \geq 3$, the differential operator L is either regular or simply irregular. Let p be the largest integer with $a_p \neq 0$, so $-\infty < p \leq p_0$, and let q be the largest integer with $b_q \neq 0$, so $-\infty < q \leq p_0$. Then $a_\kappa = 0$ for $\kappa = p+1, \dots, p_0$, and $b_\kappa = 0$ for $\kappa = q+1, \dots, p_0$. The integers p, q and the nonzero constants a_p, b_q will play major roles in the sequel.

For the case $p = q$ choose a constant $d > 0$ such that

$$|a_p|e^{-d} + |b_p|e^{-d} \leq \frac{1}{4}\min\{|a_p|, |b_p|\}, \tag{3.47}$$

and in terms of the constant d introduce the horizontal strips

$$\Gamma_0 := \{\rho = a + ib \in \mathbb{C} \mid a \geq -\pi \text{ and } |b| \leq d\},$$
$$\Gamma_1 := \{\rho = a + ib \in \mathbb{C} \mid a \leq \pi \text{ and } |b| \leq d\}.$$

Then select complex constants τ_0 and τ_1 and form the translated sectors

$$T_0 := \{\rho - \tau_0 \mid \rho \in S_0\}, \qquad T_1 := \{\rho - \tau_1 \mid \rho \in S_1\}$$

with the following properties: for the case $p = q$ we require that the sectors S_0, S_1 lie in the interiors of T_0, T_1, respectively, and that the horizontal strips Γ_0, Γ_1 lie in the interiors of T_0, T_1, respectively; for the cases $p < q$ and $p > q$ we require only that the sectors S_0, S_1 lie in the interiors of T_0, T_1, respectively. The translated sectors T_0 and T_1 have been constructed by utilizing the constants a_p, b_q determined by the differential operator L, and their construction is *independent* of the integer m. The asymptotic expansions developed in the next chapter will take place in the sectors T_0 and T_1.

Now that the integers p, q and the constants a_p, b_q have been determined and the sectors T_0, T_1 selected, the integer m can be fixed once and for all: choose any integer m with $m > n$, $m > p_0$, and $-(m - p_0 - 1) \leq p, q \leq p_0$. For this choice of m

we proceed to form the Birkhoff approximate solutions $z_k(t,\rho) = z_k(t,\rho,m)$, $k = 0,1,\dots,n-1$, the modified Birkhoff approximate solutions $y_k(t,\rho) = y_k(t,\rho,m)$, $k = 0,1,\dots,n-1$, and $x_k(t,\rho) = x_k(t,\rho,m)$, $k = 0,1,\dots,n-1$, and the approximate characteristic determinants

$$\widehat{\Delta}(\rho) = \widehat{\Delta}(\rho,m) = \det(B_i(y_k(\,\cdot\,,\rho,m))),$$

$$\widetilde{\Delta}(\rho) = \widetilde{\Delta}(\rho,m) = \det(B_i(x_k(\,\cdot\,,\rho,m)))$$

for $\rho \neq 0$ in $\mathbb{C}$. Lastly, we form the functions

$$\pi_1(\rho) = \pi_1(\rho,m) = \sum_{\kappa=-(m-p_0-1)}^{p} a_\kappa \rho^\kappa, \quad \pi_0(\rho) = \pi_0(\rho,m) = \sum_{\kappa=-(m-p_0-1)}^{q} b_\kappa \rho^\kappa$$

for $\rho \neq 0$ in $\mathbb{C}$, and the functions

$$\pi_1'(\rho) = \pi_1'(\rho,m) = \sum_{\kappa=-(m-p_0-1)}^{p_0} a_\kappa' \rho^\kappa, \quad \pi_0'(\rho) = \pi_0'(\rho,m) = \sum_{\kappa=-(m-p_0-1)}^{p_0} b_\kappa' \rho^\kappa$$

for $\rho \neq 0$ in $\mathbb{C}$. From equation (3.46) we otain the relation

$$a_\kappa' = b_\kappa(\omega_\nu)^\kappa, \qquad b_\kappa' = a_\kappa(\omega_{\nu-1})^\kappa \tag{3.48}$$

for $\kappa = p_0, p_0-1,\dots,-(m-p_0-1)$, and hence, $a_q' \neq 0$ and $a_\kappa' = 0$ for $\kappa = q+1,\dots,p_0$, and $b_p' \neq 0$ and $b_\kappa' = 0$ for $\kappa = p+1,\dots,p_0$. It follows that the functions $\pi_1'(\rho)$, $\pi_0'(\rho)$ simplify to

$$\pi_1'(\rho) = \sum_{\kappa=-(m-p_0-1)}^{q} a_\kappa' \rho^\kappa, \qquad \pi_0'(\rho) = \sum_{\kappa=-(m-p_0-1)}^{p} b_\kappa' \rho^\kappa$$

for $\rho \neq 0$ in $\mathbb{C}$. Also, in the case $p = q$ we have $|a_p'| = |b_p|$ and $|b_p'| = |a_p|$, and by equation (3.47)

$$|a_p'|e^{-d} + |b_p'|e^{-d} \leq \frac{1}{4}\min\{|a_p'|,|b_p'|\}. \tag{3.49}$$

For our future work we rewrite the representations (3.34) and (3.43) of the approximate characteristic determinants in their final forms:

$$\begin{aligned}\widehat{\Delta}(\rho) &= \pi_1(\rho)e^{i\rho} + \pi_0(\rho)\\ &\quad + \Big[\sum_{\kappa=-n(m-1)}^{-(m-p_0)} a_\kappa(m)\rho^\kappa + \widehat{\Phi}_1(\rho,m)\Big]e^{i\rho}\\ &\quad + \Big[\sum_{\kappa=-n(m-1)}^{-(m-p_0)} b_\kappa(m)\rho^\kappa + \widehat{\Phi}_0(\rho,m)\Big],\end{aligned} \tag{3.50}$$

$$\begin{aligned}\widetilde{\Delta}(\rho) &= \pi_1'(\rho)e^{-i\rho} + \pi_0'(\rho) \\ &\quad + \Big[\sum_{\kappa=-n(m-1)}^{-(m-p_0)} a_\kappa'(m)\rho^\kappa + \widetilde{\Phi}_1(\rho,m)\Big]e^{-i\rho} \\ &\quad + \Big[\sum_{\kappa=-n(m-1)}^{-(m-p_0)} b_\kappa'(m)\rho^\kappa + \widetilde{\Phi}_0(\rho,m)\Big]\end{aligned} \tag{3.51}$$

for $\rho \neq 0$ in $\mathbb{C}$.

For the case $n = 2\nu - 1$ odd, the quantities introduced in the last four paragraphs will remain fixed in the sequel. Specifically, the integer m will be held fixed with $m > n$, $m > p_0$, and $-(m - p_0 - 1) \leq p, q \leq p_0$.

CHAPTER 4

Asymptotic Expansion of Solutions

In this chapter we construct actual solutions of the differential equation (2.1) for ρ belonging to the sectors T_0 or T_1. These solutions behave asymptotically like the Birkhoff approximate solutions $z_k(t,\rho) = z_k(t,\rho,m)$, where the integer m has been fixed with $m > n$, $m > p_0$, and $-(m-p_0-1) \leq p, q \leq p_0$. Again the discussion divides naturally into the two cases n even and n odd.

4.1. Expansions for n Even

Assume that n is even: $n = 2\nu \geq 2$. We begin by examining the relationship between the complex numbers $i\rho\omega_0$, $i\rho\omega_1$, ..., $i\rho\omega_{n-1}$ for ρ belonging to the sectors S_0 or S_1, where

$$S_0\colon \text{ all } \rho = |\rho|e^{i\theta} \in \mathbb{C} \text{ with } 0 \leq \theta \leq \frac{\pi}{n},$$
$$S_1\colon \text{ all } \rho = |\rho|e^{i\theta} \in \mathbb{C} \text{ with } -\frac{\pi}{n} \leq \theta \leq 0.$$

The results in the following lemma are variations of some well-known results (see [**4**, p. 220] or [**37**, pp. 43–45]).

LEMMA 4.1. *Let n be even: $n = 2\nu$. Then there exist permutations*

$$\omega_0^0, \omega_1^0, \ldots, \omega_{n-1}^0 \quad \textit{and} \quad \omega_0^1, \omega_1^1, \ldots, \omega_{n-1}^1$$

of the nth roots of unity $\omega_0, \omega_1, \ldots, \omega_{n-1}$ such that

$$\operatorname{Re}(i\rho\omega_0^0) \leq \operatorname{Re}(i\rho\omega_1^0) \leq \cdots \leq \operatorname{Re}(i\rho\omega_{n-1}^0) \quad \textit{for all } \rho \in S_0,$$
$$\operatorname{Re}(i\rho\omega_0^1) \leq \operatorname{Re}(i\rho\omega_1^1) \leq \cdots \leq \operatorname{Re}(i\rho\omega_{n-1}^1) \quad \textit{for all } \rho \in S_1.$$

Let $\omega_0^0, \omega_1^0, \ldots, \omega_{n-1}^0$ and $\omega_0^1, \omega_1^1, \ldots, \omega_{n-1}^1$ be the permutations of the nth roots of unity $\omega_0, \omega_1, \ldots, \omega_{n-1}$ determined in the last lemma. In this section we will select constants C_0, C_1, M_0, M_1, r_0, γ_0, R_0 that can be used concurrently for the translated sectors T_0 and T_1. Indeed, for the first two constants we choose $C_0 \geq 0$ such that $|a_\alpha(t)| \leq C_0$ for $0 \leq t \leq 1$ and for $\alpha = 0, 1, \ldots, n-2$, and in terms of the sectors $T_0 = \{\rho - \tau_0 \mid \rho \in S_0\}$, $T_1 = \{\rho - \tau_1 \mid \rho \in S_1\}$, we choose $C_1 > 0$ such that

$$\left|e^{i\tau_0(\omega_i^0-\omega_j^0)(t-s)}\right| \leq C_1 \quad \text{and} \quad \left|e^{i\tau_1(\omega_i^1-\omega_j^1)(t-s)}\right| \leq C_1$$

for all $t, s \in [0,1]$ and for $i, j = 0, 1, \ldots, n-1$. At this point we fix our attention upon the sector T_0, returning later in the section to a discussion of the sector T_1.

Fix any integer k with $0 \leq k \leq n-1$, and let κ be the unique integer satisfying $0 \leq \kappa \leq n-1$ and $\omega_\kappa^0 = \omega_k$. Let us reconsider the Birkhoff approximate solution

$$z_k(t,\rho) = e^{i\rho\omega_k t}\sum_{j=0}^{m-1} z_{kj}(t)\rho^{-j} = e^{i\rho\omega_\kappa^0 t}\sum_{j=0}^{m-1} z_{kj}(t)\rho^{-j}, \qquad 0 \leq t \leq 1,$$

on the sector T_0. Our goal in the first part of this section is to construct an actual solution $v_{0k}(t,\rho)$ of the differential equation (2.1) for $\rho \in T_0$ with $|\rho|$ sufficiently large, with $v_{0k}(t,\rho)$ behaving asymptotically like $z_k(t,\rho)$ on the sector T_0 and with $v_{0k}(t,\rho)$ having nice regularity properties on T_0 relative to the t and ρ variables.

Applying Lemma 4.1, for any point $\rho \in T_0$ we have

$$\operatorname{Re}(i(\rho+\tau_0)\omega_0^0) \le \operatorname{Re}(i(\rho+\tau_0)\omega_1^0) \le \cdots \le \operatorname{Re}(i(\rho+\tau_0)\omega_{n-1}^0). \tag{4.1}$$

Now take any point $\rho \in T_0$. Then for $j = 0, 1, \ldots, \kappa$ the estimates (4.1) give

$$\operatorname{Re}(i\rho\omega_j^0) \le \operatorname{Re}(i\rho\omega_j^0 + i(\rho+\tau_0)(\omega_\kappa^0 - \omega_j^0)) = \operatorname{Re}(i\tau_0(\omega_\kappa^0 - \omega_j^0) + i\rho\omega_\kappa^0),$$

so for $0 \le s < t \le 1$ we obtain the estimates

$$\begin{aligned} \left|e^{i\rho\omega_j^0(t-s)}\right| &= e^{\operatorname{Re}(i\rho\omega_j^0(t-s))} \le e^{\operatorname{Re}(i\tau_0(\omega_\kappa^0-\omega_j^0)(t-s)+i\rho\omega_\kappa^0(t-s))} \\ &= \left|e^{i\tau_0(\omega_\kappa^0-\omega_j^0)(t-s)}\right|\left|e^{i\rho\omega_\kappa^0(t-s)}\right| \le C_1\left|e^{i\rho\omega_\kappa^0(t-s)}\right|, \end{aligned}$$

or

$$\left|e^{i\rho\omega_j^0(t-s)}\right| \le C_1\left|e^{i\rho\omega_k(t-s)}\right| \tag{4.2}$$

for $0 \le s < t \le 1$, for $\rho \in T_0$, and for $j = 0, 1, \ldots, \kappa$. A similar argument shows that

$$\left|e^{i\rho\omega_j^0(t-s)}\right| \le C_1\left|e^{i\rho\omega_k(t-s)}\right| \tag{4.3}$$

for $0 \le t < s \le 1$, for $\rho \in T_0$, and for $j = \kappa+1, \ldots, n-1$.

Let K_{01} and K_{02} be the functions defined by

$$K_{01}(t,s,\rho) := -\sum_{j=0}^{\kappa}(i\omega_j^0)e^{i\rho\omega_j^0(t-s)}, \quad K_{02}(t,s,\rho) := \sum_{j=\kappa+1}^{n-1}(i\omega_j^0)e^{i\rho\omega_j^0(t-s)}$$

for $t, s \in [0,1]$ and for $\rho \in \mathbb{C}$; let k_0 be the function defined by

$$\begin{aligned} k_0(t,s,\rho) &:= \frac{1}{n\rho^{n-1}}K_{01}(t,s,\rho), \qquad 0 \le s < t \le 1, \\ k_0(t,s,\rho) &:= \frac{1}{n\rho^{n-1}}K_{02}(t,s,\rho), \qquad 0 \le t < s \le 1, \end{aligned}$$

for $\rho \ne 0$ in $\mathbb{C}$; and let $\mathcal{K}_{0\rho}$ be the integral operator on $L^2[0,1]$ defined by

$$\mathcal{K}_{0\rho}u(t) := \int_0^1 k_0(t,s,\rho)u(s)\,ds, \qquad 0 \le t \le 1, \quad u \in L^2[0,1],$$

for each $\rho \ne 0$ in $\mathbb{C}$. From our earlier work [**34**, pp. 103–105], we know that if $u \in L^2[0,1]$ and $v = \mathcal{K}_{0\rho}u$, then $v \in H^n[0,1]$ and $(\rho^n I - \tau)v = u$, where $\tau = i^{-n}(d/dt)^n$ is the formal differential operator introduced in Section 1.1. Also, the derivatives of v are given by

$$\begin{aligned} v^{(\alpha)}(t) &= \int_0^1 \frac{\partial^\alpha k_0}{\partial t^\alpha}(t,s,\rho)u(s)\,ds \\ &= -\frac{1}{n\rho^{n-1}}\sum_{j=0}^{\kappa}(i\omega_j^0)^{\alpha+1}\rho^\alpha\int_0^t e^{i\rho\omega_j^0(t-s)}u(s)\,ds \\ &\quad + \frac{1}{n\rho^{n-1}}\sum_{j=\kappa+1}^{n-1}(i\omega_j^0)^{\alpha+1}\rho^\alpha\int_t^1 e^{i\rho\omega_j^0(t-s)}u(s)\,ds \end{aligned} \tag{4.4}$$

for $0 \le t \le 1$ and for $\alpha = 0, 1, \ldots, n-1$. We will use the integral operator $\mathcal{K}_{0\rho}$ extensively in deriving our asymptotic expansions, with the estimates (4.2) and (4.3) producing bounds for the various operators relative to the sector T_0.

Next, we begin the construction of actual solutions of the differential equation (2.1), looking for solutions of the form $u(t,\rho) = z_k(t,\rho) + \phi(t,\rho)$ where $z_k(t,\rho)$ is the Birkhoff approximate solution. From equation (2.10) we know that

$$(\rho^n I - \ell) z_k(t,\rho) = e^{i\rho\omega_k t}\eta_k(t,\rho) \tag{4.5}$$

for $0 \le t \le 1$ and $\rho \ne 0$ in $\mathbb{C}$, where the residual function $\eta_k(t,\rho)$ is given by

$$\eta_k(t,\rho) = -\frac{\rho^n}{i^n}\sum_{s=m+1}^{n+m-1}\left[\sum_{\ell+p+j=s} c_{k\ell p}(t) z_{kj}^{(\ell)}(t)\right]\rho^{-s}$$

for $0 \le t \le 1$ and $\rho \ne 0$ in $\mathbb{C}$. The function $\eta_k(t,\rho)$ depends on the fixed integers m and k. Choose a constant $M_0 \ge 0$ such that

$$|\eta_k(t,\rho)| \le M_0|\rho|^{-(m+1-n)} \tag{4.6}$$

for $0 \le t \le 1$ and for $\rho \in \mathbb{C}$ with $|\rho| \ge 1$ (recall that $m > n$). Without loss of generality we can assume that the constant M_0 is independent of the integer k.

Fix any point $\rho \ne 0$ in $\mathbb{C}$. As we determine solutions to (2.1) and to other equivalent equations, the various functions u, ϕ, ψ and constants $c_0, c_1, \ldots, c_{n-1}$ that appear will all depend upon the fixed integer k and the fixed parameter ρ. For the time being we will surpress this dependence on k and ρ in our notation. The only exceptions are in the Birkhoff approximate solution $z_k(\cdot,\rho)$ and the associated residual function $\eta_k(\cdot,\rho)$, where we continue to display both the k and ρ dependence, and in the integral operator $\mathcal{K}_{0\rho}$, where we continue to show the ρ dependence but surpress the k dependence.

Suppose $u \in H^n[0,1]$ is a solution of the differential equation (2.1), and let $\phi \in H^n[0,1]$ be the function defined by the equation $u(t) := z_k(t,\rho) + \phi(t)$ for $0 \le t \le 1$. Then by (4.5) we have

$$0 = (\rho^n I - \ell)u(t) = e^{i\rho\omega_k t}\eta_k(t,\rho) + (\rho^n I - \tau)\phi(t) - \sigma\phi(t)$$

or

$$(\rho^n I - \tau)\phi(t) = \sigma\phi(t) - e^{i\rho\omega_k t}\eta_k(t,\rho)$$

for $0 \le t \le 1$. Set $v(t) := \mathcal{K}_{0\rho}\big(\sigma\phi(t) - e^{i\rho\omega_k t}\eta_k(t,\rho)\big)$ for $0 \le t \le 1$. Then v belongs to $H^n[0,1]$ and $(\rho^n I - \tau)v(t) = \sigma\phi(t) - e^{i\rho\omega_k t}\eta_k(t,\rho)$, and hence,

$$(\rho^n I - \tau)\big(\phi(t) - v(t)\big) = 0$$

for $0 \le t \le 1$. It follows that there exist complex constants $c_0, c_1, \ldots, c_{n-1}$ such that $\phi(t) - v(t) = \sum_{j=0}^{n-1} c_j e^{i\rho\omega_j t}$, or

$$\phi(t) = \mathcal{K}_{0\rho}\sigma\phi(t) - \mathcal{K}_{0\rho}\big(e^{i\rho\omega_k t}\eta_k(t,\rho)\big) + \sum_{j=0}^{n-1} c_j e^{i\rho\omega_j t} \tag{4.7}$$

for $0 \le t \le 1$. Equation (4.7) is an integro-differential equation for the function ϕ.

Conversely, suppose $\phi \in H^n[0,1]$ is a solution of (4.7) for some constants $c_0, c_1, \ldots, c_{n-1}$. We assert that the function $u(t) := z_k(t,\rho) + \phi(t)$ is a solution of the differential equation (2.1). Clearly u belongs to $H^n[0,1]$, $(\rho^n I - \tau)\phi(t) = \sigma\phi(t) - e^{i\rho\omega_k t}\eta_k(t,\rho)$ for $0 \le t \le 1$, and by (4.5) again

$$\begin{aligned}(\rho^n I - \ell)u(t) &= e^{i\rho\omega_k t}\eta_k(t,\rho) + (\rho^n I - \tau)\phi(t) - \sigma\phi(t)\\ &= e^{i\rho\omega_k t}\eta_k(t,\rho) + \sigma\phi(t) - e^{i\rho\omega_k t}\eta_k(t,\rho) - \sigma\phi(t) = 0\end{aligned}$$

for $0 \le t \le 1$. This establishes the assertion. We conclude that finding solutions u to the differential equation (2.1) is equivalent to finding solutions ϕ to the integro-differential equation (4.7) with arbitrary constants $c_0, c_1, \dots, c_{n-1}$.

Next, consider the special case of equation (4.7) where all the constants c_j are set equal to zero:

$$\phi(t) = \mathcal{K}_{0\rho}\sigma\phi(t) - \mathcal{K}_{0\rho}\big(e^{i\rho\omega_k t}\eta_k(t,\rho)\big) \tag{4.8}$$

for $0 \le t \le 1$. Suppose $\phi \in H^n[0,1]$ is a solution of equation (4.8). Clearly $\sigma\phi(t) = \sigma\mathcal{K}_{0\rho}\sigma\phi(t) - \sigma\mathcal{K}_{0\rho}\big(e^{i\rho\omega_k t}\eta_k(t,\rho)\big)$ for $0 \le t \le 1$. Let ψ be the function defined by the equation

$$\sigma\phi(t) := e^{i\rho\omega_k t}\psi(t), \qquad 0 \le t \le 1.$$

Clearly $\psi \in H^2[0,1]$ and $\psi \in C[0,1]$, and

$$\psi(t) = e^{-i\rho\omega_k t}\sigma\mathcal{K}_{0\rho}\big(e^{i\rho\omega_k t}\psi(t)\big) - e^{-i\rho\omega_k t}\sigma\mathcal{K}_{0\rho}\big(e^{i\rho\omega_k t}\eta_k(t,\rho)\big)$$

for $0 \le t \le 1$. Writing out the last equation in detail, we get

$$\begin{aligned}\psi(t) &= \frac{1}{n\rho^{n-1}}\int_0^t \sum_{\alpha=0}^{n-2} a_\alpha(t)e^{-i\rho\omega_k(t-s)}\frac{\partial^\alpha K_{01}}{\partial t^\alpha}(t,s,\rho)\big[\psi(s) - \eta_k(s,\rho)\big]\,ds\\ &\quad + \frac{1}{n\rho^{n-1}}\int_t^1 \sum_{\alpha=0}^{n-2} a_\alpha(t)e^{-i\rho\omega_k(t-s)}\frac{\partial^\alpha K_{02}}{\partial t^\alpha}(t,s,\rho)\big[\psi(s) - \eta_k(s,\rho)\big]\,ds\end{aligned} \tag{4.9}$$

for $0 \le t \le 1$. Equation (4.9) is an integral equation for the function ψ.

Conversely, suppose $\psi \in C[0,1]$ is a solution of the integral equation (4.9). Let ϕ and u be the functions in $H^n[0,1]$ defined by

$$\begin{aligned}\phi(t) &:= \mathcal{K}_{0\rho}\big(e^{i\rho\omega_k t}[\psi(t) - \eta_k(t,\rho)]\big),\\ u(t) &:= z_k(t,\rho) + \phi(t) = z_k(t,\rho) + \mathcal{K}_{0\rho}\big(e^{i\rho\omega_k t}[\psi(t) - \eta_k(t,\rho)]\big)\end{aligned}$$

for $0 \le t \le 1$. We assert that $\sigma\phi(t) = e^{i\rho\omega_k t}\psi(t)$ for $0 \le t \le 1$. Indeed, from the definition of ϕ we have

$$\begin{aligned}\phi(t) &= \frac{1}{n\rho^{n-1}}\int_0^t K_{01}(t,s,\rho)e^{i\rho\omega_k s}[\psi(s) - \eta_k(s,\rho)]\,ds\\ &\quad + \frac{1}{n\rho^{n-1}}\int_t^1 K_{02}(t,s,\rho)e^{i\rho\omega_k s}[\psi(s) - \eta_k(s,\rho)]\,ds\end{aligned}$$

for $0 \le t \le 1$, and hence, by (4.9)

$$\begin{aligned}\sigma\phi(t) &= \frac{e^{i\rho\omega_k t}}{n\rho^{n-1}}\int_0^t \sum_{\alpha=0}^{n-2} a_\alpha(t)e^{-i\rho\omega_k(t-s)}\frac{\partial^\alpha K_{01}}{\partial t^\alpha}(t,s,\rho)\big[\psi(s) - \eta_k(s,\rho)\big]\,ds\\ &\quad + \frac{e^{i\rho\omega_k t}}{n\rho^{n-1}}\int_t^1 \sum_{\alpha=0}^{n-2} a_\alpha(t)e^{-i\rho\omega_k(t-s)}\frac{\partial^\alpha K_{02}}{\partial t^\alpha}(t,s,\rho)\big[\psi(s) - \eta_k(s,\rho)\big]\,ds\\ &= e^{i\rho\omega_k t}\,\psi(t)\end{aligned}$$

for $0 \le t \le 1$. This establishes the assertion. Also, from the definition of ϕ we have

$$\phi(t) = \mathcal{K}_{0\rho}\sigma\phi(t) - \mathcal{K}_{0\rho}\big(e^{i\rho\omega_k t}\eta_k(t,\rho)\big)$$

for $0 \le t \le 1$, and hence, ϕ is a solution of (4.8). By our earlier work it follows that the function u is a solution of the differential equation (2.1).

Summarizing, if $\psi \in C[0,1]$ is a solution of the integral equation (4.9), then the function

$$\phi(t) = \mathcal{K}_{0\rho}\big(e^{i\rho\omega_k t}[\psi(t) - \eta_k(t,\rho)]\big), \qquad 0 \le t \le 1, \tag{4.10}$$

belongs to $H^n[0,1]$ and is a solution of the integro-differential equation (4.8), $\sigma\phi(t) = e^{i\rho\omega_k t}\psi(t)$ for $0 \le t \le 1$, and the function

$$\begin{aligned} u(t) &= z_k(t,\rho) + \phi(t) \\ &= z_k(t,\rho) + \frac{e^{i\rho\omega_k t}}{n\rho^{n-1}} \int_0^t e^{-i\rho\omega_k(t-s)} K_{01}(t,s,\rho)[\psi(s) - \eta_k(s,\rho)]\,ds \\ &\quad + \frac{e^{i\rho\omega_k t}}{n\rho^{n-1}} \int_t^1 e^{-i\rho\omega_k(t-s)} K_{02}(t,s,\rho)[\psi(s) - \eta_k(s,\rho)]\,ds, \end{aligned} \tag{4.11}$$

$0 \le t \le 1$, belongs to $H^n[0,1]$ and is a solution of the differential equation (2.1). Moreover, the derivatives of u are given by

$$\begin{aligned} u^{(\alpha)}(t) &= z_k^{(\alpha)}(t,\rho) \\ &\quad + \frac{e^{i\rho\omega_k t}}{n\rho^{n-1}} \int_0^t e^{-i\rho\omega_k(t-s)} \frac{\partial^\alpha K_{01}}{\partial t^\alpha}(t,s,\rho)[\psi(s) - \eta_k(s,\rho)]\,ds \\ &\quad + \frac{e^{i\rho\omega_k t}}{n\rho^{n-1}} \int_t^1 e^{-i\rho\omega_k(t-s)} \frac{\partial^\alpha K_{02}}{\partial t^\alpha}(t,s,\rho)[\psi(s) - \eta_k(s,\rho)]\,ds \end{aligned} \tag{4.12}$$

for $0 \le t \le 1$ and for $\alpha = 0, 1, \ldots, n-1$. We conclude that finding solutions ϕ to equation (4.8) is equivalent to finding solutions ψ to equation (4.9). In view of this fact, the emphasis now shifts to solving the integral equation (4.9) for the function ψ. At this point the sector T_0 comes into play. We will show that equation (4.9) is uniquely solvable for all $\rho \in T_0$ with $|\rho|$ sufficiently large.

For fixed $\rho \ne 0$ in $\mathbb{C}$, let $\mathcal{A}_{0\rho}$ be the integral operator on $C[0,1]$ defined by

$$\mathcal{A}_{0\rho}u(t) := \int_0^1 g_0(t,s,\rho)u(s)\,ds, \qquad 0 \le t \le 1, \quad u \in C[0,1],$$

where the kernel is defined by

$$g_0(t,s,\rho) := \frac{1}{n}\sum_{\alpha=0}^{n-2} a_\alpha(t)\rho^{-(n-2)} e^{-i\rho\omega_k(t-s)} \frac{\partial^\alpha K_{01}}{\partial t^\alpha}(t,s,\rho), \quad 0 \le s < t \le 1,$$

$$g_0(t,s,\rho) := \frac{1}{n}\sum_{\alpha=0}^{n-2} a_\alpha(t)\rho^{-(n-2)} e^{-i\rho\omega_k(t-s)} \frac{\partial^\alpha K_{02}}{\partial t^\alpha}(t,s,\rho), \quad 0 \le t < s \le 1.$$

Then the integral equation (4.9) can be written in the operator form

$$\Big[I - \frac{1}{\rho}\mathcal{A}_{0\rho}\Big]\psi(t) = -\frac{1}{\rho}\mathcal{A}_{0\rho}\,\eta_k(t,\rho), \qquad 0 \le t \le 1, \tag{4.13}$$

in the setting of the Banach space $C[0,1]$.

LEMMA 4.2. *For any point $\rho \in T_0$ with $|\rho| \geq 1$,*

$$\left|\frac{\partial^\alpha K_{01}}{\partial t^\alpha}(t,s,\rho)\right| \leq C_1 n|\rho|^\alpha \left|e^{i\rho\omega_k(t-s)}\right|$$

for $0 \leq s < t \leq 1$ and for $\alpha = 0, 1, 2, \ldots$;

$$\left|\frac{\partial^\alpha K_{02}}{\partial t^\alpha}(t,s,\rho)\right| \leq C_1 n|\rho|^\alpha \left|e^{i\rho\omega_k(t-s)}\right|$$

for $0 \leq t < s \leq 1$ and for $\alpha = 0, 1, 2, \ldots$; and $|g_0(t,s,\rho)| \leq C_0 C_1 n := M_1$ for $t \neq s$ in $[0,1]$.

PROOF. Take any point $\rho \in T_0$ with $|\rho| \geq 1$. Then from (4.2) we have

$$(*) \qquad \left|\frac{\partial^\alpha K_{01}}{\partial t^\alpha}(t,s,\rho)\right| = \left|\sum_{j=0}^{\kappa}(i\omega_j^0)^{\alpha+1}\rho^\alpha e^{i\rho\omega_j^0(t-s)}\right| \leq n|\rho|^\alpha C_1\left|e^{i\rho\omega_k(t-s)}\right|$$

for $0 \leq s < t \leq 1$ and for $\alpha = 0, 1, 2, \ldots$. Similarly, by (4.3)

$$(**) \qquad \left|\frac{\partial^\alpha K_{02}}{\partial t^\alpha}(t,s,\rho)\right| = \left|\sum_{j=\kappa+1}^{n-1}(i\omega_j^0)^{\alpha+1}\rho^\alpha e^{i\rho\omega_j^0(t-s)}\right| \leq n|\rho|^\alpha C_1\left|e^{i\rho\omega_k(t-s)}\right|$$

for $0 \leq t < s \leq 1$ and for $\alpha = 0, 1, 2, \ldots$. The estimate for $|g_0(t,s,\rho)|$ follows immediately from $(*)$ and $(**)$. □

By the last lemma the norm of the operator $\mathcal{A}_{0\rho}$ satisfies the bound $\|\mathcal{A}_{0\rho}\| \leq M_1$ for all $\rho \in T_0$ with $|\rho| \geq 1$. Set $r_0 := \max\{1, M_1\} \geq 1$. Take any point $\rho \in T_0$ with $|\rho| > r_0$. Then $|\rho| > 1$, $|\rho| > M_1$, and $\|(1/\rho)\mathcal{A}_{0\rho}\| < 1$, and hence, in the operator space $\mathcal{B}(C[0,1])$ the operator $I - (1/\rho)\mathcal{A}_{0\rho}$ is invertible with its inverse given by the Neumann expansion

$$\left[I - \frac{1}{\rho}\mathcal{A}_{0\rho}\right]^{-1} = \sum_{j=0}^{\infty}\frac{1}{\rho^j}\mathcal{A}_{0\rho}^j .$$

It follows that the integral equation (4.9), or the equivalent operator equation (4.13), has a unique solution $\psi_{0k}(\,\cdot\,,\rho)$ in $C[0,1]$ given by

$$\psi_{0k}(t,\rho) = -\frac{1}{\rho}\left[I - \frac{1}{\rho}\mathcal{A}_{0\rho}\right]^{-1}\mathcal{A}_{0\rho}\,\eta_k(t,\rho) = -\sum_{j=0}^{\infty}\frac{1}{\rho^{j+1}}\mathcal{A}_{0\rho}^{j+1}\eta_k(t,\rho) \tag{4.14}$$

for $0 \leq t \leq 1$, where the series in (4.14) is converging in $C[0,1]$ under the supremum norm. In denoting the solution $\psi_{0k}(\,\cdot\,,\rho)$, we have now displayed the full dependence upon the integer k and the parameter ρ, with ρ restricted to belong to the sector T_0 with $|\rho| > r_0$.

Now use (4.11) to define the function $v_{0k}(\,\cdot\,,\rho) \in H^n[0,1]$ by

$$\begin{aligned} v_{0k}(t,\rho) &:= z_k(t,\rho) \\ &\quad + \frac{e^{i\rho\omega_k t}}{n\rho^{n-1}}\int_0^t e^{-i\rho\omega_k(t-s)}K_{01}(t,s,\rho)[\psi_{0k}(s,\rho) - \eta_k(s,\rho)]\,ds \\ &\quad + \frac{e^{i\rho\omega_k t}}{n\rho^{n-1}}\int_t^1 e^{-i\rho\omega_k(t-s)}K_{02}(t,s,\rho)[\psi_{0k}(s,\rho) - \eta_k(s,\rho)]\,ds \end{aligned} \tag{4.15}$$

for $0 \le t \le 1$ and for $\rho \in T_0$ with $|\rho| > r_0$. By our earlier work the function $v_{0k}(\,\cdot\,,\rho)$ is a solution of the differential equation (2.1), and by (4.12) the derivatives of $v_{0k}(\,\cdot\,,\rho)$ are given by

$$\begin{aligned} v_{0k}^{(\alpha)}(t,\rho) &= z_k^{(\alpha)}(t,\rho) \\ &+ \frac{e^{i\rho\omega_k t}}{n\rho^{n-1}} \int_0^t e^{-i\rho\omega_k(t-s)} \frac{\partial^\alpha K_{01}}{\partial t^\alpha}(t,s,\rho)[\psi_{0k}(s,\rho) - \eta_k(s,\rho)]\,ds \\ &+ \frac{e^{i\rho\omega_k t}}{n\rho^{n-1}} \int_t^1 e^{-i\rho\omega_k(t-s)} \frac{\partial^\alpha K_{02}}{\partial t^\alpha}(t,s,\rho)[\psi_{0k}(s,\rho) - \eta_k(s,\rho)]\,ds \end{aligned} \tag{4.16}$$

for $0 \le t \le 1$, for $\rho \in T_0$ with $|\rho| > r_0$, and for $\alpha = 0, 1, \dots, n-1$.

Let us take a closer look at the representation (4.16) of the solution $v_{0k}(\,\cdot\,,\rho)$ and its derivatives. For $\rho \ne 0$ in $\mathbb{C}$ and for $\alpha = 0, 1, \dots, n-1$, let

$$h_{0\alpha}(t,s,\rho) := \frac{1}{n\rho^{n-1}} e^{-i\rho\omega_k(t-s)} \frac{\partial^\alpha K_{01}}{\partial t^\alpha}(t,s,\rho) \quad \text{for } 0 \le s < t \le 1,$$

$$h_{0\alpha}(t,s,\rho) := \frac{1}{n\rho^{n-1}} e^{-i\rho\omega_k(t-s)} \frac{\partial^\alpha K_{02}}{\partial t^\alpha}(t,s,\rho) \quad \text{for } 0 \le t < s \le 1,$$

and let $\mathcal{B}_{0\alpha\rho}$ be the integral operator on $C[0,1]$ defined by

$$\mathcal{B}_{0\alpha\rho} u(t) := \int_0^1 h_{0\alpha}(t,s,\rho)u(s)\,ds, \qquad 0 \le t \le 1, \quad u \in C[0,1].$$

From the estimates in Lemma 4.2, we see immediately that the operator norm of $\mathcal{B}_{0\alpha\rho}$ satisfies the estimate

$$\|\mathcal{B}_{0\alpha\rho}\| \le C_1 |\rho|^{-(n-1-\alpha)}$$

for $\rho \in T_0$ with $|\rho| \ge 1$ and for $\alpha = 0, 1, \dots, n-1$. Also, since

$$\mathcal{B}_{0\alpha\rho} u(t) = e^{-i\rho\omega_k t} \frac{\partial^\alpha}{\partial t^\alpha} \mathcal{K}_{0\rho}[e^{i\rho\omega_k t} u(t)],$$

it follows that the operator $\mathcal{B}_{0\alpha\rho}$ maps $C[0,1]$ into $H^{n-\alpha}[0,1]$, and is continuous from the sup norm-structure to the $H^{n-\alpha}$-structure.

Now in terms of $\mathcal{B}_{0\alpha\rho}$ equation (4.16) can be expressed in the simpler form

$$v_{0k}^{(\alpha)}(t,\rho) = z_k^{(\alpha)}(t,\rho) + e^{i\rho\omega_k t} \mathcal{B}_{0\alpha\rho}[\psi_{0k}(t,\rho) - \eta_k(t,\rho)] \tag{4.17}$$

for $0 \le t \le 1$, for $\rho \in T_0$ with $|\rho| > r_0$, and for $\alpha = 0, 1, \dots, n-1$. Combining (4.17) and (4.14), we conclude that

$$v_{0k}^{(\alpha)}(t,\rho) = z_k^{(\alpha)}(t,\rho) - e^{i\rho\omega_k t} \sum_{j=0}^{\infty} \frac{1}{\rho^j} \mathcal{B}_{0\alpha\rho} \mathcal{A}_{0\rho}^j \eta_k(t,\rho) \tag{4.18}$$

for $0 \le t \le 1$, for $\rho \in T_0$ with $|\rho| > r_0$, and for $\alpha = 0, 1, \dots, n-1$, where the series converges in the space $H^{n-\alpha}[0,1]$ under the norm $|\ \ |_{H^{n-\alpha}}$.

For $\rho \in T_0$ with $|\rho| > r_0$ and for $\alpha = 0, 1, \dots, n-1$, set

$$E_{0k\alpha}(t,\rho) := \rho^{m-\alpha} \mathcal{B}_{0\alpha\rho}[\psi_{0k}(t,\rho) - \eta_k(t,\rho)], \qquad 0 \le t \le 1.$$

The function $E_{0k\alpha}(\,\cdot\,,\rho)$ belongs to $H^{n-\alpha}[0,1]$, and (4.17) can be rewritten as

$$v_{0k}^{(\alpha)}(t,\rho) = z_k^{(\alpha)}(t,\rho) + e^{i\rho\omega_k t} E_{0k\alpha}(t,\rho)\rho^{-m+\alpha} \tag{4.19}$$

for $0 \le t \le 1$, for $\rho \in T_0$ with $|\rho| > r_0$, and for $\alpha = 0, 1, \ldots, n-1$. To estimate the growth rate of the function $E_{0k\alpha}(\,\cdot\,, \rho)$, consider any point $\rho \in T_0$ with $|\rho| > 2r_0$. Then $\|(1/\rho)\mathcal{A}_{0\rho}\| \le M_1/|\rho| < 1/2$, and from (4.13) and (4.6) we have

$$\|\psi_{0k}(\,\cdot\,,\rho)\|_\infty \le \frac{1}{2}\,\|\psi_{0k}(\,\cdot\,,\rho)\|_\infty + \frac{1}{2}\,M_0|\rho|^{-(m+1-n)},$$

or

$$\|\psi_{0k}(\,\cdot\,,\rho)\|_\infty \le M_0|\rho|^{-(m+1-n)} \tag{4.20}$$

for $\rho \in T_0$ with $|\rho| > 2r_0$. Cf. equation (4.6). It follows that

$$\|E_{0k\alpha}(\,\cdot\,,\rho)\|_\infty \le |\rho|^{m-\alpha}\cdot C_1|\rho|^{-(n-1-\alpha)}\cdot 2M_0|\rho|^{-(m+1-n)} = 2C_1M_0,$$

or

$$\|E_{0k\alpha}(\,\cdot\,,\rho)\|_\infty \le 2C_1M_0 := \gamma_0 \qquad \text{(a constant)} \tag{4.21}$$

for $\rho \in T_0$ with $|\rho| > 2r_0$ and for $\alpha = 0, 1, \ldots, n-1$. In equations (4.19) and (4.21) we have established our asymptotic formulas for the solution $v_{0k}(\,\cdot\,,\rho)$ and its derivatives relative to the sector T_0.

Next, we apply the above construction to create a basis for the solution space of the differential equation (2.1). For $k = 0, 1, \ldots, n-1$ and $\rho \in T_0$ with $|\rho| > r_0$, let

$$v_{00}(\,\cdot\,,\rho), v_{01}(\,\cdot\,,\rho), \ldots, v_{0\,n-1}(\,\cdot\,,\rho) \in H^n[0,1]$$

be the solutions of the differential equation (2.1) constructed above. We assert that these solutions are linearly independent for $|\rho|$ sufficiently large. Indeed, from Chapter 2 the derivatives of the Birkhoff approximate solutions are given by

$$z_k^{(p)}(t,\rho) = \rho^p e^{i\rho\omega_k t}\sum_{\ell=0}^{p}\sum_{j=0}^{m-1}\alpha_{kp\ell}\,z_{kj}^{(\ell)}(t)\rho^{-\ell-j} := \rho^p e^{i\rho\omega_k t}\sum_{s=0}^{m+p-1} f_{kps}(t)\rho^{-s}$$

for $k, p = 0, 1, \ldots, n-1$, where $f_{kp0}(t) = \alpha_{kp0} = (i\omega_k)^p$. Combining this result with equation (4.19), we have

$$v_{0k}^{(\alpha)}(t,\rho) = \rho^\alpha e^{i\rho\omega_k t}\Big[(i\omega_k)^\alpha + \sum_{s=1}^{m+\alpha-1} f_{k\alpha s}(t)\rho^{-s} + E_{0k\alpha}(t,\rho)\rho^{-m}\Big]$$

for $0 \le t \le 1$, for $\rho \in T_0$ with $|\rho| > r_0$, and for $k, \alpha = 0, 1, \ldots, n-1$, where by the bound given in equation (4.21)

$$(i\omega_k)^\alpha + \sum_{s=1}^{m+\alpha-1} f_{k\alpha s}(t)\rho^{-s} + E_{0k\alpha}(t,\rho)\rho^{-m} \to (i\omega_k)^\alpha$$

uniformly on $[0,1]\times T_0$ as $|\rho| \to \infty$ for $k, \alpha = 0, 1, \ldots, n-1$. Since the Vandermonde matrix $\big((i\omega_k)^\alpha\big)$ is nonsingular, using the continuity of the determinant, we can choose a constant $R_0 \ge 2r_0$ such that the matrix

$$V(t,\rho) := \Big[(i\omega_k)^\alpha + \sum_{s=1}^{m+\alpha-1} f_{k\alpha s}(t)\rho^{-s} + E_{0k\alpha}(t,\rho)\rho^{-m}\Big]$$

is nonsingular for $0 \le t \le 1$ and for $\rho \in T_0$ with $|\rho| > R_0$. It follows that the Wronskian

$$W(v_{00}, v_{01}, \ldots, v_{0\,n-1})(t,\rho) = \rho^{n(n-1)/2}\det V(t,\rho)$$

is nonzero for $0 \le t \le 1$ and for $\rho \in T_0$ with $|\rho| > R_0$. This proves that the solutions $v_{0k}(\cdot, \rho)$, $k = 0, 1, \ldots, n-1$, are indeed linearly independent for $\rho \in T_0$ with $|\rho| > R_0$.

To develop the regularity properties of the functions $\psi_{0k}(t, \rho)$ and $v_{0k}(t, \rho)$, we will use Lemma 12.1 and Lemma 12.2 of the Appendix (Chapter 12). Again fix the integer k with $0 \le k \le n-1$. Set

$$G_0 := \{\rho \in \operatorname{Int} T_0 \mid |\rho| > R_0\},$$

an open set in the ρ plane. We know that

$$\eta_k(t, \rho) = -\frac{\rho^n}{i^n} \sum_{s=m+1}^{n+m-1} \left[\sum_{\ell+p+j=s} c_{k\ell p}(t) z_{kj}^{(\ell)}(t) \right] \rho^{-s}$$

for $0 \le t \le 1$ and $\rho \ne 0$ in $\mathbb{C}$, and hence, the function $\eta_k(t, \rho)$ is continuous on $[0, 1] \times G_0$, and $\frac{\partial}{\partial \rho} \eta_k(t, \rho)$ exists and is continuous on $[0, 1] \times G_0$. For $\rho \ne 0$ in $\mathbb{C}$ let ϕ_{01} and ϕ_{02} be the functions defined by

$$\phi_{01}(t, s, \rho) := \frac{1}{n} \sum_{\alpha=0}^{n-2} a_\alpha(t) \rho^{-(n-2)} e^{-i\rho\omega_k(t-s)} \frac{\partial^\alpha K_{01}}{\partial t^\alpha}(t, s, \rho)\, \eta_k(s, \rho),$$

$$\phi_{02}(t, s, \rho) := \frac{1}{n} \sum_{\alpha=0}^{n-2} a_\alpha(t) \rho^{-(n-2)} e^{-i\rho\omega_k(t-s)} \frac{\partial^\alpha K_{02}}{\partial t^\alpha}(t, s, \rho)\, \eta_k(s, \rho)$$

for $t, s \in [0, 1]$, and set

$$\phi_0(t, s, \rho) := \phi_{01}(t, s, \rho) \qquad \text{for } 0 \le s < t \le 1, \quad \rho \in G_0,$$
$$\phi_0(t, s, \rho) := \phi_{02}(t, s, \rho) \qquad \text{for } 0 \le t < s \le 1, \quad \rho \in G_0.$$

Clearly $\phi_0(t, s, \rho) = g_0(t, s, \rho)\eta_k(s, \rho)$ for $t \ne s$ in $[0, 1]$ and $\rho \in G_0$, and

$$\mathcal{A}_{0\rho}\eta_k(t, \rho) = \int_0^1 \phi_0(t, s, \rho)\, ds, \qquad 0 \le t \le 1, \quad \rho \in G_0.$$

From Lemma 12.1 we conclude that $\mathcal{A}_{0\rho}\eta_k(t, \rho)$ is continuous on $[0, 1] \times G_0$, and $\frac{\partial}{\partial \rho} [\mathcal{A}_{0\rho}\eta_k(t, \rho)]$ exists and is continuous on $[0, 1] \times G_0$. Proceeding by induction, for $j = 0, 1, 2, \ldots$ the function $\mathcal{A}_{0\rho}^j\eta_k(t, \rho)$ is continuous on $[0, 1] \times G_0$, and the derivative $\frac{\partial}{\partial \rho} [\mathcal{A}_{0\rho}^j\eta_k(t, \rho)]$ exists and is continuous on $[0, 1] \times G_0$.

Next, for each $\rho \in G_0$ let us examine the unique solution $\psi_{0k}(t, \rho)$ of the integral equation (4.9), where by equation (4.14)

$$\psi_{0k}(t, \rho) = -\sum_{j=1}^{\infty} \frac{1}{\rho^j} \mathcal{A}_{0\rho}^j \eta_k(t, \rho)$$

for $0 \le t \le 1$, with the series converging in $C[0, 1]$ for each $\rho \in G_0$. For the partial sums of the series,

$$S_{0N}(t, \rho) = -\sum_{j=1}^{N} \frac{1}{\rho^j} \mathcal{A}_{0\rho}^j \eta_k(t, \rho), \qquad N = 1, 2, \ldots,$$

it follows that $S_{0N}(t, \rho)$ is continuous on $[0, 1] \times G_0$, and $\frac{\partial}{\partial \rho} [S_{0N}(t, \rho)]$ exists and is continuous on $[0, 1] \times G_0$.

Since $\|\eta_k(\cdot,\rho)\|_\infty \le M_0|\rho|^{-(m+1-n)}$ for $\rho \in G_0$ by (4.6) and since $\|\mathcal{A}_{0\rho}\| \le M_1$ and $|\rho| > R_0 \ge 2M_1$ for $\rho \in G_0$, we have

$$\begin{aligned}\|\psi_{0k}(\cdot,\rho) - S_{0N}(\cdot,\rho)\|_\infty &= \Big\| \sum_{j=N+1}^{\infty} \frac{1}{\rho^j} \mathcal{A}_{0\rho}^j \eta_k(\cdot,\rho) \Big\|_\infty \\ &\le \sum_{j=N+1}^{\infty} \frac{1}{|\rho|^j} (M_1)^j \cdot M_0 |\rho|^{-(m+1-n)} \\ &\le M_0 (M_1)^{-(m+1-n)} \sum_{j=N+1}^{\infty} \frac{1}{2^{j+m+1-n}}\end{aligned}$$

for each $\rho \in G_0$, and hence,

$$\|\psi_{0k}(\cdot,\rho) - S_{0N}(\cdot,\rho)\|_\infty \le \frac{M_0(2M_1)^{-(m+1-n)}}{2^N} \tag{4.22}$$

for $\rho \in G_0$ and for $N = 1, 2, \ldots$. This result shows that the functions $S_{0N}(t,\rho)$ converge uniformly on $[0,1] \times G_0$ to $\psi_{0k}(t,\rho)$ as $N \to \infty$. From Lemma 12.2 we conclude that $\psi_{0k}(t,\rho)$ is continuous on $[0,1] \times G_0$, and the derivative $\frac{\partial}{\partial\rho}[\psi_{0k}(t,\rho)]$ exists and is continuous on $[0,1] \times G_0$. It is immediate that the function

$$\psi_{0k}(t,\rho) - \eta_k(t,\rho) = -\sum_{j=0}^{\infty} \frac{1}{\rho^j} \mathcal{A}_{0\rho}^j \eta_k(t,\rho)$$

is continuous on $[0,1] \times G_0$, and the derivative $\frac{\partial}{\partial\rho}[\psi_{0k}(t,\rho) - \eta_k(t,\rho)]$ exists and is continuous on $[0,1] \times G_0$.

Finally, consider the derivative functions $v_{0k}^{(\alpha)}(t,\rho)$ given by equation (4.16) for $0 \le t \le 1$, for $\rho \in G_0$, and for $\alpha = 0, 1, \ldots, n-1$. Again by Lemma 12.1 we see that $v_{0k}^{(\alpha)}(t,\rho)$ is continuous on $[0,1] \times G_0$, and the derivative $\frac{\partial}{\partial\rho}\left[v_{0k}^{(\alpha)}(t,\rho)\right]$ exists and is continuous on $[0,1] \times G_0$.

We summarize these results for the sector T_0 in the following theorem.

THEOREM 4.3. *Let n be even: $n = 2\nu$, let $T_0 = \{\rho - \tau_0 \mid \rho \in S_0\}$ be the translation of the sector S_0 selected in the last chapter, let m be the integer chosen in the last section with $m > n$, $m > p_0$, and $-(m - p_0 - 1) \le p \le p_0$, and for each $\rho \ne 0$ in $\mathbb{C}$ let*

$$z_k(t,\rho) = e^{i\rho\omega_k t} \sum_{j=0}^{m-1} z_{kj}(t)\rho^{-j}, \qquad k = 0, 1, \ldots, n-1,$$

be the Birkhoff approximate solutions of the differential equation (2.1) that are determined by Theorem 2.1. Then there exists a constant $R_0 > 0$ such that for $\rho \in T_0$ with $|\rho| > R_0$, there exist n functions $v_{00}(\cdot,\rho), v_{01}(\cdot,\rho), \ldots, v_{0\,n-1}(\cdot,\rho)$ in $H^n[0,1]$ that are linearly independent solutions of the differential equation (2.1), with

$$v_{0k}^{(\alpha)}(t,\rho) = z_k^{(\alpha)}(t,\rho) + e^{i\rho\omega_k t} E_{0k\alpha}(t,\rho)\rho^{-m+\alpha}$$

for $0 \le t \le 1$, for $\rho \in T_0$ with $|\rho| > R_0$, and for $k, \alpha = 0, 1, \ldots, n-1$. The function $E_{0k\alpha}(\cdot,\rho)$ belongs to $H^{n-\alpha}[0,1]$ with $|E_{0k\alpha}(t,\rho)| \le \gamma_0$ (a constant) for $0 \le t \le 1$, for $\rho \in T_0$ with $|\rho| > R_0$, and for $k, \alpha = 0, 1, \ldots, n-1$. For the open set $G_0 = \{\rho \in \operatorname{Int} T_0 \mid |\rho| > R_0\}$ and for $k, \alpha = 0, 1, \ldots, n-1$, the function $v_{0k}^{(\alpha)}(t,\rho)$ is *continuous on* $[0,1]\times G_0$, *and the derivative* $\frac{\partial}{\partial\rho}\left[v_{0k}^{(\alpha)}(t,\rho)\right]$ *exists and is continuous on* $[0,1]\times G_0$.

To construct solutions of the differential equation (2.1) on the sector T_1, we will employ the same procedure as used above for the sector T_0. Since the development is so similar, we simply sketch the results. In working on the sector T_1, we utilize the permutation $\omega_0^1, \omega_1^1, \ldots, \omega_{n-1}^1$ of $\omega_0, \omega_1, \ldots, \omega_{n-1}$ determined in Lemma 4.1. Fix any integer k with $0 \le k \le n-1$, and let κ be the unique integer satisfying $0 \le \kappa \le n-1$ and $\omega_\kappa^1 = \omega_k$. Let K_{11} and K_{12} be the functions defined by

$$K_{11}(t,s,\rho) := -\sum_{j=0}^{\kappa}(i\omega_j^1)e^{i\rho\omega_j^1(t-s)}, \quad K_{12}(t,s,\rho) := \sum_{j=\kappa+1}^{n-1}(i\omega_j^1)e^{i\rho\omega_j^1(t-s)}$$

for $t,s \in [0,1]$ and for $\rho \in \mathbb{C}$; let k_1 be the function defined by

$$k_1(t,s,\rho) := \frac{1}{n\rho^{n-1}}K_{11}(t,s,\rho), \qquad 0 \le s < t \le 1,$$

$$k_1(t,s,\rho) := \frac{1}{n\rho^{n-1}}K_{12}(t,s,\rho), \qquad 0 \le t < s \le 1,$$

for $\rho \ne 0$ in $\mathbb{C}$; and let $\mathcal{K}_{1\rho}$ be the integral operator on $L^2[0,1]$ defined by

$$\mathcal{K}_{1\rho}u(t) := \int_0^1 k_1(t,s,\rho)u(s)\,ds, \qquad 0 \le t \le 1, \quad u \in L^2[0,1],$$

for each $\rho \ne 0$ in $\mathbb{C}$. Again let $z_k(t,\rho)$ be the Birkhoff approximate solution, and let $\eta_k(t,\rho)$ be the associated residual function.

For $\rho \ne 0$ in $\mathbb{C}$ let $\mathcal{A}_{1\rho}$ be the integral operator on $C[0,1]$ defined by

$$\mathcal{A}_{1\rho}u(t) := \int_0^1 g_1(t,s,\rho)u(s)\,ds, \qquad 0 \le t \le 1, \quad u \in C[0,1],$$

where the kernel is defined by

$$g_1(t,s,\rho) := \frac{1}{n}\sum_{\alpha=0}^{n-2} a_\alpha(t)\rho^{-(n-2)}e^{-i\rho\omega_k(t-s)}\frac{\partial^\alpha K_{11}}{\partial t^\alpha}(t,s,\rho), \quad 0 \le s < t \le 1,$$

$$g_1(t,s,\rho) := \frac{1}{n}\sum_{\alpha=0}^{n-2} a_\alpha(t)\rho^{-(n-2)}e^{-i\rho\omega_k(t-s)}\frac{\partial^\alpha K_{12}}{\partial t^\alpha}(t,s,\rho), \quad 0 \le t < s \le 1.$$

Then for $\rho \in T_1$ with $|\rho| > r_0$, r_0 sufficiently large, the integral equation

$$[I - \frac{1}{\rho}\mathcal{A}_{1\rho}]\psi(t) = -\frac{1}{\rho}\mathcal{A}_{1\rho}\,\eta_k(t,\rho), \qquad 0 \le t \le 1, \tag{4.23}$$

has a unique solution $\psi_{1k}(\,\cdot\,,\rho)$ in $C[0,1]$ given by

$$\psi_{1k}(t,\rho) := -\frac{1}{\rho}\left[I - \frac{1}{\rho}\mathcal{A}_{1\rho}\right]^{-1}\mathcal{A}_{1\rho}\,\eta_k(t,\rho) = -\sum_{j=0}^{\infty}\frac{1}{\rho^{j+1}}\mathcal{A}_{1\rho}^{j+1}\eta_k(t,\rho) \tag{4.24}$$

for $0 \le t \le 1$, where the series is converging in $C[0,1]$. The function $v_{1k}(\,\cdot\,,\rho)$ in $H^n[0,1]$ defined by

$$v_{1k}(t,\rho) := z_k(t,\rho) + \mathcal{K}_{1\rho}(e^{i\rho\omega_k t}[\psi_{1k}(t,\rho) - \eta_k(t,\rho)]) \tag{4.25}$$

for $0 \le t \le 1$ and for $\rho \in T_1$ with $|\rho| > r_0$, is then a solution of the differential equation (2.1), with the derivatives of $v_{1k}(\cdot,\rho)$ given by

$$\begin{aligned} v_{1k}^{(\alpha)}(t,\rho) &= z_k^{(\alpha)}(t,\rho) \\ &\quad + \frac{e^{i\rho\omega_k t}}{n\rho^{n-1}} \int_0^t e^{-i\rho\omega_k(t-s)} \frac{\partial^\alpha K_{11}}{\partial t^\alpha}(t,s,\rho)[\psi_{1k}(s,\rho) - \eta_k(s,\rho)]\,ds \\ &\quad + \frac{e^{i\rho\omega_k t}}{n\rho^{n-1}} \int_t^1 e^{-i\rho\omega_k(t-s)} \frac{\partial^\alpha K_{12}}{\partial t^\alpha}(t,s,\rho)[\psi_{1k}(s,\rho) - \eta_k(s,\rho)]\,ds \end{aligned} \tag{4.26}$$

for $0 \le t \le 1$, for $\rho \in T_1$ with $|\rho| > r_0$, and for $\alpha = 0,1,\ldots,n-1$.

Carrying out this construction for $k = 0,1,\ldots,n-1$ and for $\rho \in T_1$ with $|\rho| > r_0$, we obtain the following basic theorem for the sector T_1.

THEOREM 4.4. *Let n be even: $n = 2\nu$, let $T_1 = \{\rho - \tau_1 \mid \rho \in S_1\}$ be the translation of the sector S_1 selected in the last chapter, let m be the integer chosen in the last section with $m > n$, $m > p_0$, and $-(m - p_0 - 1) \le p \le p_0$, and for each $\rho \ne 0$ in $\mathbb{C}$ let*

$$z_k(t,\rho) = e^{i\rho\omega_k t} \sum_{j=0}^{m-1} z_{kj}(t)\rho^{-j}, \qquad k = 0,1,\ldots,n-1,$$

be the Birkhoff approximate solutions of the differential equation (2.1) that are determined by Theorem 2.1. Then there exists a constant $R_0 > 0$ such that for $\rho \in T_1$ with $|\rho| > R_0$, there exist n functions $v_{10}(\cdot,\rho), v_{11}(\cdot,\rho),\ldots,v_{1\,n-1}(\cdot,\rho)$ in $H^n[0,1]$ that are linearly independent solutions of the differential equation (2.1), with

$$v_{1k}^{(\alpha)}(t,\rho) = z_k^{(\alpha)}(t,\rho) + e^{i\rho\omega_k t} E_{1k\alpha}(t,\rho)\rho^{-m+\alpha}$$

for $0 \le t \le 1$, for $\rho \in T_1$ with $|\rho| > R_0$, and for $k,\alpha = 0,1,\ldots,n-1$. The function $E_{1k\alpha}(\cdot,\rho)$ belongs to $H^{n-\alpha}[0,1]$ with $|E_{1k\alpha}(t,\rho)| \le \gamma_0$ (a constant) for $0 \le t \le 1$, for $\rho \in T_1$ with $|\rho| > R_0$, and for $k,\alpha = 0,1,\ldots,n-1$. For the open set $G_1 = \{\rho \in \operatorname{Int} T_1 \mid |\rho| > R_0\}$ and for $k,\alpha = 0,1,\ldots,n-1$, the function $v_{1k}^{(\alpha)}(t,\rho)$ is continuous on $[0,1] \times G_1$, and the derivative $\frac{\partial}{\partial\rho}\left[v_{1k}^{(\alpha)}(t,\rho)\right]$ exists and is continuous on $[0,1] \times G_1$.

4.2. Expansions for n Odd

In the last part of this chapter, we establish the asymptotic expansion of solutions to the differential equation (2.1) for the case n odd. The development here is identical to the case n even, and hence, we simply indicate the main features of the theory. Assume that n is odd: $n = 2\nu - 1 \ge 3$. For the sectors S_0, S_1 given by

$$\begin{aligned} S_0\colon&\ \text{all } \rho = |\rho|e^{i\theta} \in \mathbb{C} \text{ with } -\frac{\pi}{2n} \le \theta \le \frac{\pi}{2n}, \\ S_1\colon&\ \text{all } \rho = |\rho|e^{i\theta} \in \mathbb{C} \text{ with } \pi - \frac{\pi}{2n} \le \theta \le \pi + \frac{\pi}{2n}, \end{aligned}$$

we have the following analogue of Lemma 4.1.

LEMMA 4.5. *Let n be odd: $n = 2\nu - 1$. Then there exist permutations*

$$\omega_0^0, \omega_1^0, \ldots, \omega_{n-1}^0 \quad \textit{and} \quad \omega_0^1, \omega_1^1, \ldots, \omega_{n-1}^1$$

of the nth roots of unity $\omega_0, \omega_1, \dots, \omega_{n-1}$ such that

$$\begin{aligned}\operatorname{Re}(i\rho\omega_0^0) \le \operatorname{Re}(i\rho\omega_1^0) \le \cdots \le \operatorname{Re}(i\rho\omega_{n-1}^0) \quad &\textit{for all } \rho \in S_0,\\ \operatorname{Re}(i\rho\omega_0^1) \le \operatorname{Re}(i\rho\omega_1^1) \le \cdots \le \operatorname{Re}(i\rho\omega_{n-1}^1) \quad &\textit{for all } \rho \in S_1.\end{aligned}$$

Let $\omega_0^0, \omega_1^0, \dots, \omega_{n-1}^0$ and $\omega_0^1, \omega_1^1, \dots, \omega_{n-1}^1$ be the permutations of the nth roots of unity $\omega_0, \omega_1, \dots, \omega_{n-1}$ given in Lemma 4.5. We begin our expansion of solutions for ρ belonging to the translated sector T_0. Fix any integer k with $0 \le k \le n-1$, and let κ be the unique integer satisfying $0 \le \kappa \le n-1$ and $\omega_\kappa^0 = \omega_k$. Let K_{01} and K_{02} be the functions defined by

$$K_{01}(t,s,\rho) := -\sum_{j=0}^{\kappa} (i\omega_j^0) e^{i\rho\omega_j^0(t-s)}, \quad K_{02}(t,s,\rho) := \sum_{j=\kappa+1}^{n-1} (i\omega_j^0) e^{i\rho\omega_j^0(t-s)}$$

for $t, s \in [0,1]$ and for $\rho \in \mathbb{C}$; let k_0 be the function defined by

$$\begin{aligned} k_0(t,s,\rho) &:= \frac{1}{n\rho^{n-1}} K_{01}(t,s,\rho), \qquad 0 \le s < t \le 1,\\ k_0(t,s,\rho) &:= \frac{1}{n\rho^{n-1}} K_{02}(t,s,\rho), \qquad 0 \le t < s \le 1,\end{aligned}$$

for $\rho \ne 0$ in $\mathbb{C}$; and let $\mathcal{K}_{0\rho}$ be the integral operator on $L^2[0,1]$ defined by

$$\mathcal{K}_{0\rho} u(t) := \int_0^1 k_0(t,s,\rho) u(s)\, ds, \qquad 0 \le t \le 1, \quad u \in L^2[0,1],$$

for each $\rho \ne 0$ in $\mathbb{C}$. The operator $\mathcal{K}_{0\rho}$ has the same properties as its earlier version for the case n even. For the fixed integer k, we form the Birkhoff approximate solution $z_k(t,\rho)$ and the corresponding residual function $\eta_k(t,\rho)$, with

$$(\rho^n I - \ell) z_k(t,\rho) = e^{i\rho\omega_k t} \eta_k(t,\rho) \tag{4.27}$$

for $0 \le t \le 1$ and $\rho \ne 0$ in $\mathbb{C}$, and with

$$|\eta_k(t,\rho)| \le M_0 |\rho|^{-(m+1-n)} \tag{4.28}$$

for $0 \le t \le 1$ and for $\rho \in \mathbb{C}$ with $|\rho| \ge 1$.

Fix any point $\rho \ne 0$ in $\mathbb{C}$. A function $u(t)$ belonging to $H^n[0,1]$ is a solution of the differential equation (2.1) if and only if the function $\phi(t) = u(t) - z_k(t,\rho)$ is a solution of the integro-differential equation

$$\phi(t) = \mathcal{K}_{0\rho}\sigma\phi(t) - \mathcal{K}_{0\rho}\big(e^{i\rho\omega_k t}\eta_k(t,\rho)\big) + \sum_{j=0}^{n-1} c_j e^{i\rho\omega_j t}, \qquad 0 \le t \le 1, \tag{4.29}$$

for some constants $c_0, c_1, \dots, c_{n-1}$. Consider the special case of equation (4.29) where all the constants c_j are equal to zero:

$$\phi(t) = \mathcal{K}_{0\rho}\sigma\phi(t) - \mathcal{K}_{0\rho}\big(e^{i\rho\omega_k t}\eta_k(t,\rho)\big), \qquad 0 \le t \le 1. \tag{4.30}$$

If $\phi(t)$ is a function in $H^n[0,1]$ that is a solution of equation (4.30), then the function $\psi(t) = e^{-i\rho\omega_k t}\sigma\phi(t)$ belonging to $C[0,1]$ is a solution of the integral equation

(4.31)
$$\psi(t) = \frac{1}{n\rho^{n-1}} \int_0^t \sum_{\alpha=0}^{n-2} a_\alpha(t) e^{-i\rho\omega_k(t-s)} \frac{\partial^\alpha K_{01}}{\partial t^\alpha}(t,s,\rho)\big[\psi(s) - \eta_k(s,\rho)\big]\,ds$$
$$+ \frac{1}{n\rho^{n-1}} \int_t^1 \sum_{\alpha=0}^{n-2} a_\alpha(t) e^{-i\rho\omega_k(t-s)} \frac{\partial^\alpha K_{02}}{\partial t^\alpha}(t,s,\rho)\big[\psi(s) - \eta_k(s,\rho)\big]\,ds,$$

$0 \le t \le 1$. Conversely, if $\psi(t)$ is a function in $C[0,1]$ that is a solution of the integral equation (4.31), then the function $\phi(t) = \mathcal{K}_{0\rho}\big(e^{i\rho\omega_k t}[\psi(t) - \eta_k(t,\rho)]\big)$ is a solution of equation (4.30), and the function

$$u(t) = z_k(t,\rho) + \phi(t) = z_k(t,\rho) + \mathcal{K}_{0\rho}\big(e^{i\rho\omega_k t}[\psi(t) - \eta_k(t,\rho)]\big)$$
$$(4.32) \qquad = z_k(t,\rho) + \frac{e^{i\rho\omega_k t}}{n\rho^{n-1}} \int_0^t e^{-i\rho\omega_k(t-s)} K_{01}(t,s,\rho)[\psi(s) - \eta_k(s,\rho)]\,ds$$
$$+ \frac{e^{i\rho\omega_k t}}{n\rho^{n-1}} \int_t^1 e^{-i\rho\omega_k(t-s)} K_{02}(t,s,\rho)[\psi(s) - \eta_k(s,\rho)]\,ds,$$

$0 \le t \le 1$, is a solution of the differential equation (2.1), where the derivatives of $u(t)$ are given by

$$u^{(\alpha)}(t) = z_k^{(\alpha)}(t,\rho)$$
$$(4.33) \qquad + \frac{e^{i\rho\omega_k t}}{n\rho^{n-1}} \int_0^t e^{-i\rho\omega_k(t-s)} \frac{\partial^\alpha K_{01}}{\partial t^\alpha}(t,s,\rho)[\psi(s) - \eta_k(s,\rho)]\,ds$$
$$+ \frac{e^{i\rho\omega_k t}}{n\rho^{n-1}} \int_t^1 e^{-i\rho\omega_k(t-s)} \frac{\partial^\alpha K_{02}}{\partial t^\alpha}(t,s,\rho)[\psi(s) - \eta_k(s,\rho)]\,ds$$

for $0 \le t \le 1$ and for $\alpha = 0, 1, \ldots, n-1$. It is in solving equation (4.31) that the sector T_0 comes into play.

For each $\rho \neq 0$ in $\mathbb{C}$, let $\mathcal{A}_{0\rho}$ be the integral operator on $C[0,1]$ defined by

$$\mathcal{A}_{0\rho} u(t) := \int_0^1 g_0(t,s,\rho) u(s)\,ds, \qquad 0 \le t \le 1, \quad u \in C[0,1],$$

where the kernel is defined by

$$g_0(t,s,\rho) := \frac{1}{n} \sum_{\alpha=0}^{n-2} a_\alpha(t) \rho^{-(n-2)} e^{-i\rho\omega_k(t-s)} \frac{\partial^\alpha K_{01}}{\partial t^\alpha}(t,s,\rho), \quad 0 \le s < t \le 1,$$
$$g_0(t,s,\rho) := \frac{1}{n} \sum_{\alpha=0}^{n-2} a_\alpha(t) \rho^{-(n-2)} e^{-i\rho\omega_k(t-s)} \frac{\partial^\alpha K_{02}}{\partial t^\alpha}(t,s,\rho), \quad 0 \le t < s \le 1.$$

Then the integral equation (4.31) can be written in the operator form

$$(4.34) \qquad [I - \frac{1}{\rho}\mathcal{A}_{0\rho}]\psi(t) = -\frac{1}{\rho}\mathcal{A}_{0\rho}\,\eta_k(t,\rho), \qquad 0 \le t \le 1,$$

in the setting of the Banach space $C[0,1]$. It follows that for $\rho \in T_0$ with $|\rho| > r_0$, r_0 sufficiently large, the integral equation (4.31), or the equivalent operator equation

(4.34), has a unique solution $\psi_{0k}(\cdot,\rho)$ in $C[0,1]$ given by

$$\psi_{0k}(t,\rho) := -\frac{1}{\rho}\Big[I - \frac{1}{\rho}\mathcal{A}_{0\rho}\Big]^{-1}\mathcal{A}_{0\rho}\,\eta_k(t,\rho) = -\sum_{j=0}^{\infty}\frac{1}{\rho^{j+1}}\mathcal{A}_{0\rho}^{j+1}\eta_k(t,\rho) \tag{4.35}$$

for $0 \le t \le 1$, where the series in (4.35) is converging in $C[0,1]$ under the supremum norm. Now use (4.32) to define the function $v_{0k}(\cdot,\rho) \in H^n[0,1]$ by

$$\begin{aligned} v_{0k}(t,\rho) := {} & z_k(t,\rho) \\ & + \frac{e^{i\rho\omega_k t}}{n\rho^{n-1}}\int_0^t e^{-i\rho\omega_k(t-s)}K_{01}(t,s,\rho)[\psi_{0k}(s,\rho)-\eta_k(s,\rho)]\,ds \\ & + \frac{e^{i\rho\omega_k t}}{n\rho^{n-1}}\int_t^1 e^{-i\rho\omega_k(t-s)}K_{02}(t,s,\rho)[\psi_{0k}(s,\rho)-\eta_k(s,\rho)]\,ds \end{aligned} \tag{4.36}$$

for $0 \le t \le 1$ and for $\rho \in T_0$ with $|\rho| > r_0$. The function $v_{0k}(\cdot,\rho)$ is a solution of the differential equation (2.1), and the derivatives of $v_{0k}(\cdot,\rho)$ are given by

$$\begin{aligned} v_{0k}^{(\alpha)}(t,\rho) = {} & z_k^{(\alpha)}(t,\rho) \\ & + \frac{e^{i\rho\omega_k t}}{n\rho^{n-1}}\int_0^t e^{-i\rho\omega_k(t-s)}\frac{\partial^\alpha K_{01}}{\partial t^\alpha}(t,s,\rho)[\psi_{0k}(s,\rho)-\eta_k(s,\rho)]\,ds \\ & + \frac{e^{i\rho\omega_k t}}{n\rho^{n-1}}\int_t^1 e^{-i\rho\omega_k(t-s)}\frac{\partial^\alpha K_{02}}{\partial t^\alpha}(t,s,\rho)[\psi_{0k}(s,\rho)-\eta_k(s,\rho)]\,ds \end{aligned} \tag{4.37}$$

for $0 \le t \le 1$, for $\rho \in T_0$ with $|\rho| > r_0$, and for $\alpha = 0,1,\ldots,n-1$.

Repeating the above construction for $k = 0,1,\ldots,n-1$ and for $\rho \in T_0$ with $|\rho| > r_0$, we produce solutions

$$v_{00}(\cdot,\rho), v_{01}(\cdot,\rho), \ldots, v_{0\,n-1}(\cdot,\rho) \in H^n[0,1]$$

of the differential equation (2.1). These solutions are linearly independent for $\rho \in T_0$ with $|\rho|$ sufficiently large, and the regularity properties of the solutions $v_{0k}(t,\rho)$ then follow from the lemmas in the Appendix (Chapter 12).

These results for the sector T_0 are collected below in a basic theorem.

THEOREM 4.6. *Let n be odd: $n = 2\nu - 1$, let $T_0 = \{\rho - \tau_0 \mid \rho \in S_0\}$ be the translation of the sector S_0 selected in the last chapter, let m be the integer chosen in the last chapter with $m > n$, $m > p_0$, and $-(m - p_0 - 1) \le p, q \le p_0$, and for each $\rho \ne 0$ in $\mathbb{C}$ let*

$$z_k(t,\rho) = e^{i\rho\omega_k t}\sum_{j=0}^{m-1} z_{kj}(t)\rho^{-j}, \qquad k = 0,1,\ldots,n-1,$$

be the Birkhoff approximate solutions of the differential equation (2.1) that are determined by Theorem 2.1. Then there exists a constant $R_0 > 0$ such that for $\rho \in T_0$ with $|\rho| > R_0$, there exist n functions $v_{00}(\cdot,\rho), v_{01}(\cdot,\rho),\ldots,v_{0\,n-1}(\cdot,\rho)$ in $H^n[0,1]$ that are linearly independent solutions of the differential equation (2.1), with

$$v_{0k}^{(\alpha)}(t,\rho) = z_k^{(\alpha)}(t,\rho) + e^{i\rho\omega_k t}E_{0k\alpha}(t,\rho)\rho^{-m+\alpha}$$

for $0 \le t \le 1$, for $\rho \in T_0$ with $|\rho| > R_0$, and for $k,\alpha = 0,1,\ldots,n-1$. The function $E_{0k\alpha}(\cdot,\rho)$ belongs to $H^{n-\alpha}[0,1]$ with $|E_{0k\alpha}(t,\rho)| \le \gamma_0$ (a constant) for $0 \le t \le 1$, for $\rho \in T_0$ with $|\rho| > R_0$, and for $k,\alpha = 0,1,\ldots,n-1$. For the open set $G_0 = \{\rho \in \operatorname{Int} T_0 \mid |\rho| > R_0\}$ and for $k,\alpha = 0,1,\ldots,n-1$, the function $v_{0k}^{(\alpha)}(t,\rho)$ is *continuous on* $[0,1]\times G_0$, *and the derivative* $\frac{\partial}{\partial\rho}\left[v_{0k}^{(\alpha)}(t,\rho)\right]$ *exists and is continuous on* $[0,1]\times G_0$.

To obtain solutions of the differential equation (2.1) for ρ belonging to the sector T_1, we use the same argument as used earlier for the case n even. Indeed, fix any integer k with $0\le k\le n-1$, and let κ be the unique integer satisfying $0\le\kappa\le n-1$ and $\omega_\kappa^1=\omega_k$. The key to this work is to introduce the integral operators $\mathcal{K}_{1\rho}$ and $\mathcal{A}_{1\rho}$. First, let K_{11} and K_{12} be the functions defined by

$$K_{11}(t,s,\rho):=-\sum_{j=0}^{\kappa}(i\omega_j^1)e^{i\rho\omega_j^1(t-s)},\quad K_{12}(t,s,\rho):=\sum_{j=\kappa+1}^{n-1}(i\omega_j^1)e^{i\rho\omega_j^1(t-s)}$$

for $t,s\in[0,1]$ and for $\rho\in\mathbb{C}$; let k_1 be the function defined by

$$k_1(t,s,\rho):=\frac{1}{n\rho^{n-1}}K_{11}(t,s,\rho),\qquad 0\le s<t\le 1,$$
$$k_1(t,s,\rho):=\frac{1}{n\rho^{n-1}}K_{12}(t,s,\rho),\qquad 0\le t<s\le 1,$$

for $\rho\ne 0$ in $\mathbb{C}$; and let $\mathcal{K}_{1\rho}$ be the integral operator on $L^2[0,1]$ defined by

$$\mathcal{K}_{1\rho}u(t):=\int_0^1 k_1(t,s,\rho)u(s)\,ds,\qquad 0\le t\le 1,\quad u\in L^2[0,1],$$

for each $\rho\ne 0$ in $\mathbb{C}$. Second, for each $\rho\ne 0$ in $\mathbb{C}$, let $\mathcal{A}_{1\rho}$ be the integral operator on $C[0,1]$ defined by

$$\mathcal{A}_{1\rho}u(t):=\int_0^1 g_1(t,s,\rho)u(s)\,ds,\qquad 0\le t\le 1,\quad u\in C[0,1],$$

where the kernel is defined by

$$g_1(t,s,\rho):=\frac{1}{n}\sum_{\alpha=0}^{n-2}a_\alpha(t)\rho^{-(n-2)}e^{-i\rho\omega_k(t-s)}\frac{\partial^\alpha K_{11}}{\partial t^\alpha}(t,s,\rho),\quad 0\le s<t\le 1,$$
$$g_1(t,s,\rho):=\frac{1}{n}\sum_{\alpha=0}^{n-2}a_\alpha(t)\rho^{-(n-2)}e^{-i\rho\omega_k(t-s)}\frac{\partial^\alpha K_{12}}{\partial t^\alpha}(t,s,\rho),\quad 0\le t<s\le 1.$$

Then for $\rho\in T_1$ with $|\rho|>r_0$, r_0 sufficiently large, the integral equation

$$[I-\frac{1}{\rho}\mathcal{A}_{1\rho}]\psi(t)=-\frac{1}{\rho}\mathcal{A}_{1\rho}\eta_k(t,\rho),\qquad 0\le t\le 1,\tag{4.38}$$

has a unique solution $\psi_{1k}(\,\cdot\,,\rho)$ in $C[0,1]$ given by

$$\psi_{1k}(t,\rho):=-\frac{1}{\rho}\left[I-\frac{1}{\rho}\mathcal{A}_{1\rho}\right]^{-1}\mathcal{A}_{1\rho}\eta_k(t,\rho)=-\sum_{j=0}^{\infty}\frac{1}{\rho^{j+1}}\mathcal{A}_{1\rho}^{j+1}\eta_k(t,\rho)\tag{4.39}$$

for $0\le t\le 1$, where the series is converging in $C[0,1]$. The function $v_{1k}(\,\cdot\,,\rho)$ in $H^n[0,1]$ defined by

$$v_{1k}(t,\rho):=z_k(t,\rho)+\mathcal{K}_{1\rho}\big(e^{i\rho\omega_\kappa t}[\psi_{1k}(t,\rho)-\eta_k(t,\rho)]\big)\tag{4.40}$$

for $0 \le t \le 1$ and for $\rho \in T_1$ with $|\rho| > r_0$, is then a solution of the differential equation (2.1), with the derivatives of $v_{1k}(\,\cdot\,,\rho)$ given by

$$
\begin{aligned}
v_{1k}^{(\alpha)}(t,\rho) &= z_k^{(\alpha)}(t,\rho) \\
&\quad + \frac{e^{i\rho\omega_k t}}{n\rho^{n-1}} \int_0^t e^{-i\rho\omega_k(t-s)} \frac{\partial^\alpha K_{11}}{\partial t^\alpha}(t,s,\rho)[\psi_{1k}(s,\rho) - \eta_k(s,\rho)]\,ds \\
&\quad + \frac{e^{i\rho\omega_k t}}{n\rho^{n-1}} \int_t^1 e^{-i\rho\omega_k(t-s)} \frac{\partial^\alpha K_{12}}{\partial t^\alpha}(t,s,\rho)[\psi_{1k}(s,\rho) - \eta_k(s,\rho)]\,ds
\end{aligned}
\tag{4.41}
$$

for $0 \le t \le 1$, for $\rho \in T_1$ with $|\rho| > r_0$, and for $\alpha = 0, 1, \dots, n-1$.

Carrying out this construction for $k = 0, 1, \dots, n-1$ and for $\rho \in T_1$ with $|\rho| > r_0$, we are lead to the following basic theorem for the sector T_1.

THEOREM 4.7. *Let n be odd: $n = 2\nu - 1$, let $T_1 = \{\rho - \tau_1 \mid \rho \in S_1\}$ be the translation of the sector S_1 selected in the last chapter, let m be the integer chosen in the last chapter with $m > n$, $m > p_0$, and $-(m - p_0 - 1) \le p, q \le p_0$, and for each $\rho \ne 0$ in $\mathbb{C}$ let*

$$z_k(t,\rho) = e^{i\rho\omega_k t} \sum_{j=0}^{m-1} z_{kj}(t)\rho^{-j}, \qquad k = 0, 1, \dots, n-1,$$

be the Birkhoff approximate solutions of the differential equation (2.1) that are determined by Theorem 2.1. Then there exists a constant $R_0 > 0$ such that for $\rho \in T_1$ with $|\rho| > R_0$, there exist n functions $v_{10}(\,\cdot\,,\rho), v_{11}(\,\cdot\,,\rho), \dots, v_{1\,n-1}(\,\cdot\,,\rho)$ in $H^n[0,1]$ that are linearly independent solutions of the differential equation (2.1), with

$$v_{1k}^{(\alpha)}(t,\rho) = z_k^{(\alpha)}(t,\rho) + e^{i\rho\omega_k t} E_{1k\alpha}(t,\rho)\rho^{-m+\alpha}$$

for $0 \le t \le 1$, for $\rho \in T_1$ with $|\rho| > R_0$, and for $k, \alpha = 0, 1, \dots, n-1$. The function $E_{1k\alpha}(\,\cdot\,,\rho)$ belongs to $H^{n-\alpha}[0,1]$ with $|E_{1k\alpha}(t,\rho)| \le \gamma_0$ (a constant) for $0 \le t \le 1$, for $\rho \in T_1$ with $|\rho| > R_0$, and for $k, \alpha = 0, 1, \dots, n-1$. For the open set $G_1 = \{\rho \in \operatorname{Int} T_1 \mid |\rho| > R_0\}$ and for $k, \alpha = 0, 1, \dots, n-1$, the function $v_{1k}^{(\alpha)}(t,\rho)$ is continuous on $[0,1] \times G_1$, and the derivative $\frac{\partial}{\partial\rho}\left[v_{1k}^{(\alpha)}(t,\rho)\right]$ exists and is continuous on $[0,1] \times G_1$.

CHAPTER 5

The Characteristic Determinant

Now that we have constructed independent solutions of the differential equation (2.1) on the sectors T_0 and T_1, the next step is to develop the characteristic determinant on T_0 and T_1.

5.1. The Characteristic Determinant for n Even

Assume that n is even: $n = 2\nu \geq 2$. Consider the sectors

$$S_0: \text{ all } \rho = |\rho| e^{i\theta} \in \mathbb{C} \text{ with } 0 \leq \theta \leq \frac{\pi}{n},$$
$$S_1: \text{ all } \rho = |\rho| e^{i\theta} \in \mathbb{C} \text{ with } -\frac{\pi}{n} \leq \theta \leq 0,$$

and the corresponding translated sectors T_0, T_1. For each point $\rho \in T_0$ with $|\rho| > R_0$ let $v_{00}(\cdot, \rho), v_{01}(\cdot, \rho), \ldots, v_{0\,n-1}(\cdot, \rho)$ be the n linearly independent solutions of the differential equation (2.1) constructed in Theorem 4.3. For fixed ρ the function $v_{0k}(\cdot, \rho)$ belongs to $H^n[0,1]$, and $v_{0k}(\cdot, \rho)$ and its derivatives have the asymptotic expansions

$$v_{0k}^{(\eta)}(t, \rho) = z_k^{(\eta)}(t, \rho) + e^{i\rho\omega_k t} E_{0k\eta}(t, \rho) \rho^{-m+\eta} \tag{5.1}$$

for $0 \leq t \leq 1$, for $\rho \in T_0$ with $|\rho| > R_0$, and for $k, \eta = 0, 1, \ldots, n-1$. In this representation the function $z_k(t, \rho) = e^{i\rho\omega_k t} \sum_{j=0}^{m-1} z_{kj}(t) \rho^{-j}$ is the Birkhoff approximate solution of (2.1) constructed in Chapter 2, and the function $E_{0k\eta}(\cdot, \rho)$ belongs to $H^{n-\eta}[0,1]$ and satisfies the bound $|E_{0k\eta}(t, \rho)| \leq \gamma_0$ for $0 \leq t \leq 1$, for $\rho \in T_0$ with $|\rho| > R_0$, and for $k, \eta = 0, 1, \ldots, n-1$.

Similarly, for each $\rho \in T_1$ with $|\rho| > R_0$ let $v_{10}(\cdot, \rho), v_{11}(\cdot, \rho), \ldots, v_{1\,n-1}(\cdot, \rho)$ be the n linearly independent solutions of the differential equation (2.1) constructed in Theorem 4.4. Again the function $v_{1k}(\cdot, \rho)$ belongs to $H^n[0,1]$ with

$$v_{1k}^{(\eta)}(t, \rho) = z_k^{(\eta)}(t, \rho) + e^{i\rho\omega_k t} E_{1k\eta}(t, \rho) \rho^{-m+\eta} \tag{5.2}$$

for $0 \leq t \leq 1$, for $\rho \in T_1$ with $|\rho| > R_0$, and for $k, \eta = 0, 1, \ldots, n-1$. Here the function $E_{1k\eta}(\cdot, \rho)$ belongs to the space $H^{n-\eta}[0,1]$ and satisfies the bound $|E_{1k\eta}(t, \rho)| \leq \gamma_0$ for $0 \leq t \leq 1$, for $\rho \in T_1$ with $|\rho| > R_0$, and for $k, \eta = 0, 1, \ldots, n-1$.

We will use these functions to construct the characteristic determinant of the differential operator L. The first set is used for the construction on the sector T_0, and the second for the construction on T_1. For the sector T_0 we develop the theory in detail. Since the development for the sector T_1 is similar, we simply sketch the theory for T_1.

In Chapter 4 we showed that

$$(5.3)\qquad \begin{aligned} v_{0k}^{(\eta)}(t,\rho) &= \rho^{\eta} e^{i\rho\omega_k t}\Big[(i\omega_k)^{\eta} + \sum_{\ell=1}^{m+\eta-1} f_{k\eta\ell}(t)\rho^{-\ell} + E_{0k\eta}(t,\rho)\rho^{-m}\Big] \\ &:= \rho^{\eta} e^{i\rho\omega_k t}\big[(i\omega_k)^{\eta} + F_{0k\eta}(t,\rho)\big] \end{aligned}$$

for $0 \le t \le 1$, for $\rho \in T_0$ with $|\rho| > R_0$, and for $k, \eta = 0, 1, \dots, n-1$. The function $F_{0k\eta}(\,\cdot\,,\rho)$ belongs to $H^{n-\eta}[0,1]$ with $F_{0k\eta}(t,\rho) \to 0$ uniformly on $[0,1] \times T_0$ as $|\rho| \to \infty$ for $k, \eta = 0, 1, \dots, n-1$. Choose a constant $R_1 \ge R_0$ such that

$$(5.4)\qquad |F_{0k\eta}(t,\rho)| \le 1$$

for $0 \le t \le 1$, for $\rho \in T_0$ with $|\rho| > R_1$, and for $k, \eta = 0, 1, \dots, n-1$.

Relative to the sector T_0, we form the *modified solutions* of the differential equation (2.1):

$$u_{0k}(t,\rho) := v_{0k}(t,\rho) = y_k(t,\rho) + e^{i\rho\omega_k t} E_{0k0}(t,\rho)\rho^{-m}, \qquad k = 0, 1, \dots, \nu-1,$$

$$u_{0k}(t,\rho) := e^{-i\rho\omega_k} v_{0k}(t,\rho) = y_k(t,\rho) + e^{i\rho\omega_k (t-1)} E_{0k0}(t,\rho)\rho^{-m}, \qquad k = \nu, \dots, n-1,$$

for $0 \le t \le 1$ and for $\rho \in T_0$ with $|\rho| > R_0$. Cf. the definitions of the functions $y_k(\,\cdot\,,\rho)$ given at the beginning of Chapter 3. Clearly these functions form a basis for the solution space of the differential equation (2.1) for $\rho \in T_0$ with $|\rho| > R_0$. Using (5.1) and (5.3), the derivatives of the solutions $u_{0k}(\,\cdot\,,\rho)$ can be expressed as

$$(5.5)\qquad \begin{aligned} u_{0k}^{(\eta)}(t,\rho) &= y_k^{(\eta)}(t,\rho) + e^{i\rho\omega_k t} E_{0k\eta}(t,\rho)\rho^{-m+\eta} \\ &= \rho^{\eta} e^{i\rho\omega_k t}\big[(i\omega_k)^{\eta} + F_{0k\eta}(t,\rho)\big], \qquad k = 0, 1, \dots, \nu-1, \\ u_{0k}^{(\eta)}(t,\rho) &= y_k^{(\eta)}(t,\rho) + e^{i\rho\omega_k (t-1)} E_{0k\eta}(t,\rho)\rho^{-m+\eta} \\ &= \rho^{\eta} e^{i\rho\omega_k (t-1)}\big[(i\omega_k)^{\eta} + F_{0k\eta}(t,\rho)\big], \quad k = \nu, \dots, n-1, \end{aligned}$$

for $0 \le t \le 1$, for $\rho \in T_0$ with $|\rho| > R_0$, and for $\eta = 0, 1, \dots, n-1$. Applying the regularity results of Theorem 4.3, for fixed $t \in [0,1]$ the functions $u_{0k}^{(\eta)}(t,\rho)$ and $F_{0k\eta}(t,\rho)$ are analytic functions of the ρ variable on the open set

$$G_0 = \{\rho \in \operatorname{Int} T_0 \mid |\rho| > R_0\}.$$

It follows that the boundary data $u_{0k}^{(\eta)}(0,\rho)$, $u_{0k}^{(\eta)}(1,\rho)$, $k, \eta = 0, 1, \dots, n-1$, consists of functions of ρ that are analytic on G_0.

Next, we establish various bounds and growth rates relative to the sector T_0 for the functions appearing above. Fix a real number σ_0 with $0 < \sigma_0 < \pi/10$, set $\alpha := \sin(\sigma_0/n) > 0$, and then form the sector

$$\Sigma\colon\ \text{all } \rho = |\rho| e^{i\theta} \in \mathbb{C} \text{ with } -\frac{2\pi}{n} + \frac{\sigma_0}{n} \le \theta \le \frac{2\pi}{n} - \frac{\sigma_0}{n}$$

in the ρ plane. The reason for the constant $\pi/10$ is that eventually we will apply the basic completeness theorem [**34**, p. 80] in which we need five rays with the angles between adjacent rays being less than $\pi/2$ (see Chapter 9). Clearly $-2\pi/n + \sigma_0/n < -\pi/n < \pi/n < 2\pi/n - \sigma_0/n$, and hence, any ρ in $T_0 \cup T_1$ with $|\rho|$ sufficiently large lies in the sector Σ. Without loss of generality we can assume that the constant $R_0 > 0$

chosen earlier (see Theorem 4.3 and Theorem 4.4) has the additional property that $\rho \in T_0 \cup T_1$ with $|\rho| > R_0$ implies $\rho \in \Sigma$ and can also assume that $R_1 = R_0$.

Take any point $\rho = a + ib \in \Sigma$. Then

$$\left|e^{i\rho\omega_0 t}\right| = \left|e^{i\rho t}\right| = e^{-bt}, \qquad 0 \le t \le 1, \tag{5.6}$$

$$\left|e^{i\rho\omega_\nu (t-1)}\right| = \left|e^{i\rho(1-t)}\right| = e^{-b(1-t)}, \qquad 0 \le t \le 1. \tag{5.7}$$

Clearly these two exponential functions are unbounded on the sector Σ. For $k = 1, \dots, \nu - 1$ we have $\sigma_0/n < 2\pi/n - \sigma_0/n$ and

$$\frac{\pi}{2} + \left(-\frac{2\pi}{n} + \frac{\sigma_0}{n}\right) + \frac{2\pi}{n} \le \arg(i\rho\omega_k t) \le \frac{\pi}{2} + \left(\frac{2\pi}{n} - \frac{\sigma_0}{n}\right) + \frac{2\pi(\nu-1)}{n},$$

and hence, $\pi/2 + \sigma_0/n \le \arg(i\rho\omega_k t) \le 3\pi/2 - \sigma_0/n$. It follows that

$$\begin{aligned}\operatorname{Re}(i\rho\omega_k t) &= |i\rho\omega_k t| \cos[\arg(i\rho\omega_k t)] \\ &\le t|\rho| \cos\left(\frac{\pi}{2} + \frac{\sigma_0}{n}\right) = -t|\rho| \sin\frac{\sigma_0}{n}\end{aligned}$$

and $\left|e^{i\rho\omega_k t}\right| = e^{\operatorname{Re}(i\rho\omega_k t)} \le e^{-t\alpha|\rho|} \le 1$, or

$$|e^{i\rho\omega_k t}| \le e^{-t\alpha|\rho|} \le 1, \qquad 0 \le t \le 1, \tag{5.8}$$

for $\rho = a + ib \in \Sigma$ and for $k = 1, \dots, \nu - 1$. Similarly,

$$|e^{i\rho\omega_k (t-1)}| \le e^{-(1-t)\alpha|\rho|} \le 1, \qquad 0 \le t \le 1, \tag{5.9}$$

for $\rho = a + ib \in \Sigma$ and for $k = \nu + 1, \dots, n - 1$.

From the estimates (5.6)–(5.9) it is immediate that

$$\left|e^{i\rho\omega_0}\right| = \left|e^{i\rho}\right| = e^{-b}, \tag{5.10}$$

$$\left|e^{-i\rho\omega_\nu}\right| = \left|e^{i\rho}\right| = e^{-b}, \tag{5.11}$$

$$\left|e^{i\rho\omega_k}\right| \le e^{-\alpha|\rho|} \le 1, \qquad k = 1, \dots, \nu - 1, \tag{5.12}$$

$$\left|e^{-i\rho\omega_k}\right| \le e^{-\alpha|\rho|} \le 1, \qquad k = \nu + 1, \dots, n - 1, \tag{5.13}$$

for all $\rho = a + ib \in \Sigma$. Thus, the exponentials $e^{i\rho\omega_0} = e^{-i\rho\omega_\nu} = e^{i\rho}$ are unbounded on Σ as $b \to -\infty$, while the exponentials $e^{i\rho\omega_k}$, $1 \le k \le \nu - 1$, and $e^{-i\rho\omega_k}$, $\nu + 1 \le k \le n - 1$, go to 0 very rapidly on Σ as $|\rho| \to \infty$. Note that the estimates (5.6)–(5.13) are also valid for $\rho \in T_0 \cup T_1$ with $|\rho| > R_0$, and in particular, they are valid for $\rho \in S_0 \cup S_1$ with $|\rho| > R_0$ or for $\rho \in G_0$ or $\rho \in G_1$.

Applying the estimates (5.6)–(5.9) and (5.4) to the representations (5.5) with $\eta = 0$, it follows that

$$|u_{00}(t, \rho)| \le 2e^{-bt}, \tag{5.14}$$

$$|u_{0\nu}(t, \rho)| \le 2e^{-b(1-t)}, \tag{5.15}$$

$$|u_{0k}(t, \rho)| \le 2e^{-t\alpha|\rho|} \le 2, \qquad k = 1, \dots, \nu - 1, \tag{5.16}$$

$$|u_{0k}(t, \rho)| \le 2e^{-(1-t)\alpha|\rho|} \le 2, \qquad k = \nu + 1, \dots, n - 1, \tag{5.17}$$

for $0 \le t \le 1$ and for $\rho = a + ib \in T_0$ with $|\rho| > R_0$. In particular, for $\rho = a + ib$ belonging to the sector S_0 (where $b \ge 0$) with $|\rho| > R_0$, we have

$$|u_{0k}(t, \rho)| \le 2, \qquad 0 \le t \le 1, \quad k = 0, 1, \dots, n - 1. \tag{5.18}$$

These bounds will be used in Chapter 6 to determine the growth rate of the Green's function and the resolvent.

Using the modified solutions $u_{0k}(t,\rho)$, $k=0,1,\dots,n-1$, we form the functions

$$M_{0ik}(\rho) := B_i(u_{0k}(\,\cdot\,,\rho)) = \sum_{\eta=0}^{m_i} \alpha_{i\eta} u_{0k}^{(\eta)}(0,\rho) + \sum_{\eta=0}^{m_i} \beta_{i\eta} u_{0k}^{(\eta)}(1,\rho)$$

for $\rho \in T_0$ with $|\rho| > R_0$ and for $i=1,\dots,n$ and $k=0,1,\dots,n-1$. Clearly these functions are analytic on the open set G_0. For $i=1,\dots,n$ and $k=0,1,\dots,\nu-1$ define

$$\widetilde{P}_{0ik}(\rho) := \sum_{\eta=0}^{m_i} \alpha_{i\eta} E_{0k\eta}(0,\rho)\rho^{-m+\eta}, \quad \widetilde{Q}_{0ik}(\rho) := \sum_{\eta=0}^{m_i} \beta_{i\eta} E_{0k\eta}(1,\rho)\rho^{-m+\eta}$$

for $\rho \in T_0$ with $|\rho| > R_0$; for $i=1,\dots,n$ and $k=\nu,\dots,n-1$ define

$$\widetilde{P}_{0ik}(\rho) := \sum_{\eta=0}^{m_i} \beta_{i\eta} E_{0k\eta}(1,\rho)\rho^{-m+\eta}, \quad \widetilde{Q}_{0ik}(\rho) := \sum_{\eta=0}^{m_i} \alpha_{i\eta} E_{0k\eta}(0,\rho)\rho^{-m+\eta}$$

for $\rho \in T_0$ with $|\rho| > R_0$; and in terms of these functions and the functions $\widehat{P}_{ik}(\rho)$, $\widehat{Q}_{ik}(\rho)$ appearing in equations (3.1) and (3.2), for $i=1,\dots,n$ and $k=0,1,\dots,n-1$ define

$$P_{0ik}(\rho) := \widehat{P}_{ik}(\rho) + \widetilde{P}_{0ik}(\rho), \quad Q_{0ik}(\rho) := \widehat{Q}_{ik}(\rho) + \widetilde{Q}_{0ik}(\rho)$$

for $\rho \in T_0$ with $|\rho| > R_0$. Clearly these functions are analytic on the open set G_0. We can then express the functions M_{0ik} as follows: for $i=1,\dots,n$ and $k=0,1,\dots,\nu-1$

$$\begin{aligned} M_{0ik}(\rho) &= \sum_{\eta=0}^{m_i} \alpha_{i\eta}\, [y_k^{(\eta)}(0,\rho) + E_{0k\eta}(0,\rho)\rho^{-m+\eta}] \\ &\quad + \sum_{\eta=0}^{m_i} \beta_{i\eta}\, [y_k^{(\eta)}(1,\rho) + e^{i\rho\omega_k} E_{0k\eta}(1,\rho)\rho^{-m+\eta}] \\ &= \widehat{P}_{ik}(\rho) + \widetilde{P}_{0ik}(\rho) + \widehat{Q}_{ik}(\rho)e^{i\rho\omega_k} + \widetilde{Q}_{0ik}(\rho)e^{i\rho\omega_k}, \end{aligned}$$

while for $i=1,\dots,n$ and $k=\nu,\dots,n-1$

$$\begin{aligned} M_{0ik}(\rho) &= \sum_{\eta=0}^{m_i} \alpha_{i\eta}\, [y_k^{(\eta)}(0,\rho) + e^{-i\rho\omega_k} E_{0k\eta}(0,\rho)\rho^{-m+\eta}] \\ &\quad + \sum_{\eta=0}^{m_i} \beta_{i\eta}\, [y_k^{(\eta)}(1,\rho) + E_{0k\eta}(1,\rho)\rho^{-m+\eta}] \\ &= \widehat{Q}_{ik}(\rho)e^{-i\rho\omega_k} + \widetilde{Q}_{0ik}(\rho)e^{-i\rho\omega_k} + \widehat{P}_{ik}(\rho) + \widetilde{P}_{0ik}(\rho). \end{aligned}$$

Thus, for $i=1,\dots,n$ and $k=0,1,\dots,\nu-1$ we have

$$\begin{aligned} M_{0ik}(\rho) &= [\widehat{P}_{ik}(\rho) + \widetilde{P}_{0ik}(\rho)] + [\widehat{Q}_{ik}(\rho) + \widetilde{Q}_{0ik}(\rho)]e^{i\rho\omega_k} \\ &= P_{0ik}(\rho) + Q_{0ik}(\rho)e^{i\rho\omega_k} \end{aligned} \tag{5.19}$$

for $\rho \in T_0$ with $|\rho| > R_0$, and for $i=1,\dots,n$ and $k=\nu,\dots,n-1$ we have

$$\begin{aligned} M_{0ik}(\rho) &= [\widehat{P}_{ik}(\rho) + \widetilde{P}_{0ik}(\rho)] + [\widehat{Q}_{ik}(\rho) + \widetilde{Q}_{0ik}(\rho)]e^{-i\rho\omega_k} \\ &= P_{0ik}(\rho) + Q_{0ik}(\rho)e^{-i\rho\omega_k} \end{aligned} \tag{5.20}$$

for $\rho \in T_0$ with $|\rho| > R_0$.

The *characteristic determinant* of the differential operator L relative to the sector T_0 is the analytic function Δ_0 defined by

$$\Delta_0(\rho) := \det(B_i(u_{0k}(\,\cdot\,,\rho))) = \det(M_{0ik}(\rho)) \quad \text{for } \rho \in G_0.$$

For any complex number $\lambda = \rho^n$ with $\rho \in G_0$, we know that λ is an eigenvalue of L if and only if $\Delta_0(\rho) = 0$, and hence, in Chapter 7 we will proceed to compute the zeros of Δ_0 in the sector T_0. Applying (5.19) and (5.20), we can express the characteristic determinant in the form

(5.21)
$$\Delta_0(\rho) =$$
$$\det\begin{pmatrix} P_{010}(\rho)+Q_{010}(\rho)e^{i\rho} & \overbrace{P_{01k}(\rho)+Q_{01k}(\rho)e^{i\rho\omega_k}}^{1\le k\le \nu-1} & P_{01\nu}(\rho)+Q_{01\nu}(\rho)e^{i\rho} & \overbrace{P_{01k}(\rho)+Q_{01k}(\rho)e^{-i\rho\omega_k}}^{\nu+1\le k\le n-1} \\ \vdots & \vdots & \vdots & \vdots \\ P_{0n0}(\rho)+Q_{0n0}(\rho)e^{i\rho} & P_{0nk}(\rho)+Q_{0nk}(\rho)e^{i\rho\omega_k} & P_{0n\nu}(\rho)+Q_{0n\nu}(\rho)e^{i\rho} & P_{0nk}(\rho)+Q_{0nk}(\rho)e^{-i\rho\omega_k} \end{pmatrix}$$

for $\rho \in G_0$. Cf. equation (3.10) for the approximate characteristic determinant.

For the functions $\widehat{P}_{ik}(\rho)$, $\widehat{Q}_{ik}(\rho)$, we see from either (3.1), (3.2) or (3.7), (3.8) that the powers of ρ appearing in the sums are $\rho^{m_i}, \rho^{m_i-1}, \ldots, \rho^{-(m-1)}$. Hence, there exists a constant $\gamma_0 > 0$ such that

$$|\widehat{P}_{ik}(\rho)| \le \gamma_0|\rho|^{m_i}, \qquad |\widehat{Q}_{ik}(\rho)| \le \gamma_0|\rho|^{m_i} \tag{5.22}$$

for $\rho \in \mathbb{C}$ with $|\rho| \ge 1$ and for $i = 1, \ldots, n$ and $k = 0, 1, \ldots, n-1$. In the definitions of the functions $\widetilde{P}_{0ik}(\rho)$, $\widetilde{Q}_{0ik}(\rho)$, we see that the functions $E_{0k\eta}(0,\rho)$, $E_{0k\eta}(1,\rho)$ are bounded for $\rho \in T_0$ with $|\rho| > R_0$ by Theorem 4.3, and that the powers of ρ appearing in the sums are $\rho^{-(m-m_i)}, \rho^{-(m-m_i+1)}, \ldots, \rho^{-m}$. Thus, there exists a constant $\gamma_1 > 0$ such that

$$|\widetilde{P}_{0ik}(\rho)| \le \gamma_1|\rho|^{-(m-m_i)}, \qquad |\widetilde{Q}_{0ik}(\rho)| \le \gamma_1|\rho|^{-(m-m_i)} \tag{5.23}$$

for $\rho \in T_0$ with $|\rho| > R_0$ and for $i = 1, \ldots, n$ and $k = 0, 1, \ldots, n-1$.

For $i = 1, \ldots, n$ define

$$\begin{aligned} \widetilde{F}_{0ik}(\rho) &:= \widetilde{P}_{0ik}(\rho) + [\widehat{Q}_{ik}(\rho) + \widetilde{Q}_{0ik}(\rho)]e^{i\rho\omega_k}, \qquad k = 1, \ldots, \nu-1, \\ \widetilde{F}_{0ik}(\rho) &:= \widetilde{P}_{0ik}(\rho) + [\widehat{Q}_{ik}(\rho) + \widetilde{Q}_{0ik}(\rho)]e^{-i\rho\omega_k}, \quad k = \nu+1, \ldots, n-1, \end{aligned}$$

for $\rho \in T_0$ with $|\rho| > R_0$. Clearly these functions are analytic on the open set G_0. From the estimates (5.12), (5.13) and (5.22), (5.23) we have

$$\begin{aligned} |\widetilde{F}_{0ik}(\rho)| &\le \gamma_1|\rho|^{-(m-m_i)} + [\gamma_0|\rho|^{m_i} + \gamma_1|\rho|^{-(m-m_i)}]\, e^{-\alpha|\rho|} \\ &\le \gamma_2|\rho|^{-(m-m_i)} \end{aligned} \tag{5.24}$$

for $\rho \in T_0$ with $|\rho| > R_0$, for $i = 1, \ldots, n$, and for $k = 1, \ldots, \nu-1$ and $k = \nu+1, \ldots, n-1$. In terms of these functions we can rewrite the representation

(5.21) of the characteristic determinant in the form

(5.25)

$$\Delta_0(\rho) =$$

$$\det\begin{pmatrix} P_{010}(\rho)+Q_{010}(\rho)e^{i\rho} & \overset{1\le k\le \nu-1}{\widehat{P}_{1k}(\rho)+\widetilde{F}_{01k}(\rho)} & P_{01\nu}(\rho)+Q_{01\nu}(\rho)e^{i\rho} & \overset{\nu+1\le k\le n-1}{\widehat{P}_{1k}(\rho)+\widetilde{F}_{01k}(\rho)} \\ \vdots & \vdots & \vdots & \vdots \\ P_{0n0}(\rho)+Q_{0n0}(\rho)e^{i\rho} & \widehat{P}_{nk}(\rho)+\widetilde{F}_{0nk}(\rho) & P_{0n\nu}(\rho)+Q_{0n\nu}(\rho)e^{i\rho} & \widehat{P}_{nk}(\rho)+\widetilde{F}_{0nk}(\rho) \end{pmatrix}$$

for $\rho \in G_0$.

We now proceed to expand the determinant for $\Delta_0(\rho)$ that appears in equation (5.25). These expansions parallel the ones used earlier in equations (3.12)–(3.13) for the approximate characteristic determinant $\widehat{\Delta}(\rho)$, and in fact, the functions $\widehat{\pi}_i(\rho)$, $i = 0, 1, 2$, that were introduced in Chapter 3 will also appear in these new expansions for $\Delta_0(\rho)$.

Indeed, suppose we expand the determinant in (5.25) using the linearity of the determinant in the 0th and νth columns:

$$\Delta_0(\rho) = D_{02}(\rho)e^{2i\rho} + D_{01}(\rho)e^{i\rho} + D_{00}(\rho) \tag{5.26}$$

for $\rho \in G_0$, where

$$D_{02}(\rho) := \det\begin{pmatrix} \widehat{Q}_{10}(\rho)+\widetilde{Q}_{010}(\rho) & \overset{1\le k\le \nu-1}{\widehat{P}_{1k}(\rho)+\widetilde{F}_{01k}(\rho)} & \widehat{Q}_{1\nu}(\rho)+\widetilde{Q}_{01\nu}(\rho) & \overset{\nu+1\le k\le n-1}{\widehat{P}_{1k}(\rho)+\widetilde{F}_{01k}(\rho)} \\ \vdots & \vdots & \vdots & \vdots \\ \widehat{Q}_{n0}(\rho)+\widetilde{Q}_{0n0}(\rho) & \widehat{P}_{nk}(\rho)+\widetilde{F}_{0nk}(\rho) & \widehat{Q}_{n\nu}(\rho)+\widetilde{Q}_{0n\nu}(\rho) & \widehat{P}_{nk}(\rho)+\widetilde{F}_{0nk}(\rho) \end{pmatrix},$$

$$D_{01}(\rho) := \det\begin{pmatrix} \widehat{P}_{10}(\rho)+\widetilde{P}_{010}(\rho) & \overset{1\le k\le \nu-1}{\widehat{P}_{1k}(\rho)+\widetilde{F}_{01k}(\rho)} & \widehat{Q}_{1\nu}(\rho)+\widetilde{Q}_{01\nu}(\rho) & \overset{\nu+1\le k\le n-1}{\widehat{P}_{1k}(\rho)+\widetilde{F}_{01k}(\rho)} \\ \vdots & \vdots & \vdots & \vdots \\ \widehat{P}_{n0}(\rho)+\widetilde{P}_{0n0}(\rho) & \widehat{P}_{nk}(\rho)+\widetilde{F}_{0nk}(\rho) & \widehat{Q}_{n\nu}(\rho)+\widetilde{Q}_{0n\nu}(\rho) & \widehat{P}_{nk}(\rho)+\widetilde{F}_{0nk}(\rho) \end{pmatrix}$$

$$+ \det\begin{pmatrix} \widehat{Q}_{10}(\rho)+\widetilde{Q}_{010}(\rho) & \overset{1\le k\le \nu-1}{\widehat{P}_{1k}(\rho)+\widetilde{F}_{01k}(\rho)} & \widehat{P}_{1\nu}(\rho)+\widetilde{P}_{01\nu}(\rho) & \overset{\nu+1\le k\le n-1}{\widehat{P}_{1k}(\rho)+\widetilde{F}_{01k}(\rho)} \\ \vdots & \vdots & \vdots & \vdots \\ \widehat{Q}_{n0}(\rho)+\widetilde{Q}_{0n0}(\rho) & \widehat{P}_{nk}(\rho)+\widetilde{F}_{0nk}(\rho) & \widehat{P}_{n\nu}(\rho)+\widetilde{P}_{0n\nu}(\rho) & \widehat{P}_{nk}(\rho)+\widetilde{F}_{0nk}(\rho) \end{pmatrix},$$

and

$$D_{00}(\rho) := \det\begin{pmatrix} \widehat{P}_{10}(\rho)+\widetilde{P}_{010}(\rho) & \overset{1\le k\le \nu-1}{\widehat{P}_{1k}(\rho)+\widetilde{F}_{01k}(\rho)} & \widehat{P}_{1\nu}(\rho)+\widetilde{P}_{01\nu}(\rho) & \overset{\nu+1\le k\le n-1}{\widehat{P}_{1k}(\rho)+\widetilde{F}_{01k}(\rho)} \\ \vdots & \vdots & \vdots & \vdots \\ \widehat{P}_{n0}(\rho)+\widetilde{P}_{0n0}(\rho) & \widehat{P}_{nk}(\rho)+\widetilde{F}_{0nk}(\rho) & \widehat{P}_{n\nu}(\rho)+\widetilde{P}_{0n\nu}(\rho) & \widehat{P}_{nk}(\rho)+\widetilde{F}_{0nk}(\rho) \end{pmatrix}$$

for $\rho \in G_0$. Clearly the functions $D_{0i}(\rho)$, $i = 0, 1, 2$, are analytic on the open set G_0. In these representations we will treat the "hat terms" as principal terms and the "tilde terms" as small perturbation terms. See (5.22)–(5.24).

Now consider the function $D_{02}(\rho)$. If we expand $D_{02}(\rho)$ using the linearity of the determinant in its n columns, then $D_{02}(\rho)$ becomes the sum of 2^n determinants, starting with the determinant

$$\widehat{\pi}_2(\rho) = \det\begin{pmatrix} \widehat{Q}_{10}(\rho) & \widehat{P}_{11}(\rho) & \cdots & \widehat{P}_{1\nu-1}(\rho) & \widehat{Q}_{1\nu}(\rho) & \widehat{P}_{1\nu+1}(\rho) & \cdots & \widehat{P}_{1n-1}(\rho) \\ \vdots & \vdots & & \vdots & \vdots & \vdots & & \vdots \\ \widehat{Q}_{n0}(\rho) & \widehat{P}_{n1}(\rho) & \cdots & \widehat{P}_{n\nu-1}(\rho) & \widehat{Q}_{n\nu}(\rho) & \widehat{P}_{n\nu+1}(\rho) & \cdots & \widehat{P}_{nn-1}(\rho) \end{pmatrix},$$

which is precisely the function introduced in Chapter 3. Each of the remaining $2^n - 1$ determinants contains at least one column consisting of the functions $\widetilde{Q}_{0i0}(\rho)$, $i = 1, \ldots, n$, or of the functions $\widetilde{F}_{0ik}(\rho)$, $i = 1, \ldots, n$, or of the functions $\widetilde{Q}_{0i\nu}(\rho)$, $i = 1, \ldots, n$. When such a determinant is expanded, it becomes the sum of $n!$ products each having modulus less than or equal to

$$\gamma|\rho|^{m_1}|\rho|^{m_2}\cdots|\rho|^{m_{i-1}}|\rho|^{-(m-m_i)}|\rho|^{m_{i+1}}\cdots|\rho|^{m_n} = \gamma|\rho|^{-(m-p_0)}$$

by virtue of the estimates (5.22)–(5.24). It follows that the function $D_{02}(\rho)$ can be expressed as

$$D_{02}(\rho) = \widehat{\pi}_2(\rho) + \widetilde{\Phi}_{02}(\rho)$$

for $\rho \in G_0$, where the function $\widetilde{\Phi}_{02}(\rho)$ is analytic on the open set G_0 and satisfies the estimate

$$|\widetilde{\Phi}_{02}(\rho)| \leq \gamma_3|\rho|^{-(m-p_0)}$$

for $\rho \in G_0$. Similarly, we can express $D_{01}(\rho)$ and $D_{00}(\rho)$ as

$$D_{01}(\rho) = \widehat{\pi}_1(\rho) + \widetilde{\Phi}_{01}(\rho), \qquad D_{00}(\rho) = \widehat{\pi}_0(\rho) + \widetilde{\Phi}_{00}(\rho)$$

for $\rho \in G_0$, where the functions $\widetilde{\Phi}_{01}(\rho)$, $\widetilde{\Phi}_{00}(\rho)$ are analytic on the open set G_0 with

$$|\widetilde{\Phi}_{01}(\rho)| \leq \gamma_3|\rho|^{-(m-p_0)}, \qquad |\widetilde{\Phi}_{00}(\rho)| \leq \gamma_3|\rho|^{-(m-p_0)}$$

for $\rho \in G_0$.

Combining the above results, we obtain our principal representation of the characteristic determinant Δ_0 relative to the sector T_0:

$$\Delta_0(\rho) = \widehat{\pi}_2(\rho)e^{2i\rho} + \widehat{\pi}_1(\rho)e^{i\rho} + \widehat{\pi}_0(\rho) + \widetilde{\Phi}_{02}(\rho)e^{2i\rho} + \widetilde{\Phi}_{01}(\rho)e^{i\rho} + \widetilde{\Phi}_{00}(\rho) \tag{5.27}$$

for $\rho \in G_0$, where the functions $\widehat{\pi}_i(\rho)$, $i = 0, 1, 2$, are analytic for $\rho \neq 0$ in $\mathbb{C}$, and the functions $\widetilde{\Phi}_{0i}(\rho)$, $i = 0, 1, 2$, are analytic on the open set G_0 with

$$|\widetilde{\Phi}_{0i}(\rho)| \leq \gamma_3|\rho|^{-(m-p_0)}, \qquad i = 0, 1, 2, \tag{5.28}$$

for $\rho \in G_0$. The functions $\widehat{\pi}_i(\rho)$, $i = 0, 1, 2$, are the functions introduced in Chapter 3 in our formation of the approximate characteristic determinant $\widehat{\Delta}(\rho) = \widehat{\Delta}(\rho, m)$; they are determined completely by the Birkhoff approximate solutions. The functions $\widetilde{\Phi}_{0i}(\rho)$, $i = 0, 1, 2$, contain all the perturbation terms that are produced in constructing the actual solutions of the differential equation (2.1).

Let us recall some of the results of Chapter 3 where we constructed the approximate characteristic determinant and classified the differential operator L. First, we are assuming that n is even and that the differential operator L is either regular or simply irregular. This yields the integer p with $-\infty < p \leq p_0$, $a_p \neq 0$ and $c_p \neq 0$, and $a_\kappa = c_\kappa = 0$ for $\kappa = p+1, \ldots, p_0$. Second, the integer q is defined as follows: if $b_\kappa = 0$ for $\kappa = p+1, \ldots, p_0$, then $q = p$ and $b_q = b_p$ can be either zero or nonzero; if $b_\kappa \neq 0$ for some κ with $p+1 \leq \kappa \leq p_0$, then q is the largest such integer and b_q is nonzero. In either case we have $p \leq q \leq p_0$ and $b_\kappa = 0$ for $\kappa = q+1, \ldots, p_0$.

Third, the translated sectors T_0 and T_1 are formed subject to the condition (3.31). Fourth, the integer m is fixed with $m > n$, $m > p_0$, and $-(m-p_0-1) \leq p \leq p_0$, and then the corresponding Birkhoff approximate solutions $z_k(t,\rho)$, $k = 0, 1, \ldots, n-1$, are formed, and the modified functions $y_k(t,\rho)$, $k = 0, 1, \ldots, n-1$, are determined. Fifth, the functions $\pi_i(\rho)$, $i = 0, 1, 2$, are defined by

$$(5.29)\quad \pi_2(\rho) = \sum_{\kappa=-(m-p_0-1)}^{p} a_\kappa \rho^\kappa, \quad \pi_1(\rho) = \sum_{\kappa=-(m-p_0-1)}^{q} b_\kappa \rho^\kappa, \quad \pi_0(\rho) = \sum_{\kappa=-(m-p_0-1)}^{p} c_\kappa \rho^\kappa$$

for $\rho \neq 0$ in $\mathbb{C}$. From equations (3.24), (3.27), and (3.28) it is immediate that

$$(5.30)\quad \begin{gathered} \widehat{\pi}_2(\rho) = \pi_2(\rho) + \sum_{\kappa=-n(m-1)}^{-(m-p_0)} a_\kappa(m)\rho^\kappa, \quad \widehat{\pi}_1(\rho) = \pi_1(\rho) + \sum_{\kappa=-n(m-1)}^{-(m-p_0)} b_\kappa(m)\rho^\kappa, \\ \widehat{\pi}_0(\rho) = \pi_0(\rho) + \sum_{\kappa=-n(m-1)}^{-(m-p_0)} c_\kappa(m)\rho^\kappa \end{gathered}$$

for $\rho \neq 0$ in $\mathbb{C}$.

Finally, in terms of (5.27) and (5.30) we define the functions

$$\begin{gathered} \Phi_{02}(\rho) := \sum_{\kappa=-n(m-1)}^{-(m-p_0)} a_\kappa(m)\rho^\kappa + \widetilde{\Phi}_{02}(\rho), \quad \Phi_{01}(\rho) := \sum_{\kappa=-n(m-1)}^{-(m-p_0)} b_\kappa(m)\rho^\kappa + \widetilde{\Phi}_{01}(\rho), \\ \Phi_{00}(\rho) := \sum_{\kappa=-n(m-1)}^{-(m-p_0)} c_\kappa(m)\rho^\kappa + \widetilde{\Phi}_{00}(\rho) \end{gathered}$$

for $\rho \in G_0$. These functions are clearly analytic on the open set G_0, and by using them we can rewrite the representation (5.27) in the simpler form

$$(5.31)\quad \Delta_0(\rho) = \pi_2(\rho)e^{2i\rho} + \pi_1(\rho)e^{i\rho} + \pi_0(\rho) + \Phi_{02}(\rho)e^{2i\rho} + \Phi_{01}(\rho)e^{i\rho} + \Phi_{00}(\rho)$$

for $\rho \in G_0$. In equation (5.31) the functions $\pi_i(\rho)$, $i = 0, 1, 2$, are given by (5.29), they are analytic for $\rho \neq 0$ in $\mathbb{C}$, and they are determined exclusively by the Birkhoff approximate solutions and the boundary values $B_1, \ldots, B_n$; the functions $\Phi_{0i}(\rho)$, $i = 0, 1, 2$, are analytic on the open set G_0 and satisfy the growth rates

$$(5.32)\quad |\Phi_{0i}(\rho)| \leq \gamma_4 |\rho|^{-(m-p_0)}, \qquad i = 0, 1, 2,$$

for $\rho \in G_0$. The representation (5.31) is our working form for the characteristic determinant Δ_0 relative to the sector T_0. Compare the representation (5.31) for the characteristic determinant Δ_0 to the representation (3.32) for the approximate characteristic determinant $\widehat{\Delta}$. In Chapter 7 we will determine the zeros of Δ_0 in the open set G_0, and hence, determine eigenvalues for the differential operator L.

The above results are summarized in the following theorem. **Here we assume the conditions set forth in Chapter 3:** (i) $n = 2\nu$ is even; (ii) the differential operator L is either regular or simply irregular; (iii) the integers p and q have been determined with $-\infty < p \leq q \leq p_0$ and with $a_p \neq 0$, $c_p \neq 0$, and $a_\kappa = c_\kappa = 0$ for $\kappa = p+1, \ldots, p_0$ and $b_\kappa = 0$ for $\kappa = q+1, \ldots, p_0$; (iv) the translated sectors T_0 and T_1 have been chosen; (v) the integer m has been fixed with $m > n$, $m > p_0$, and $-(m - p_0 - 1) \leq p \leq p_0$; and (vi) the functions $\pi_i(\rho)$, $i = 0, 1, 2$, have been determined as per Chapter 3 or equation (5.29).

THEOREM 5.1. *Let n be even: $n = 2\nu$. Under the assumptions* (i)–(vi), *let $\pi_i(\rho)$, $i = 0, 1, 2$, be the functions defined in Section 3.2 and in equation* (5.29), *let $v_{0k}(\cdot, \rho)$, $k = 0, 1, \ldots, n-1$, be the linearly independent solutions of the differential equation* (2.1) *constructed in Theorem 4.3 for $\rho \in T_0$ with $|\rho| > R_0$, let $u_{0k}(\cdot, \rho)$, $k = 0, 1, \ldots, n-1$, be the modified solutions of* (2.1) *defined above for $\rho \in T_0$ with $|\rho| > R_0$, and let Δ_0 be the characteristic determinant of the differential operator L given by*

$$\Delta_0(\rho) = \det(B_i(u_{0k}(\cdot, \rho))) \quad \textit{for } \rho \in G_0,$$

where $G_0 = \{\rho \in \operatorname{Int} T_0 \mid |\rho| > R_0\}$. Then Δ_0 is analytic on the open set G_0, and Δ_0 has the representation

$$\Delta_0(\rho) = \pi_2(\rho)e^{2i\rho} + \pi_1(\rho)e^{i\rho} + \pi_0(\rho) + \Phi_{02}(\rho)e^{2i\rho} + \Phi_{01}(\rho)e^{i\rho} + \Phi_{00}(\rho)$$

for $\rho \in G_0$, where the functions $\Phi_{0i}(\rho)$, $i = 0, 1, 2$, are analytic on G_0 and satisfy the estimates $|\Phi_{0i}(\rho)| \leq \gamma|\rho|^{-(m-p_0)}$ for $\rho \in G_0$ and for $i = 0, 1, 2$.

Next, for the case $n = 2\nu$ even, we form the characteristic determinant on the sector T_1. The starting point for the discussion is the set of functions $v_{1k}(\cdot, \rho)$, $k = 0, 1, \ldots, n-1$, which form a basis for the solution space of the differential equation (2.1) for $\rho \in T_1$ with $|\rho| > R_0$. In terms of these functions we introduce the *modified solutions* of differential equation (2.1) relative to the sector T_1:

$$\begin{aligned}
u_{10}(t, \rho) &:= e^{-i\rho\omega_0} v_{10}(t, \rho), && \\
u_{1k}(t, \rho) &:= v_{1k}(t, \rho), && k = 1, \ldots, \nu - 1, \\
u_{1\nu}(t, \rho) &:= v_{1\nu}(t, \rho), && \\
u_{1k}(t, \rho) &:= e^{-i\rho\omega_k} v_{1k}(t, \rho), && k = \nu + 1, \ldots, n-1,
\end{aligned}$$

for $0 \leq t \leq 1$ and for $\rho \in T_1$ with $|\rho| > R_0$. These functions also form a basis for the solution space of the differential equation (2.1). From the regularity results of Theorem 4.4, for fixed $t \in [0, 1]$ each of the functions $u_{1k}^{(\eta)}(t, \rho)$, $k, \eta = 0, 1, \ldots, n-1$, is an analytic function of the ρ variable on the open set

$$G_1 = \{\rho \in \operatorname{Int} T_1 \mid |\rho| > R_0\},$$

and the boundary data $u_{1k}^{(\eta)}(0, \rho)$, $u_{1k}^{(\eta)}(1, \rho)$, $k, \eta = 0, 1, \ldots, n-1$, consists of functions of ρ that are analytic on G_1.

For these solutions we obtain the estimates

$$|u_{10}(t, \rho)| \leq 2e^{b(1-t)}, \tag{5.33}$$

$$|u_{1\nu}(t, \rho)| \leq 2e^{bt}, \tag{5.34}$$

$$|u_{1k}(t, \rho)| \leq 2e^{-t\alpha|\rho|} \leq 2, \qquad k = 1, \ldots, \nu - 1, \tag{5.35}$$

$$|u_{1k}(t, \rho)| \leq 2e^{-(1-t)\alpha|\rho|} \leq 2, \qquad k = \nu + 1, \ldots, n - 1, \tag{5.36}$$

for $0 \leq t \leq 1$ and for $\rho = a + ib \in T_1$ with $|\rho| > R_0$. In particular, for $\rho = a + ib$ belonging to the sector S_1 (where $b \leq 0$) with $|\rho| > R_0$, we have

$$|u_{1k}(t, \rho)| \leq 2, \qquad 0 \leq t \leq 1, \qquad k = 0, 1, \ldots, n - 1. \tag{5.37}$$

These bounds will be used in Chapter 6 to determine the growth rate of the Green's function and the resolvent.

The *characteristic determinant* of the differential operator L relative to the sector T_1 is the analytic function Δ_1 defined by

$$\Delta_1(\rho) := \det(B_i(u_{1k}(\cdot,\rho))) \quad \text{for } \rho \in G_1.$$

For any complex number $\lambda = \rho^n$ with $\rho \in G_1$, we know that λ is an eigenvalue of L if and only if $\Delta_1(\rho) = 0$. In Chapter 7 we will compute the zeros of Δ_1 in the sector T_1. When the determinant for Δ_1 is expanded using linearity in the 0th and νth columns, we obtain the representation

$$\begin{aligned}\Delta_1(\rho) &= \pi_2(\rho) + \pi_1(\rho)e^{-i\rho} + \pi_0(\rho)e^{-2i\rho} \\ &\quad + \Phi_{12}(\rho) + \Phi_{11}(\rho)e^{-i\rho} + \Phi_{10}(\rho)e^{-2i\rho}\end{aligned} \tag{5.38}$$

for $\rho \in G_1$. In this representation the functions $\pi_i(\rho)$, $i = 0, 1, 2$, are given by (5.29), and they are analytic for $\rho \neq 0$ in $\mathbb{C}$; the functions $\Phi_{1i}(\rho)$, $i = 0, 1, 2$, are analytic on the open set G_1 and satisfy the growth rates

$$|\Phi_{1i}(\rho)| \leq \gamma_0 |\rho|^{-(m-p_0)}, \qquad i = 0, 1, 2, \tag{5.39}$$

for $\rho \in G_1$. The representation (5.38) is our working form for the characteristic determinant Δ_1 relative to the sector T_1.

Let us summarize the above results for the characteristic determinant Δ_1 in a theorem. Again we assume conditions (i)–(vi) that precede Theorem 5.1.

THEOREM 5.2. *Let n be even: $n = 2\nu$. Under the assumptions* (i)–(vi), *let $\pi_i(\rho)$, $i = 0, 1, 2$, be the functions defined in Section 3.2 and in equation* (5.29), *let $v_{1k}(\cdot,\rho)$, $k = 0, 1, \ldots, n-1$, be the linearly independent solutions of the differential equation* (2.1) *constructed in Theorem 4.4 for $\rho \in T_1$ with $|\rho| > R_0$, let $u_{1k}(\cdot,\rho)$, $k = 0, 1, \ldots, n-1$, be the modified solutions of* (2.1) *defined above for $\rho \in T_1$ with $|\rho| > R_0$, and let Δ_1 be the characteristic determinant of the differential operator L given by*

$$\Delta_1(\rho) = \det(B_i(u_{1k}(\cdot,\rho))) \quad \textit{for } \rho \in G_1,$$

where $G_1 = \{\rho \in \operatorname{Int} T_1 \mid |\rho| > R_0\}$. Then Δ_1 is analytic on the open set G_1, and Δ_1 has the representation

$$\begin{aligned}\Delta_1(\rho) &= \pi_2(\rho) + \pi_1(\rho)e^{-i\rho} + \pi_0(\rho)e^{-2i\rho} \\ &\quad + \Phi_{12}(\rho) + \Phi_{11}(\rho)e^{-i\rho} + \Phi_{10}(\rho)e^{-2i\rho}\end{aligned}$$

for $\rho \in G_1$, where the functions $\Phi_{1i}(\rho)$, $i = 0, 1, 2$, are analytic on G_1 and satisfy the estimates $|\Phi_{1i}(\rho)| \leq \gamma|\rho|^{-(m-p_0)}$ for $\rho \in G_1$ and for $i = 0, 1, 2$.

5.2. The Characteristic Determinant for n Odd

Assume that n is odd: $n = 2\nu - 1 \geq 3$, and consider the sectors

$$S_0\colon \text{ all } \rho = |\rho|e^{i\theta} \in \mathbb{C} \text{ with } -\frac{\pi}{2n} \leq \theta \leq \frac{\pi}{2n},$$

$$S_1\colon \text{ all } \rho = |\rho|e^{i\theta} \in \mathbb{C} \text{ with } \pi - \frac{\pi}{2n} \leq \theta \leq \pi + \frac{\pi}{2n},$$

and the corresponding translated sectors T_0, T_1. The development of the characteristic determinants for the case n odd is almost identical to the development for the case n even. Consequently, we will indicate only the highlights of this theory.

For each $\rho \in T_0$ with $|\rho| > R_0$ let $v_{00}(\cdot,\rho), v_{01}(\cdot,\rho), \ldots, v_{0\,n-1}(\cdot,\rho)$ be the n linearly independent solutions of the differential equation (2.1) constructed in

Theorem 4.6. The function $v_{0k}(\cdot,\rho)$ belongs to $H^n[0,1]$, and $v_{0k}(\cdot,\rho)$ and its derivatives have the asymptotic expansions

$$v_{0k}^{(\eta)}(t,\rho) = z_k^{(\eta)}(t,\rho) + e^{i\rho\omega_k t}E_{0k\eta}(t,\rho)\rho^{-m+\eta} \tag{5.40}$$

for $0 \le t \le 1$, for $\rho \in T_0$ with $|\rho| > R_0$, and for $k,\eta = 0,1,\dots,n-1$. Similarly, for each $\rho \in T_1$ with $|\rho| > R_0$ let $v_{10}(\cdot,\rho), v_{11}(\cdot,\rho),\dots,v_{1\,n-1}(\cdot,\rho)$ be the n linearly independent solutions of the differential equation (2.1) constructed in Theorem 4.7. Again each function $v_{1k}(\cdot,\rho)$ belongs to $H^n[0,1]$ with

$$v_{1k}^{(\eta)}(t,\rho) = z_k^{(\eta)}(t,\rho) + e^{i\rho\omega_k t}E_{1k\eta}(t,\rho)\rho^{-m+\eta} \tag{5.41}$$

for $0 \le t \le 1$, for $\rho \in T_1$ with $|\rho| > R_0$, and for $k,\eta = 0,1,\dots,n-1$. In these expansions the function $z_k(t,\rho)$ is the Birkhoff approximate solution constructed in Chapter 2.

For the sector T_0 we form the *modified solutions* of the differential equation (2.1):

$$\begin{aligned} u_{0k}(t,\rho) &:= v_{0k}(t,\rho), & k &= 0,1,\dots,\nu-1,\\ u_{0k}(t,\rho) &:= e^{-i\rho\omega_k} v_{0k}(t,\rho), & k &= \nu,\dots,n-1, \end{aligned}$$

for $0 \le t \le 1$ and for $\rho \in T_0$ with $|\rho| > R_0$. These modified solutions form a basis for the solution space of the differential equation (2.1) for $\rho \in T_0$ with $|\rho| > R_0$. Applying Theorem 4.6, for fixed $t \in [0,1]$ each of the functions $u_{0k}^{(\eta)}(t,\rho)$, $k,\eta = 0,1,\dots,n-1$, is an analytic function of the ρ variable on the open set

$$G_0 = \{\rho \in \operatorname{Int} T_0 \mid |\rho| > R_0\}.$$

Hence, the boundary data $u_{0k}^{(\eta)}(0,\rho)$, $u_{0k}^{(\eta)}(1,\rho)$, $k,\eta = 0,1,\dots,n-1$, consists of functions of ρ that are analytic on G_0.

Next, fix σ_0 with $0 < \sigma_0 < \pi/10$ and set $\alpha := \sin(\sigma_0/n) > 0$. Arguing as in the even order case, we obtain the following estimates:

$$|u_{00}(t,\rho)| \le 2e^{-bt}, \tag{5.42}$$

$$|u_{0k}(t,\rho)| \le 2e^{-t\alpha|\rho|} \le 2, \qquad k = 1,\dots,\nu-1, \tag{5.43}$$

$$|u_{0k}(t,\rho)| \le 2e^{-(1-t)\alpha|\rho|} \le 2, \qquad k = \nu,\dots,n-1, \tag{5.44}$$

for $0 \le t \le 1$ and for $\rho = a + ib \in T_0$ with $|\rho| > R_0$. In particular, for $\rho = a + ib$ belonging to the sector S_0 with $|\rho| > R_0$ and $b \ge 0$, we have

$$|u_{0k}(t,\rho)| \le 2, \qquad 0 \le t \le 1, \quad k = 0,1,\dots,n-1. \tag{5.45}$$

The *characteristic determinant* of the differential operator L relative to the sector T_0 is the analytic function Δ_0 defined by

$$\Delta_0(\rho) := \det(B_i(u_{0k}(\cdot,\rho))) \quad \text{for } \rho \in G_0.$$

For any complex number $\lambda = \rho^n$ with $\rho \in G_0$, we know that λ is an eigenvalue of L if and only if $\Delta_0(\rho) = 0$. When the determinant for Δ_0 is expanded using linearity in the 0th column, we obtain the representation

$$\Delta_0(\rho) = \pi_1(\rho)e^{i\rho} + \pi_0(\rho) + \Phi_{01}(\rho)e^{i\rho} + \Phi_{00}(\rho) \tag{5.46}$$

for $\rho \in G_0$. In equation (5.46) the functions $\pi_i(\rho)$, $i = 0, 1$, are the functions defined earlier in Chapter 3 for the case n odd, viz.,

$$(5.47) \qquad \pi_1(\rho) = \sum_{\kappa=-(m-p_0-1)}^{p} a_\kappa \rho^\kappa, \quad \pi_0(\rho) = \sum_{\kappa=-(m-p_0-1)}^{q} b_\kappa \rho^\kappa$$

for $\rho \neq 0$ in $\mathbb{C}$, these functions being analytic for $\rho \neq 0$ in $\mathbb{C}$; the functions $\Phi_{0i}(\rho)$, $i = 0, 1$, are analytic on the open set G_0 and satisfy the growth rates

$$(5.48) \qquad |\Phi_{0i}(\rho)| \leq \gamma_0 |\rho|^{-(m-p_0)}, \qquad i = 0, 1,$$

for $\rho \in G_0$.

The representation (5.46) is our working form for the characteristic determinant Δ_0 relative to the sector T_0. Compare the representation (5.46) for the characteristic determinant Δ_0 to the representation (3.50) for the approximate characteristic determinant $\widehat{\Delta}$.

The above results are summarized in the following theorem. **We assume the conditions set forth in Chapter 3:** (i) $n = 2\nu - 1$ is odd; (ii) the differential operator L is either regular or simply irregular; (iii) the integers p and q have been determined with $-\infty < p, q \leq p_0$ and with $a_p \neq 0$, $b_q \neq 0$, and $a_\kappa = 0$ for $\kappa = p+1, \ldots, p_0$ and $b_\kappa = 0$ for $\kappa = q+1, \ldots, p_0$; (iv) the translated sectors T_0 and T_1 have been chosen; (v) the integer m has been fixed with $m > n$, $m > p_0$, and $-(m - p_0 - 1) \leq p, q \leq p_0$; and (vi) the functions $\pi_i(\rho)$, $i = 0, 1$, have been determined as per Chapter 3 or equation (5.47).

THEOREM 5.3. *Let n be odd: $n = 2\nu - 1$. Under the assumptions* (i)–(vi), *let $\pi_i(\rho)$, $i = 0, 1$, be the functions defined in Section 3.3 and in equation* (5.47), *let $v_{0k}(\cdot, \rho)$, $k = 0, 1, \ldots, n-1$, be the linearly independent solutions of the differential equation* (2.1) *constructed in Theorem 4.6 for $\rho \in T_0$ with $|\rho| > R_0$, let $u_{0k}(\cdot, \rho)$, $k = 0, 1, \ldots, n-1$, be the modified solutions of* (2.1) *defined above for $\rho \in T_0$ with $|\rho| > R_0$, and let Δ_0 be the characteristic determinant of the differential operator L given by*

$$\Delta_0(\rho) = \det(B_i(u_{0k}(\cdot, \rho))) \quad \textit{for } \rho \in G_0,$$

where $G_0 = \{\rho \in \operatorname{Int} T_0 \mid |\rho| > R_0\}$. Then Δ_0 is analytic on the open set G_0, and Δ_0 has the representation

$$\Delta_0(\rho) = \pi_1(\rho)e^{i\rho} + \pi_0(\rho) + \Phi_{01}(\rho)e^{i\rho} + \Phi_{00}(\rho)$$

for $\rho \in G_0$, where the functions $\Phi_{0i}(\rho)$, $i = 0, 1$, are analytic on G_0 and satisfy the estimates $|\Phi_{0i}(\rho)| \leq \gamma |\rho|^{-(m-p_0)}$ for $\rho \in G_0$ and for $i = 0, 1$.

In the final part of this section we form the characteristic determinant on the sector T_1. The starting point for the discussion is the set of functions $v_{1k}(\cdot, \rho)$, $k = 0, 1, \ldots, n-1$, which form a basis for the solution space of the differential equation (2.1) for $\rho \in T_1$ with $|\rho| > R_0$. In terms of these functions we introduce the *modified solutions* of differential equation (2.1) relative to the sector T_1:

$$u_{1k}(t, \rho) := e^{-i\rho\omega_k} v_{1k}(t, \rho), \qquad k = 0, 1, \ldots, \nu - 1,$$
$$u_{1k}(t, \rho) := v_{1k}(t, \rho), \qquad k = \nu, \ldots, n-1,$$

for $0 \leq t \leq 1$ and for $\rho \in T_1$ with $|\rho| > R_0$. These functions also form a basis for the solution space of the differential equation (2.1). From the regularity results of

Theorem 4.7, for fixed $t \in [0,1]$ each of the functions $u_{1k}^{(\eta)}(t,\rho)$, $k,\eta = 0,1,\dots,n-1$, is an analytic function of the ρ variable on the open set

$$G_1 = \{\rho \in \operatorname{Int} T_1 \mid |\rho| > R_0\}.$$

Thus, the boundary data $u_{1k}^{(\eta)}(0,\rho)$, $u_{1k}^{(\eta)}(1,\rho)$, $k,\eta = 0,1,\dots,n-1$, consists of functions of ρ that are analytic on G_1.

For these solutions we obtain the estimates

$$|u_{10}(t,\rho)| \le 2e^{-b(t-1)}, \tag{5.49}$$

$$|u_{1k}(t,\rho)| \le 2e^{-(1-t)\alpha|\rho|} \le 2, \qquad k = 1,\dots,\nu-1, \tag{5.50}$$

$$|u_{1k}(t,\rho)| \le 2e^{-t\alpha|\rho|} \le 2, \qquad k = \nu,\dots,n-1, \tag{5.51}$$

for $0 \le t \le 1$ and for $\rho = a+ib \in T_1$ with $|\rho| > R_0$. In particular, for $\rho = a+ib$ belonging to the sector S_1 with $|\rho| > R_0$ and $b \le 0$, we have

$$|u_{1k}(t,\rho)| \le 2, \qquad 0 \le t \le 1, \quad k = 0,1,\dots,n-1. \tag{5.52}$$

The *characteristic determinant* of the differential operator L relative to the sector T_1 is the analytic function Δ_1 defined by

$$\Delta_1(\rho) := \det(B_i(u_{1k}(\,\cdot\,,\rho))) \quad \text{for } \rho \in G_1.$$

A complex number $\lambda = \rho^n$ with $\rho \in G_1$ is an eigenvalue of L if and only if $\Delta_1(\rho) = 0$. If we expand the determinant for Δ_1 using linearity in the 0th column, then we obtain the representation

$$\Delta_1(\rho) = \pi_1'(\rho)e^{-i\rho} + \pi_0'(\rho) + \Phi_{11}(\rho)e^{-i\rho} + \Phi_{10}(\rho) \tag{5.53}$$

for $\rho \in G_1$. In equation (5.53) the functions $\pi_i'(\rho)$, $i = 0,1$, are the functions defined earlier in Chapter 3 for the case n odd, viz.,

$$\pi_1'(\rho) = \sum_{\kappa=-(m-p_0-1)}^{q} a_\kappa' \rho^\kappa, \quad \pi_0'(\rho) = \sum_{\kappa=-(m-p_0-1)}^{p} b_\kappa' \rho^\kappa \tag{5.54}$$

for $\rho \ne 0$ in $\mathbb{C}$, these functions being analytic for $\rho \ne 0$ in $\mathbb{C}$; the functions $\Phi_{1i}(\rho)$, $i = 0,1$, are analytic on the open set G_1 and satisfy the growth rates

$$|\Phi_{1i}(\rho)| \le \gamma_4 |\rho|^{-(m-p_0)}, \qquad i = 0,1, \tag{5.55}$$

for $\rho \in G_1$. Recall that the functions $\pi_i(\rho)$, $i = 0,1$, appearing in the representation (5.46) of the characteristic determinant $\Delta_0(\rho)$ are related to the functions $\pi_i'(\rho)$, $i = 0,1$, appearing in the representation (5.53) of the characteristic determinant $\Delta_1(\rho)$ by equation (3.46), namely

$$\pi_1'(\rho) = \pi_0(\rho\omega_\nu), \qquad \pi_0'(\rho) = \pi_1(\rho\omega_{\nu-1}) \tag{5.56}$$

for $\rho \ne 0$ in $\mathbb{C}$.

The representation (5.53) is our working form for the characteristic determinant Δ_1 relative to the sector T_1. Compare the representation (5.53) for the characteristic determinant Δ_1 to the representation (3.51) for the approximate characteristic determinant $\widetilde{\Delta}$.

The above results are summarized in the following theorem. Again we assume conditions (i)–(vi) that precede Theorem 5.3.

THEOREM 5.4. *Let n be odd: $n = 2\nu - 1$. Under the assumptions* (i)–(vi), *let $\pi_i'(\rho)$, $i = 0, 1$, be the functions defined in Section 3.3 and in equation* (5.54) *and satisfying the relation* (5.56), *let $v_{1k}(\cdot, \rho)$, $k = 0, 1, \dots, n-1$, be the linearly independent solutions of the differential equation* (2.1) *constructed in Theorem 4.7 for $\rho \in T_1$ with $|\rho| > R_0$, let $u_{1k}(\cdot, \rho)$, $k = 0, 1, \dots, n-1$, be the modified solutions of* (2.1) *defined above for $\rho \in T_1$ with $|\rho| > R_0$, and let Δ_1 be the characteristic determinant of the differential operator L given by*

$$\Delta_1(\rho) = \det(B_i(u_{1k}(\cdot, \rho))) \quad \textit{for } \rho \in G_1,$$

where $G_1 = \{\rho \in \operatorname{Int} T_1 \mid |\rho| > R_0\}$. Then Δ_1 is analytic on the open set G_1, and Δ_1 has the representation

$$\Delta_1(\rho) = \pi_1'(\rho)e^{-i\rho} + \pi_0'(\rho) + \Phi_{11}(\rho)e^{-i\rho} + \Phi_{10}(\rho)$$

for $\rho \in G_1$, where the functions $\Phi_{1i}(\rho)$, $i = 0, 1$, are analytic on G_1 and satisfy the estimates $|\Phi_{1i}(\rho)| \le \gamma|\rho|^{-(m-p_0)}$ for $\rho \in G_1$ and for $i = 0, 1$.

5.3. Special Case: $n = 2$

To illustrate these ideas, let us examine the special case $n = 2$, where the differential operator L is determined by the formal differential operator

$$\ell = -\left(\frac{d}{dt}\right)^2 + q(t), \qquad q(t) = a_0(t),$$

and by the boundary values

$$\begin{aligned} B_1(u) &= a_1u'(0) + b_1u'(1) + a_0u(0) + b_0u(1), \\ B_2(u) &= c_1u'(0) + d_1u'(1) + c_0u(0) + d_0u(1). \end{aligned}$$

The boundary coefficient matrix

$$A = \begin{pmatrix} a_1 & b_1 & a_0 & b_0 \\ c_1 & d_1 & c_0 & d_0 \end{pmatrix}$$

is assumed to be in reduced row echelon form with rank 2, and $p_0 = m_1 + m_2$ can have the values 2, 1, or 0. Let $A_{ij}, 1 \le i < j \le 4$, denote the determinant of the 2×2 submatrix of A obtained by retaining the ith and jth columns. The six parameters A_{ij} play very prominent roles in the characteristic determinants of L.

Fix the integer m at the value $m = 3$, so we certainly satisfy the conditions $m > n$ and $m > p_0$. Theorems 5.1 and 5.2 are applicable provided the integer p satisfies the condition $-(2 - p_0) \le p \le p_0$ with $a_p \ne 0$. These theorems can not be used if L is simply irregular with $-\infty < p < -(2 - p_0)$ or if L is degenerate irregular. We will see that these latter cases are quite rare, but can nonetheless occur.

In Section 2.2 the Birkhoff approximate solutions $z_0(t, \rho)$, $z_1(t, \rho)$ were calculated for $m = 3$:

$$\begin{aligned} z_0(t, \rho) &= e^{i\rho t}\Big\{1 + \frac{1}{2i}\,Q(t)\rho^{-1} + \Big[\frac{1}{4}\,q(t) - \frac{1}{4}\,q(0) - \frac{1}{8}\,Q(t)^2\Big]\rho^{-2}\Big\}, \\ z_1(t, \rho) &= e^{-i\rho t}\Big\{1 - \frac{1}{2i}\,Q(t)\rho^{-1} + \Big[\frac{1}{4}\,q(t) - \frac{1}{4}\,q(0) - \frac{1}{8}\,Q(t)^2\Big]\rho^{-2}\Big\} \end{aligned}$$

for $0 \le t \le 1$ and for $\rho \ne 0$ in $\mathbb{C}$, where $Q(t) = \int_0^t q(\xi)\,d\xi$ for $0 \le t \le 1$. The corresponding modified approximate solutions are

$$(5.57)\qquad \begin{aligned} y_0(t,\rho) &= e^{i\rho t}\Big\{1 + \frac{1}{2i}\,Q(t)\rho^{-1} + \Big[\frac{1}{4}\,q(t) - \frac{1}{4}\,q(0) - \frac{1}{8}\,Q(t)^2\Big]\rho^{-2}\Big\},\\ y_1(t,\rho) &= e^{-i\rho(t-1)}\Big\{1 - \frac{1}{2i}\,Q(t)\rho^{-1} + \Big[\frac{1}{4}\,q(t) - \frac{1}{4}\,q(0) - \frac{1}{8}\,Q(t)^2\Big]\rho^{-2}\Big\} \end{aligned}$$

for $0 \le t \le 1$ and for $\rho \ne 0$ in $\mathbb{C}$. We can then form the functions

$$(5.58)\qquad B_i(y_k(\,\cdot\,,\rho)) = \widehat{P}_{ik}(\rho) + \widehat{Q}_{ik}(\rho)e^{i\rho}$$

for $\rho \ne 0$ in $\mathbb{C}$ and for $i = 1, 2,\ k = 0, 1$, and hence, obtain the approximate characteristic determinant:

$$(5.59)\qquad \begin{aligned} \widehat{\Delta}(\rho) &= \det\begin{pmatrix} \widehat{P}_{10}(\rho) + \widehat{Q}_{10}(\rho)e^{i\rho} & \widehat{P}_{11}(\rho) + \widehat{Q}_{11}(\rho)e^{i\rho} \\ \widehat{P}_{20}(\rho) + \widehat{Q}_{20}(\rho)e^{i\rho} & \widehat{P}_{21}(\rho) + \widehat{Q}_{21}(\rho)e^{i\rho} \end{pmatrix} \\ &= \widehat{\pi}_2(\rho)e^{2i\rho} + \widehat{\pi}_1(\rho)e^{i\rho} + \widehat{\pi}_0(\rho) \end{aligned}$$

for $\rho \ne 0$ in $\mathbb{C}$. Cf. equation (3.13). The functions $\widehat{\pi}_i(\rho)$, $i = 0, 1, 2$, are given by

$$(5.60)\qquad \begin{aligned} \widehat{\pi}_2(\rho) &= \det\begin{pmatrix} \widehat{Q}_{10}(\rho) & \widehat{Q}_{11}(\rho) \\ \widehat{Q}_{20}(\rho) & \widehat{Q}_{21}(\rho) \end{pmatrix}, \\ \widehat{\pi}_0(\rho) &= -\widehat{\pi}_2(-\rho) = \det\begin{pmatrix} \widehat{P}_{10}(\rho) & \widehat{P}_{11}(\rho) \\ \widehat{P}_{20}(\rho) & \widehat{P}_{21}(\rho) \end{pmatrix}, \end{aligned}$$

$$(5.61)\qquad \widehat{\pi}_1(\rho) = \det\begin{pmatrix} \widehat{P}_{10}(\rho) & \widehat{Q}_{11}(\rho) \\ \widehat{P}_{20}(\rho) & \widehat{Q}_{21}(\rho) \end{pmatrix} + \det\begin{pmatrix} \widehat{Q}_{10}(\rho) & \widehat{P}_{11}(\rho) \\ \widehat{Q}_{20}(\rho) & \widehat{P}_{21}(\rho) \end{pmatrix}$$

for $\rho \ne 0$ in $\mathbb{C}$.

From equations (3.24), (3.27), and (3.28) we have

$$\begin{aligned} \widehat{\pi}_2(\rho) &= \sum_{\kappa=-(2-p_0)}^{p_0} a_\kappa \rho^\kappa + \sum_{\kappa=-4}^{-(3-p_0)} a_\kappa(3)\rho^\kappa, \\ \widehat{\pi}_1(\rho) &= \sum_{\kappa=-(2-p_0)}^{p_0} b_\kappa \rho^\kappa + \sum_{\kappa=-4}^{-(3-p_0)} b_\kappa(3)\rho^\kappa, \\ \widehat{\pi}_0(\rho) &= \sum_{\kappa=-(2-p_0)}^{p_0} c_\kappa \rho^\kappa + \sum_{\kappa=-4}^{-(3-p_0)} c_\kappa(3)\rho^\kappa \end{aligned}$$

for $\rho \ne 0$ in $\mathbb{C}$, with

$$(5.62)\qquad \pi_2(\rho) = \sum_{\kappa=-(2-p_0)}^{p_0} a_\kappa\rho^\kappa, \quad \pi_1(\rho) = \sum_{\kappa=-(2-p_0)}^{p_0} b_\kappa\rho^\kappa, \quad \pi_0(\rho) = \sum_{\kappa=-(2-p_0)}^{p_0} c_\kappa\rho^\kappa$$

for $\rho \ne 0$ in $\mathbb{C}$. Therefore, the powers of ρ appearing in the functions $\pi_i(\rho)$, $i = 0, 1, 2$, are $p_0 = 2{:}\,\rho^2, \rho^1, \rho^0$; $p_0 = 1{:}\,\rho^1, \rho^0, \rho^{-1}$; $p_0 = 0{:}\,\rho^0, \rho^{-1}, \rho^{-2}$. It follows that in calculating the functions $\widehat{\pi}_i(\rho)$, $i = 0, 1, 2$, if we compute the coefficients of the powers $\rho^2, \rho^1, \rho^0, \rho^{-1}, \rho^{-2}$ explicitly, then the corresponding functions $\pi_i(\rho)$, $i = 0, 1, 2$, can be read off.

Next, we calculate the functions appearing in (5.59). Substituting (5.57) into (5.58), the functions $\widehat{P}_{ik}(\rho)$, $\widehat{Q}_{ik}(\rho)$ are first determined. Then using (5.60) and (5.61), the functions $\widehat{\pi}_i(\rho)$, $i = 0, 1, 2$, are computed. After a lengthy calculation we arrive at the results:

(5.63)

$$\begin{aligned}\widehat{\pi}_2(\rho) = {} & -A_{12}\rho^2 + \Big\{\frac{i}{2}A_{12}Q(1) + i(A_{14}+A_{23})\Big\}\rho \\ & + \Big\{A_{12}\Big[\frac{1}{4}q(1) + \frac{3}{4}q(0) + \frac{1}{8}Q(1)^2\Big] + \frac{1}{2}(A_{14}+A_{23})Q(1) - A_{34}\Big\}\rho^0 \\ & + \Big\{-A_{12}\Big[\frac{i}{4}q'(0) + \frac{i}{4}q(0)Q(1) - \frac{i}{4}q'(1) + \frac{i}{4}q(1)Q(1)\Big] \\ & + iA_{14}\Big[\frac{1}{4}q(1) - \frac{3}{4}q(0) - \frac{1}{8}Q(1)^2\Big] \\ & - iA_{23}\Big[\frac{1}{4}q(1) + \frac{1}{4}q(0) + \frac{1}{8}Q(1)^2\Big] + \frac{i}{2}A_{34}Q(1)\Big\}\rho^{-1} \\ & + \Big\{-A_{12}\Big[\frac{1}{8}q'(0)Q(1) + \frac{1}{8}q(0)q(1) + \frac{1}{8}q(0)^2 + \frac{1}{16}q(0)Q(1)^2\Big] \\ & + iA_{14}\Big[\frac{i}{4}q'(0) + \frac{i}{4}q(0)Q(1)\Big] - iA_{23}\Big[\frac{i}{4}q'(1) - \frac{i}{4}q(1)Q(1)\Big] \\ & - A_{34}\Big[\frac{1}{4}q(1) - \frac{1}{4}q(0) - \frac{1}{8}Q(1)^2\Big]\Big\}\rho^{-2} + \rho^{-3}, \rho^{-4} \text{ terms,}\end{aligned}$$

(5.64) $$\widehat{\pi}_1(\rho) = 2i(A_{13}+A_{24})\rho - i(A_{13}+A_{24})q(0)\rho^{-1} + \rho^{-3}, \rho^{-4} \text{ terms,}$$

(5.65)

$$\begin{aligned}\widehat{\pi}_0(\rho) = {} & A_{12}\rho^2 + \Big\{\frac{i}{2}A_{12}Q(1) + i(A_{14}+A_{23})\Big\}\rho \\ & - \Big\{A_{12}\Big[\frac{1}{4}q(1) + \frac{3}{4}q(0) + \frac{1}{8}Q(1)^2\Big] + \frac{1}{2}(A_{14}+A_{23})Q(1) - A_{34}\Big\}\rho^0 \\ & + \Big\{-A_{12}\Big[\frac{i}{4}q'(0) + \frac{i}{4}q(0)Q(1) - \frac{i}{4}q'(1) + \frac{i}{4}q(1)Q(1)\Big] \\ & + iA_{14}\Big[\frac{1}{4}q(1) - \frac{3}{4}q(0) - \frac{1}{8}Q(1)^2\Big] \\ & - iA_{23}\Big[\frac{1}{4}q(1) + \frac{1}{4}q(0) + \frac{1}{8}Q(1)^2\Big] + \frac{i}{2}A_{34}Q(1)\Big\}\rho^{-1} \\ & + \Big\{A_{12}\Big[\frac{1}{8}q'(0)Q(1) + \frac{1}{8}q(0)q(1) + \frac{1}{8}q(0)^2 + \frac{1}{16}q(0)Q(1)^2\Big] \\ & - iA_{14}\Big[\frac{i}{4}q'(0) + \frac{i}{4}q(0)Q(1)\Big] + iA_{23}\Big[\frac{i}{4}q'(1) - \frac{i}{4}q(1)Q(1)\Big] \\ & + A_{34}\Big[\frac{1}{4}q(1) - \frac{1}{4}q(0) - \frac{1}{8}Q(1)^2\Big]\Big\}\rho^{-2} + \rho^{-3}, \rho^{-4} \text{ terms}\end{aligned}$$

for $\rho \neq 0$ in $\mathbb{C}$. Equations (5.63)–(5.65) uniquely determine the key functions $\pi_i(\rho)$, $i = 0, 1, 2$, and then equations (5.31) and (5.38) lead immediately to the asymptotic expansions of the characteristic determinants Δ_0 and Δ_1.

Let us look at the possible forms of the characteristic determinants. We will follow the classification scheme given in the four part series [**30, 31, 32, 33**]. See [**30**, p. 280].

Case 1. $A_{12} \neq 0$. For this case the boundary coefficient matrix A must have the form

$$\begin{pmatrix} 1 & 0 & a_0 & b_0 \\ 0 & 1 & c_0 & d_0 \end{pmatrix},$$

and hence, $A_{12} = 1$, $m_1 = m_2 = 1$, and $p_0 = 2$. From (5.63)–(5.65) we see that $p = q = 2$, $a_2 = -c_2 = -1$ (see equation (3.30)), and $b_2 = 0$. In Case 1 the differential operator L is regular. From (5.31) and (5.38) the characteristic determinants have the asymptotic expansions

$$\begin{aligned} \Delta_0(\rho) &= [-\rho^2 + O(\rho)]e^{2i\rho} + O(\rho)e^{i\rho} + [\rho^2 + O(\rho)] && \text{for } \rho \in G_0, \\ \Delta_1(\rho) &= [-\rho^2 + O(\rho)] + O(\rho)e^{-i\rho} + [\rho^2 + O(\rho)]e^{-2i\rho} && \text{for } \rho \in G_1. \end{aligned} \tag{5.66}$$

Case 2. $A_{12} = 0$, $A_{14} + A_{23} \neq 0$. Here there are three possible forms for the boundary coefficient matrix A:

$$\begin{pmatrix} 1 & b_1 & 0 & b_0 \\ 0 & 0 & 1 & d_0 \end{pmatrix}, \quad \begin{pmatrix} 1 & b_1 & a_0 & 0 \\ 0 & 0 & 0 & 1 \end{pmatrix}, \quad \begin{pmatrix} 0 & 1 & 0 & b_0 \\ 0 & 0 & 1 & d_0 \end{pmatrix},$$

and for each of these cases it is clear that $p_0 = 1$. The forms

$$\begin{pmatrix} 0 & 1 & a_0 & 0 \\ 0 & 0 & 0 & 1 \end{pmatrix}, \quad \begin{pmatrix} 0 & 0 & 1 & 0 \\ 0 & 0 & 0 & 1 \end{pmatrix}$$

are not possible because of the condition $A_{14} + A_{23} \neq 0$. Also, we have $p = q = 1$, $a_1 = c_1 = i(A_{14}+A_{23})$, and $b_1 = 2i(A_{13}+A_{24})$. The differential operator L is again regular in this case. For the characteristic determinants we have the asymptotic expansions

$$\begin{aligned} \Delta_0(\rho) &= [i(A_{14} + A_{23})\rho + O(1)]e^{2i\rho} + [2i(A_{13} + A_{24})\rho + O(1)]e^{i\rho} \\ &\qquad + [i(A_{14} + A_{23})\rho + O(1)] \qquad \text{for } \rho \in G_0, \\ \Delta_1(\rho) &= [i(A_{14} + A_{23})\rho + O(1)] + [2i(A_{13} + A_{24})\rho + O(1)]e^{-i\rho} \\ &\qquad + [i(A_{14} + A_{23})\rho + O(1)]e^{-2i\rho} \quad \text{for } \rho \in G_1. \end{aligned} \tag{5.67}$$

Case 3. $A_{12} = 0$, $A_{14}+A_{23} = 0$, $A_{34} \neq 0$, $A_{13}+A_{24} = 0$. The boundary coefficient matrix A must have either the form

$$\begin{pmatrix} 1 & b_1 & 0 & b_0 \\ 0 & 0 & 1 & d_0 \end{pmatrix} \quad \text{or} \quad \begin{pmatrix} 0 & 0 & 1 & 0 \\ 0 & 0 & 0 & 1 \end{pmatrix},$$

where $p_0 = 1$ in the first case and $p_0 = 0$ in the second case, which corresponds to Dirichlet boundary conditions. From (5.63)–(5.65) we see that $p = q = 0$, $a_0 = -c_0 = -A_{34}$, and $b_0 = 0$. Thus, the differential operator L where $p_0 = 1$ is simply irregular, while the differential operator L where $p_0 = 0$ (Dirichlet) is regular. For the characteristic determinants we obtain the asymptotic expansions

(5.68)

$$\begin{aligned} \Delta_0(\rho) &= [-A_{34} + O(\rho^{-1})]e^{2i\rho} + O(\rho^{-1})e^{i\rho} + [A_{34} + O(\rho^{-1})] && \text{for } \rho \in G_0, \\ \Delta_1(\rho) &= [-A_{34} + O(\rho^{-1})] + O(\rho^{-1})e^{-i\rho} + [A_{34} + O(\rho^{-1})]e^{-2i\rho} && \text{for } \rho \in G_1. \end{aligned}$$

Case 4. $A_{12} = 0$, $A_{14} + A_{23} = 0$, $A_{34} \neq 0$, $A_{13} + A_{24} \neq 0$. For the fourth case the boundary coefficient matrix A has one of the two forms

$$\begin{pmatrix} 1 & b_1 & 0 & b_0 \\ 0 & 0 & 1 & d_0 \end{pmatrix} \quad \text{or} \quad \begin{pmatrix} 0 & 1 & a_0 & 0 \\ 0 & 0 & 0 & 1 \end{pmatrix},$$

so $p_0 = 1$. From (5.63)–(5.65) we see that $p = 0$, $q = 1$, $a_0 = -c_0 = -A_{34}$, and $b_1 = 2i(A_{13} + A_{24})$, and hence, the differential operator L is always simply irregular in this case. For the characteristic determinants we have the representations

$$(5.69)\qquad \begin{aligned} \Delta_0(\rho) &= [-A_{34} + O(\rho^{-1})]e^{2i\rho} + [2i(A_{13} + A_{24})\rho + O(\rho^{-1})]e^{i\rho} \\ &\quad + [A_{34} + O(\rho^{-1})] \qquad \text{for } \rho \in G_0, \\ \Delta_1(\rho) &= [-A_{34} + O(\rho^{-1})] + [2i(A_{13} + A_{24})\rho + O(\rho^{-1})]e^{-i\rho} \\ &\quad + [A_{34} + O(\rho^{-1})]e^{-2i\rho} \quad \text{for } \rho \in G_1. \end{aligned}$$

Case 5. $A_{12} = 0$, $A_{14}+A_{23} = 0$, $A_{34} = 0$. In this last case the boundary coefficient matrix A must have the same form as in Case 4:

$$\begin{pmatrix} 1 & b_1 & 0 & b_0 \\ 0 & 0 & 1 & d_0 \end{pmatrix} \quad \text{or} \quad \begin{pmatrix} 0 & 1 & a_0 & 0 \\ 0 & 0 & 0 & 1 \end{pmatrix},$$

and hence, $p_0 = 1$. The functions $\pi_i(\rho)$, $i = 0, 1, 2$, are extracted from (5.63)–(5.65) by keeping only the powers $\rho^1, \rho^0, \rho^{-1}$. Thus,

$$\begin{aligned} \pi_2(\rho) &= \Big\{ iA_{14}\Big[\frac{1}{4}\, q(1) - \frac{3}{4}\, q(0) - \frac{1}{8}\, Q(1)^2 \Big] \\ &\qquad - iA_{23}\Big[\frac{1}{4}\, q(1) + \frac{1}{4}\, q(0) + \frac{1}{8}\, Q(1)^2 \Big] \Big\} \rho^{-1}, \\ \pi_1(\rho) &= 2i(A_{13} + A_{24})\rho - i(A_{13} + A_{24})q(0)\rho^{-1}, \\ \pi_0(\rho) &= \Big\{ iA_{14}\Big[\frac{1}{4}\, q(1) - \frac{3}{4}\, q(0) - \frac{1}{8}\, Q(1)^2 \Big] \\ &\qquad - iA_{23}\Big[\frac{1}{4}\, q(1) + \frac{1}{4}\, q(0) + \frac{1}{8}\, Q(1)^2 \Big] \Big\} \rho^{-1} \end{aligned}$$

for $\rho \neq 0$ in $\mathbb{C}$, and for the constants we have $a_1 = c_1 = 0$, $a_0 = -c_0 = 0$,

$$\begin{aligned} a_{-1} = c_{-1} &= iA_{14}\Big[\frac{1}{4}\, q(1) - \frac{3}{4}\, q(0) - \frac{1}{8}\, Q(1)^2 \Big] \\ &\qquad - iA_{23}\Big[\frac{1}{4}\, q(1) + \frac{1}{4}\, q(0) + \frac{1}{8}\, Q(1)^2 \Big], \end{aligned}$$

$$b_1 = 2i(A_{13} + A_{24}), \quad b_0 = 0, \quad b_{-1} = -i(A_{13} + A_{24})q(0).$$

This shows that the differential operator L is irregular. More precisely, if $a_{-1} \neq 0$, then $p = -1$, $q = 1$ or $q = -1$, and L is simply irregular. Example 1.11 in Chapter 1 is a model of this case. On the other hand, if $a_{-1} = 0$, then we can not determine whether L is simply irregular or degenerate irregular. To get a definitive classification, we must use a larger integer m, thus enlarging our "window" for viewing the constants a_κ, b_κ, c_κ. In the simply irregular case where $a_{-1} \neq 0$, the characteristic determinants have the form

$$(5.70)\qquad \begin{aligned} \Delta_0(\rho) &= [a_{-1}\rho^{-1} + O(\rho^{-2})]e^{2i\rho} + [b_1\rho + b_{-1}\rho^{-1} + O(\rho^{-2})]e^{i\rho} \\ &\quad + [c_{-1}\rho^{-1} + O(\rho^{-2})] \qquad \text{for } \rho \in G_0, \\ \Delta_1(\rho) &= [a_{-1}\rho^{-1} + O(\rho^{-2})] + [b_1\rho + b_{-1}\rho^{-1} + O(\rho^{-2})]e^{-i\rho} \\ &\quad + [c_{-1}\rho^{-1} + O(\rho^{-2})]e^{-2i\rho} \quad \text{for } \rho \in G_1. \end{aligned}$$

CHAPTER 6

The Green's Function

For both n even and n odd, if we start with any point $\lambda = \rho^n$ in $\mathbb{C}$ with $\rho \in G_0$ and $\Delta_0(\rho) \neq 0$ or with $\rho \in G_1$ and $\Delta_1(\rho) \neq 0$, then λ belongs to the resolvent set $\rho(L)$ and the resolvent $R_\lambda(L) = (\lambda I - L)^{-1}$ exists as an integral operator on $L^2[0,1]$ with the Green's function $G(t,s;\lambda)$ as its kernel:

$$R_\lambda(L)u(t) = \int_0^1 G(t,s;\lambda)u(s)\,ds, \qquad 0 \leq t \leq 1, \quad u \in L^2[0,1].$$

In this chapter we construct important representations of the resolvent and the Green's function, and then use these representations to derive their growth rates for ρ belonging to the sectors S_0 and S_1. Our representations are first developed for ρ belonging to the open set G_0, and then analogous representations are established for ρ in the open set G_1. Again the discussion is divided into the cases n even and n odd.

6.1. The Green's Function for n Even

Assume that n is even: $n = 2\nu \geq 2$. For the even order case recall that the sectors S_0 and S_1 are given by

$$S_0\colon\ \text{all } \rho = |\rho|e^{i\theta} \in \mathbb{C} \text{ with } 0 \leq \theta \leq \frac{\pi}{n},$$
$$S_1\colon\ \text{all } \rho = |\rho|e^{i\theta} \in \mathbb{C} \text{ with } -\frac{\pi}{n} \leq \theta \leq 0,$$

with T_0 and T_1 the corresponding translated sectors. For each $\rho \in T_0$ with $|\rho| > R_0$, let us consider the basis $v_{00}(\,\cdot\,,\rho), v_{01}(\,\cdot\,,\rho), \ldots, v_{0\,n-1}(\,\cdot\,,\rho)$ for the solution space of the differential equation (2.1) determined in Theorem 4.3. In Chapter 5 we showed that

$$v_{0k}^{(\alpha)}(t,\rho) = \rho^\alpha e^{i\rho\omega_k t}\big[(i\omega_k)^\alpha + F_{0k\alpha}(t,\rho)\big] \tag{6.1}$$

for $0 \leq t \leq 1$, for $\rho \in T_0$ with $|\rho| > R_0$, and for $k, \alpha = 0, 1, \ldots, n-1$, where the function $F_{0k\alpha}(\,\cdot\,,\rho)$ belongs to $H^{n-\alpha}[0,1]$ with $F_{0k\alpha}(t,\rho) \to 0$ uniformly on $[0,1] \times T_0$ as $|\rho| \to \infty$ and with

$$|F_{0k\alpha}(t,\rho)| \leq 1 \tag{6.2}$$

for $0 \leq t \leq 1$, for $\rho \in T_0$ with $|\rho| > R_0$, and for $k, \alpha = 0, 1, \ldots, n-1$. Also, for $\rho \in T_0$ with $|\rho| > R_0$ we formed the modified solutions of the differential equation (2.1):

$$u_{0k}(t,\rho) = v_{0k}(t,\rho), \qquad k = 0, 1, \ldots, \nu - 1,$$
$$u_{0k}(t,\rho) = e^{-i\rho\omega_k} v_{0k}(t,\rho), \qquad k = \nu, \ldots, n-1,$$

for $0 \le t \le 1$. These functions have the representations

$$
(6.3) \qquad \begin{aligned} u_{0k}^{(\alpha)}(t,\rho) &= \rho^{\alpha} e^{i\rho\omega_k t}\big[(i\omega_k)^{\alpha} + F_{0k\alpha}(t,\rho)\big], && k = 0,1,\dots,\nu-1, \\ u_{0k}^{(\alpha)}(t,\rho) &= \rho^{\alpha} e^{i\rho\omega_k (t-1)}\big[(i\omega_k)^{\alpha} + F_{0k\alpha}(t,\rho)\big], && k = \nu,\dots,n-1, \end{aligned}
$$

for $0 \le t \le 1$, for $\rho \in T_0$ with $|\rho| > R_0$, and for $\alpha = 0,1,\dots,n-1$. From equation (5.18) we have the bounds

$$
(6.4) \qquad |u_{0k}(t,\rho)| \le 2, \qquad 0 \le t \le 1, \quad k = 0,1,\dots,n-1,
$$

for $\rho = a + ib$ belonging to the sector S_0 (where $b \ge 0$) with $|\rho| > R_0$.

Let L_0 be the nth order differential operator in $L^2[0,1]$ defined by

$$
\mathcal{D}(L_0) = \{u \in H^n[0,1] \mid u^{(n-i)}(0) = 0,\ i = 1,\dots,n\}, \qquad L_0 u = \ell u.
$$

Clearly the resolvent set $\rho(L_0)$ is equal to $\mathbb{C}$ due to the initial value conditions at $t = 0$. We begin by computing the Green's function $g(t,s;\lambda)$ of the differential operator $\lambda I - L_0$, and then use it to compute the Green's function $G(t,s;\lambda)$ of the differential operator $\lambda I - L$. The algorithm in Example III.3.18 of [**28**] will be employed.

For $\lambda = \rho^n$ in $\mathbb{C}$ and $\rho \in T_0$ with $|\rho| > R_0$, our earlier work has shown that the Green's function $g(t,s;\lambda)$ is given by

$$
(6.5) \qquad \begin{aligned} g(t,s;\lambda) &= \sum_{k=0}^{n-1} v_{0k}(t,\rho)\eta_{0k}(s,\rho), && 0 \le s < t \le 1, \\ g(t,s;\lambda) &= 0, && 0 \le t < s \le 1, \end{aligned}
$$

where the functions $\eta_{0k}(\,\cdot\,,\rho)$, $k = 0,1,\dots,n-1$, belong to $H^n[0,1]$ and are determined by the linear system

$$
(6.6) \qquad \sum_{k=0}^{n-1} v_{0k}^{(\alpha)}(s,\rho)\eta_{0k}(s,\rho) = -\delta_{\alpha\, n-1} i^n, \qquad \alpha = 0,1,\dots,n-1,
$$

for $0 \le s \le 1$. Using equation (6.1), the system (6.6) can be rewritten in the form

$$
(6.7) \qquad \begin{aligned} \sum_{k=0}^{n-1} \big[(i\omega_k)^{\alpha} + F_{0k\alpha}(s,\rho)\big] e^{i\rho\omega_k s}\eta_{0k}(s,\rho) &= -\delta_{\alpha\, n-1} i^n \rho^{-(n-1)}, \\ \alpha &= 0,1,\dots,n-1, \end{aligned}
$$

for $0 \le s \le 1$. We will treat (6.7) as a linear system for the unknowns $e^{i\rho\omega_k s}\eta_{0k}(s,\rho)$, $k = 0,1,\dots,n-1$, with $n \times n$ coefficient matrix $A_0(s,\rho) := \big((i\omega_k)^{\alpha} + F_{0k\alpha}(s,\rho)\big)$. Previously in Chapter 4 this matrix was shown to be nonsingular for $0 \le s \le 1$ and for $\rho \in T_0$ with $|\rho| > R_0$.

For the Vandermonde matrix

$$
V := \big((i\omega_k)^{\alpha}\big) = \begin{pmatrix} 1 & 1 & \cdots & 1 \\ i\omega_0 & i\omega_1 & \cdots & i\omega_{n-1} \\ \vdots & \vdots & & \vdots \\ (i\omega_0)^{n-1} & (i\omega_1)^{n-1} & \cdots & (i\omega_{n-1})^{n-1} \end{pmatrix},
$$

the inverse is given by

$$V^{-1} = \frac{1}{n i^{n-1}} \left(i^{n-k-1} \omega_\alpha^{n-k} \right) = \left(\frac{1}{n i^{k}} \omega_\alpha^{n-k} \right)$$

$$= \frac{1}{n i^{n-1}} \begin{pmatrix} i^{n-1}\omega_0^n & i^{n-2}\omega_0^{n-1} & \cdots & i\omega_0^2 & \omega_0 \\ i^{n-1}\omega_1^n & i^{n-2}\omega_1^{n-1} & \cdots & i\omega_1^2 & \omega_1 \\ \vdots & \vdots & & \vdots & \vdots \\ i^{n-1}\omega_{n-1}^n & i^{n-2}\omega_{n-1}^{n-1} & \cdots & i\omega_{n-1}^2 & \omega_{n-1} \end{pmatrix}.$$

Let us write the inverse of $A_0(s, \rho)$ in the form

$$A_0(s, \rho)^{-1} := \left(\frac{1}{n i^{k}} \omega_\alpha^{n-k} \left[1 + G_{0k\alpha}(s, \rho) \right] \right)$$

for $0 \le s \le 1$ and for $\rho \in T_0$ with $|\rho| > R_0$. Then from (6.7) it follows that

$$\begin{pmatrix} e^{i\rho\omega_0 s}\eta_{00}(s, \rho) \\ e^{i\rho\omega_1 s}\eta_{01}(s, \rho) \\ \vdots \\ e^{i\rho\omega_{n-1} s}\eta_{0\,n-1}(s, \rho) \end{pmatrix} = A_0(s, \rho)^{-1} \begin{pmatrix} 0 \\ \vdots \\ 0 \\ -i^{n}\rho^{-(n-1)} \end{pmatrix}$$

$$= \frac{-i}{n\rho^{n-1}} \begin{pmatrix} \omega_0[1 + G_{0\,n-1\,0}(s, \rho)] \\ \omega_1[1 + G_{0\,n-1\,1}(s, \rho)] \\ \vdots \\ \omega_{n-1}[1 + G_{0\,n-1\,n-1}(s, \rho)] \end{pmatrix},$$

and hence,

$$\eta_{0k}(s, \rho) = -\frac{i\omega_k}{n\rho^{n-1}} e^{-i\rho\omega_k s}[1 + G_{0\,n-1\,k}(s, \rho)] \tag{6.8}$$

for $0 \le s \le 1$, for $\rho \in T_0$ with $|\rho| > R_0$, and for $k = 0, 1, \ldots, n-1$. Since the functions $\eta_{0k}(\,\cdot\,, \rho)$ belong to $H^n[0, 1]$, the functions $G_{0\,n-1\,k}(\,\cdot\,, \rho)$ also belong to $H^n[0, 1]$.

We assert that the functions $G_{0k\alpha}(s, \rho) \to 0$ uniformly on $[0, 1] \times T_0$ as $|\rho| \to \infty$ for $k, \alpha = 0, 1, \ldots, n-1$. Take any $\epsilon > 0$. Using the continuity of matrix inversion and the fact that the Vandermonde matrix V is nonsingular, choose a number $\delta > 0$ such that if $W = (w_{\alpha k})$ is an $n \times n$ matrix with $|(i\omega_k)^\alpha - w_{\alpha k}| < \delta$ for $\alpha, k = 0, 1, \ldots, n-1$, then W is nonsingular, and the inverse $W^{-1} = (x_{\alpha k})$ satisfies

$$\left| \frac{1}{n i^{k}} \omega_\alpha^{n-k} - x_{\alpha k} \right| < \epsilon \qquad \text{for } \alpha, k = 0, 1, \ldots, n-1.$$

Choose a constant $R_\delta \ge R_0$ such that $|F_{0k\alpha}(s, \rho)| < \delta$ for $0 \le s \le 1$ and for $\rho \in T_0$ with $|\rho| > R_\delta$, for $\alpha, k = 0, 1, \ldots, n-1$. Then for $0 \le s \le 1$ and for $\rho \in T_0$ with $|\rho| > R_\delta$, we have

$$|(i\omega_k)^\alpha - [(i\omega_k)^\alpha + F_{0k\alpha}(s, \rho)]| = |F_{0k\alpha}(s, \rho)| < \delta$$

for $\alpha, k = 0, 1, \ldots, n-1$, and hence, from the above

$$\left| \frac{1}{n i^{k}} \omega_\alpha^{n-k} - \frac{1}{n i^{k}} \omega_\alpha^{n-k} \left[1 + G_{0k\alpha}(s, \rho) \right] \right| = \frac{1}{n} |G_{0k\alpha}(s, \rho)| < \epsilon$$

for $0 \le s \le 1$ and for $\rho \in T_0$ with $|\rho| > R_\delta$, for $k, \alpha = 0, 1, \ldots, n-1$. This establishes the assertion. Based on this assertion, without loss of generality we can assume that the constant R_0 chosen earlier also yields the bound

$$|G_{0k\alpha}(s,\rho)| \le 1 \tag{6.9}$$

for $0 \le s \le 1$, for $\rho \in T_0$ with $|\rho| > R_0$, and for $k, \alpha = 0, 1, \dots, n-1$.

Summarizing, for $\lambda = \rho^n$ in $\mathbb{C}$ and $\rho \in T_0$ with $|\rho| > R_0$, the Green's function for the differential operator $\lambda I - L_0$ is given by

$$\begin{aligned} g(t,s;\lambda) &= -\frac{1}{n\rho^{n-1}} \sum_{k=0}^{n-1} (i\omega_k) e^{i\rho\omega_k(t-s)} [1 + F_{0k0}(t,\rho)][1 + G_{0\,n-1\,k}(s,\rho)], \\ & \qquad\qquad 0 \le s < t \le 1, \\ g(t,s;\lambda) &= 0, \qquad\qquad 0 \le t < s \le 1, \end{aligned} \tag{6.10}$$

where the functions $F_{0k0}(\,\cdot\,,\rho)$ and $G_{0\,n-1\,k}(\,\cdot\,,\rho)$ belong to $H^n[0,1]$ and satisfy the bounds given in equations (6.2) and (6.9).

Next, we rewrite (6.5) or (6.10) in the form

$$g(t,s;\lambda) = k_0(t,s;\rho) + \ell_0(t,s;\rho)$$

for $t \ne s$ in $[0,1]$ and for $\lambda = \rho^n$ in $\mathbb{C}$ and $\rho \in T_0$ with $|\rho| > R_0$, where

$$\begin{aligned} k_0(t,s;\rho) &:= \sum_{k=0}^{\nu-1} v_{0k}(t,\rho)\eta_{0k}(s,\rho), \qquad 0 \le s < t \le 1, \\ k_0(t,s;\rho) &:= -\sum_{k=\nu}^{n-1} v_{0k}(t,\rho)\eta_{0k}(s,\rho), \qquad 0 \le t < s \le 1, \end{aligned} \tag{6.11}$$

and

$$\ell_0(t,s;\rho) := \sum_{k=\nu}^{n-1} v_{0k}(t,\rho)\eta_{0k}(s,\rho), \qquad 0 \le t, s \le 1.$$

The Green's function $g(t,s;\lambda)$ is the kernel of the integral operator $R_\lambda(L_0)$, which assigns to each u in $L^2[0,1]$ the unique solution $z \in \mathcal{D}(L_0)$ of the initial value problem $(\lambda I - L_0)z = u$. Also, the function $\ell_0(t,s;\rho)$ is the kernel of an integral operator which maps $L^2[0,1]$ into the solution space of the differential equation $(\rho^n I - \ell)u = 0$.

For each $\rho \in T_0$ with $|\rho| > R_0$, let $K_{0\rho}$ be the integral operator on $L^2[0,1]$ defined by

$$K_{0\rho}u(t) := \int_0^1 k_0(t,s;\rho)u(s)\,ds, \qquad 0 \le t \le 1,$$

for $u \in L^2[0,1]$. From the above remarks it follows that if $u \in L^2[0,1]$ and $v = K_{0\rho}u$, then v belongs to $H^n[0,1]$ and $(\rho^n I - \ell)v = u$. Using induction with equation (6.6), we see immediately that the derivatives of $v = K_{0\rho}u$ satisfy the equations

$$\begin{aligned} v^{(j)}(t) &= \int_0^1 \frac{\partial^j k_0}{\partial t^j}(t,s;\rho)u(s)\,ds \\ &= \sum_{k=0}^{\nu-1} v_{0k}^{(j)}(t,\rho) \int_0^t \eta_{0k}(s,\rho)u(s)\,ds - \sum_{k=\nu}^{n-1} v_{0k}^{(j)}(t,\rho) \int_t^1 \eta_{0k}(s,\rho)u(s)\,ds \end{aligned} \tag{6.12}$$

for $0 \le t \le 1$ and for $j = 0, 1, \dots, n-1$, valid for each $\rho \in T_0$ with $|\rho| > R_0$.

Using (6.1) and (6.8), for $\rho \in T_0$ with $|\rho| > R_0$ the kernel $k_0(t, s; \rho)$ can be expressed as

$$(6.13)\quad \begin{aligned} k_0(t,s;\rho) &= -\frac{1}{n\rho^{n-1}} \sum_{k=0}^{\nu-1} (i\omega_k) e^{i\rho\omega_k(t-s)} [1 + F_{0k0}(t,\rho)][1 + G_{0\,n-1\,k}(s,\rho)], \\ & \qquad\qquad 0 \le s < t \le 1, \\ k_0(t,s;\rho) &= \frac{1}{n\rho^{n-1}} \sum_{k=\nu}^{n-1} (i\omega_k) e^{i\rho\omega_k(t-s)} [1 + F_{0k0}(t,\rho)][1 + G_{0\,n-1\,k}(s,\rho)], \\ & \qquad\qquad 0 \le t < s \le 1. \end{aligned}$$

Now take any point $\rho = a + ib \in T_0$ with $|\rho| > R_0$. If t, s are real numbers with $0 \le s < t \le 1$, then $0 < t - s \le 1$, and by (5.8) we obtain the estimates

$$(6.14)\qquad \left|e^{i\rho\omega_0(t-s)}\right| = e^{-b(t-s)},$$

$$(6.15)\qquad \left|e^{i\rho\omega_k(t-s)}\right| \le e^{-(t-s)\alpha|\rho|} \le 1, \qquad k = 1, \dots, \nu - 1;$$

for real numbers t, s with $0 \le t < s \le 1$, we have $0 \le 1 + t - s < 1$, and by (5.9)

$$(6.16)\qquad \left|e^{i\rho\omega_\nu(t-s)}\right| = e^{-b(s-t)},$$

$$(6.17)\qquad \left|e^{i\rho\omega_k(t-s)}\right| = \left|e^{i\rho\omega_k(1+t-s-1)}\right| \le e^{-(s-t)\alpha|\rho|} \le 1, \quad k = \nu + 1, \dots, n - 1.$$

In particular, if $\rho = a + ib \in S_0$ with $|\rho| > R_0$, then these estimates give

$$(6.18)\qquad \left|e^{i\rho\omega_k(t-s)}\right| \le 1, \qquad 0 \le s < t \le 1, \quad k = 0, 1, \dots, \nu - 1,$$

$$(6.19)\qquad \left|e^{i\rho\omega_k(t-s)}\right| \le 1, \qquad 0 \le t < s \le 1, \quad k = \nu, \dots, n - 1.$$

Applying (6.18), (6.19) and (6.2), (6.9) to the representation (6.13), it follows that

$$(6.20)\qquad |k_0(t, s; \rho)| \le \frac{2}{|\rho|^{n-1}}$$

for $t \ne s$ in $[0, 1]$ and for $\rho \in S_0$ with $|\rho| > R_0$.

Finally, fix any point $\lambda = \rho^n$ in $\mathbb{C}$ with $\rho \in G_0$, and assume that $\Delta_0(\rho) \ne 0$, so the point λ belongs to the resolvent set $\rho(L)$. Using the integral operator $K_{0\rho}$, we can establish our representations of the resolvent $R_\lambda(L)$ and the associated Green's function $G(t, s; \lambda)$. Indeed, take any function $u \in L^2[0, 1]$, and set

$$v = K_{0\rho} u \qquad \text{and} \qquad w = R_\lambda(L) u.$$

Clearly the functions v and w belong to $H^n[0, 1]$, and

$$(\lambda I - \ell) v = u = (\lambda I - \ell) w.$$

Thus, there exist constants $c_0, c_1, \dots, c_{n-1}$ (depending on ρ) such that

$$R_\lambda(L) u(t) = w(t) = v(t) + \sum_{k=0}^{n-1} c_k u_{0k}(t, \rho), \qquad 0 \le t \le 1.$$

The functions $u_{0k}(\,\cdot\,, \rho)$, $k = 0, 1, \dots, n - 1$, are the modified solutions of the differential equation (2.1) introduced earlier. They form a basis for the solution space of (2.1), and the characteristic determinant $\Delta_0(\rho)$ is defined in terms of them.

Applying the boundary value B_i to both sides of the last equation, we obtain the linear system

$$\sum_{k=0}^{n-1} M_{0ik}(\rho)c_k = -B_i(v), \qquad i = 1, \dots, n, \tag{6.21}$$

for the constants $c_0, c_1, \dots, c_{n-1}$, where as in Chapter 5

$$M_{0ik}(\rho) = B_i(u_{0k}(\,\cdot\,, \rho)) = \sum_{j=0}^{m_i} \alpha_{ij} u_{0k}^{(j)}(0, \rho) + \sum_{j=0}^{m_i} \beta_{ij} u_{0k}^{(j)}(1, \rho)$$

for $i = 1, \dots, n$, $k = 0, 1, \dots, n-1$. The $n \times n$ coefficient matrix $(M_{0ik}(\rho))$ in (6.21) is nonsingular because $\det(M_{0ik}(\rho)) = \Delta_0(\rho) \neq 0$.

Fix an index i with $1 \leq i \leq n$, and consider the quantity $B_i(v) = B_i(K_{0\rho} u)$. From equation (6.12) and equations (6.1), (6.8), we have

$$\begin{aligned} v^{(j)}(0) = \frac{1}{n\rho^{n-1}} &\sum_{k=\nu}^{n-1} (i\omega_k)\rho^j [(i\omega_k)^j + F_{0kj}(0, \rho)] \\ &\times \int_0^1 e^{-i\rho\omega_k s}[1 + G_{0\,n-1\,k}(s, \rho)]u(s)\,ds \end{aligned}$$

and

$$\begin{aligned} v^{(j)}(1) = -\frac{1}{n\rho^{n-1}} &\sum_{k=0}^{\nu-1} (i\omega_k)\rho^j [(i\omega_k)^j + F_{0kj}(1, \rho)] \\ &\times \int_0^1 e^{i\rho\omega_k(1-s)}[1 + G_{0\,n-1\,k}(s, \rho)]u(s)\,ds \end{aligned}$$

for $j = 0, 1, \dots, n-1$, and hence,

$$\begin{aligned} B_i(v) &= \sum_{j=0}^{m_i} \alpha_{ij} v^{(j)}(0) + \sum_{j=0}^{m_i} \beta_{ij} v^{(j)}(1) \\ &= \frac{1}{n\rho^{n-1}} \sum_{k=0}^{\nu-1} \Big\{ \sum_{j=0}^{m_i} (-1)\beta_{ij}(i\omega_k)\rho^j [(i\omega_k)^j + F_{0kj}(1, \rho)] \Big\} \\ &\qquad \times \int_0^1 e^{i\rho\omega_k(1-s)}[1 + G_{0\,n-1\,k}(s, \rho)]u(s)\,ds \\ &\quad + \frac{1}{n\rho^{n-1}} \sum_{k=\nu}^{n-1} \Big\{ \sum_{j=0}^{m_i} \alpha_{ij}(i\omega_k)\rho^j [(i\omega_k)^j + F_{0kj}(0, \rho)] \Big\} \\ &\qquad \times \int_0^1 e^{-i\rho\omega_k s}[1 + G_{0\,n-1\,k}(s, \rho)]u(s)\,ds \end{aligned} \tag{6.22}$$

for $i = 1, \dots, n$. For $i = 1, \dots, n$ and $k = 0, 1, \dots, \nu - 1$ define

$$\mathcal{T}_{0ik}(\rho) := \sum_{j=0}^{m_i} (-1)\beta_{ij}(i\omega_k)\rho^j [(i\omega_k)^j + F_{0kj}(1, \rho)]$$

for $\rho \in T_0$ with $|\rho| > R_0$, and for $i = 1, \dots, n$ and $k = \nu, \dots, n-1$ define

$$\mathcal{T}_{0ik}(\rho) := \sum_{j=0}^{m_i} \alpha_{ij}(i\omega_k)\rho^j [(i\omega_k)^j + F_{0kj}(0, \rho)]$$

for $\rho \in T_0$ with $|\rho| > R_0$; and for $k = 0, 1, \ldots, \nu - 1$ define

$$U_{0k}(s, \rho) := e^{i\rho\omega_k(1-s)}[1 + G_{0\,n-1\,k}(s, \rho)], \qquad 0 \le s \le 1,$$

for $\rho \in T_0$ with $|\rho| > R_0$, and for $k = \nu, \ldots, n-1$ define

$$U_{0k}(s, \rho) := e^{-i\rho\omega_k s}[1 + G_{0\,n-1\,k}(s, \rho)], \qquad 0 \le s \le 1,$$

for $\rho \in T_0$ with $|\rho| > R_0$. Then we can express (6.22) in the simpler form

$$B_i(v) = \frac{1}{n\rho^{n-1}} \sum_{k=0}^{n-1} \mathcal{T}_{0ik}(\rho) \int_0^1 U_{0k}(s, \rho) u(s)\, ds \tag{6.23}$$

for $i = 1, \ldots, n$, where the functions $\mathcal{T}_{0ik}(\rho)$ are analytic functions of ρ on G_0, and for fixed ρ in T_0 with $|\rho| > R_0$ the functions $U_{0k}(\,\cdot\,, \rho)$ belong to $H^n[0, 1]$.

In terms of the matrix $(M_{0ik}(\rho))$, let $\widetilde{M}_{0ik}(\rho)$ denote the cofactor of the entry $M_{0ik}(\rho)$:

$\widetilde{M}_{0ik}(\rho) = (-1)^{i+k+1}$ times the determinant of the $(n-1) \times (n-1)$ submatrix of $(M_{0j\ell}(\rho))$ obtained by deleting the ith row and kth column.

Clearly the cofactors $\widetilde{M}_{0ik}(\rho)$, $i = 1, \ldots, n$, $k = 0, 1, \ldots, n-1$, are analytic functions in the ρ variable on the open set G_0. These cofactors arise naturally when we solve for the constants $c_0, c_1, \ldots, c_{n-1}$ in the linear system (6.21) by means of Cramer's rule:

$$c_k = \frac{-1}{\Delta_0(\rho)} \sum_{j=1}^{n} \widetilde{M}_{0jk}(\rho) B_j(v),$$

or

$$c_k = \frac{-1}{n\rho^{n-1}\Delta_0(\rho)} \sum_{l=0}^{n-1} \sum_{j=1}^{n} \widetilde{M}_{0jk}(\rho) \mathcal{T}_{0jl}(\rho) \int_0^1 U_{0l}(s, \rho) u(s)\, ds \tag{6.24}$$

for $k = 0, 1, \ldots, n-1$.

Combining the above results, we conclude that

$$\begin{aligned} R_\lambda(L)u(t) &= K_{0\rho}u(t) \\ &\quad - \frac{1}{n\rho^{n-1}\Delta_0(\rho)} \sum_{k,l=0}^{n-1} \sum_{j=1}^{n} \widetilde{M}_{0jk}(\rho) \mathcal{T}_{0jl}(\rho) u_{0k}(t, \rho) \int_0^1 U_{0l}(s, \rho) u(s)\, ds \end{aligned} \tag{6.25}$$

for $0 \le t \le 1$ and for $u \in L^2[0, 1]$, where (6.25) is valid for $\lambda = \rho^n$ in $\mathbb{C}$ with $\rho \in G_0$ and with $\Delta_0(\rho) \ne 0$. Also, from (6.25) the associated Green's function is given by

$$\begin{aligned} G(t, s; \lambda) &= k_0(t, s; \rho) \\ &\quad - \frac{1}{n\rho^{n-1}\Delta_0(\rho)} \sum_{k,l=0}^{n-1} \sum_{j=1}^{n} \widetilde{M}_{0jk}(\rho) \mathcal{T}_{0jl}(\rho) u_{0k}(t, \rho) U_{0l}(s, \rho) \end{aligned} \tag{6.26}$$

for $t \ne s$ in $[0, 1]$, where (6.26) is valid for $\lambda = \rho^n$ in $\mathbb{C}$ with $\rho \in G_0$ and with $\Delta_0(\rho) \ne 0$.

To effectively use the representations (6.25) and (6.26), we must determine bounds or growth rates for the various functions appearing there. For the basis functions $u_{0k}(t, \rho)$ we have already established the necessary bounds in equation (6.4). For the kernel $k_0(t, s; \rho)$ the required growth rate is given in equation (6.20).

Let us proceed to calculate bounds and growth rates for the functions $\mathcal{T}_{0ik}(\rho)$, $U_{0k}(s,\rho)$, and $\widetilde{M}_{0ik}(\rho)$.

First, consider the functions $\mathcal{T}_{0ik}(\rho)$. From their definitions and the estimate (6.2), it is immediate that

$$|\mathcal{T}_{0ik}(\rho)| \leq \gamma_1 |\rho|^{m_i} \tag{6.27}$$

for $\rho \in T_0$ with $|\rho| > R_0$ and for $i = 1, \ldots, n$, $k = 0, 1, \ldots, n-1$. Second, for the functions $U_{0k}(s,\rho)$, from their definitions and the earlier estimates (5.8), (5.9), together with (6.9), we obtain the estimates (replace t by $1-s$)

$$|U_{00}(s,\rho)| \leq 2e^{-b(1-s)}, \tag{6.28}$$

$$|U_{0\nu}(s,\rho)| \leq 2e^{-bs}, \tag{6.29}$$

$$|U_{0k}(s,\rho)| \leq 2e^{-(1-s)\alpha|\rho|} \leq 2, \qquad k = 1, \ldots, \nu - 1, \tag{6.30}$$

$$|U_{0k}(s,\rho)| \leq 2e^{-s\alpha|\rho|} \leq 2, \qquad k = \nu + 1, \ldots, n-1, \tag{6.31}$$

for $0 \leq s \leq 1$ and for $\rho = a + ib \in T_0$ with $|\rho| > R_0$. In particular, for $\rho = a + ib$ belonging to the sector S_0 (where $b \geq 0$) with $|\rho| > R_0$, we have

$$|U_{0k}(s,\rho)| \leq 2, \qquad 0 \leq s \leq 1, \quad k = 0, 1, \ldots, n-1. \tag{6.32}$$

Third, fix indices i and k with $1 \leq i \leq n$ and $0 \leq k \leq n-1$, and let us consider the cofactor $\widetilde{M}_{0ik}(\rho)$. It is formed by taking $(-1)^{i+k+1}$ times the determinant of the $(n-1) \times (n-1)$ matrix obtained by deleting the ith row and the kth column of the matrix

$$\begin{array}{cccc} & {\scriptstyle 1\leq k\leq \nu-1} & & {\scriptstyle \nu+1\leq k\leq n-1} \\ \end{array}$$
$$\begin{pmatrix} P_{010}(\rho)+Q_{010}(\rho)e^{i\rho} & \widehat{P}_{1k}(\rho)+\widetilde{F}_{01k}(\rho) & P_{01\nu}(\rho)+Q_{01\nu}(\rho)e^{i\rho} & \widehat{P}_{1k}(\rho)+\widetilde{F}_{01k}(\rho) \\ \vdots & \vdots & \vdots & \vdots \\ P_{0n0}(\rho)+Q_{0n0}(\rho)e^{i\rho} & \widehat{P}_{nk}(\rho)+\widetilde{F}_{0nk}(\rho) & P_{0n\nu}(\rho)+Q_{0n\nu}(\rho)e^{i\rho} & \widehat{P}_{nk}(\rho)+\widetilde{F}_{0nk}(\rho) \end{pmatrix}.$$

Suppose we proceed to expand the determinant for $\widetilde{M}_{0ik}(\rho)$ in the same manner as $\Delta_0(\rho)$ was expanded in Chapter 5. See equation (5.25). If $1 \leq k \leq \nu - 1$ or $\nu + 1 \leq k \leq n-1$, then the determinant of the $(n-1) \times (n-1)$ submatrix that leads to $\widetilde{M}_{0ik}(\rho)$ is first expanded using linearity in both the 0th and νth columns; this yields the terms $e^{2i\rho}$, $e^{i\rho}$, 1. On the other hand, if $k = 0$, then the initial expansion takes place in only the νth column; this gives the terms $e^{i\rho}$, 1. And if $k = \nu$, then the expansion initially is in the 0th column, leading to the terms $e^{i\rho}$, 1.

Next, all the $(n-1) \times (n-1)$ determinants are expanded using linearity in all $n-1$ columns. This expansion produces the representation

$$\begin{aligned} \widetilde{M}_{0ik}(\rho) &= \tilde{\pi}''_{ik}(\rho)e^{2i\rho} + \tilde{\pi}'_{ik}(\rho)e^{i\rho} + \tilde{\pi}_{ik}(\rho) \\ &\quad + \tilde{\phi}''_{0ik}(\rho)e^{2i\rho} + \tilde{\phi}'_{0ik}(\rho)e^{i\rho} + \tilde{\phi}_{0ik}(\rho) \end{aligned} \tag{6.33}$$

for $\rho \in G_0$. In this equation the functions $\tilde{\pi}''_{ik}$, $\tilde{\pi}'_{ik}$, $\tilde{\pi}_{ik}$ are formed from the functions $\widehat{P}_{ik}$, $\widehat{Q}_{ik}$ introduced in Chapter 3 (see equations (3.1) and (3.2)); they are analytic for $\rho \neq 0$ in $\mathbb{C}$, and each one has the simple form

$$\sum_{j=-(n-1)(m-1)}^{p_0 - m_i} A_{ikj}\rho^j, \qquad \rho \neq 0 \text{ in } \mathbb{C}.$$

Cf. (3.3); (3.20), (3.25), (3.26); and (5.30). All the perturbation terms are contained in the functions $\tilde{\phi}''_{0ik}$, $\tilde{\phi}'_{0ik}$, $\tilde{\phi}_{0ik}$; they are analytic for ρ in the open set G_0, and satisfy the growth rates

$$\begin{gathered}|\tilde{\phi}''_{0ik}(\rho)| \leq \gamma_2|\rho|^{-(m-p_0+m_i)}, \qquad |\tilde{\phi}'_{0ik}(\rho)| \leq \gamma_2|\rho|^{-(m-p_0+m_i)}, \\ |\tilde{\phi}_{0ik}(\rho)| \leq \gamma_2|\rho|^{-(m-p_0+m_i)}\end{gathered} \tag{6.34}$$

for $\rho \in G_0$. In the special cases $k = 0$ or $k = \nu$, then $\tilde{\pi}''_{ik}(\rho) = 0$ and $\tilde{\phi}''_{0ik}(\rho) = 0$.

Take any point $\rho = a + ib$ in the sector S_0 with $|\rho| > R_0$. Clearly $b \geq 0$, $|e^{i\rho}| = e^{-b} \leq 1$, and $|e^{2i\rho}| \leq 1$, and hence, by (6.33) and (6.34)

$$|\widetilde{M}_{0ik}(\rho)| \leq 3\gamma_3|\rho|^{p_0-m_i} + 3\gamma_2|\rho|^{-(m-p_0+m_i)} \leq \gamma_4|\rho|^{p_0-m_i} \tag{6.35}$$

for $i = 1, \ldots, n$, $k = 0, 1, \ldots, n-1$. Combining this result with (6.27), we obtain the estimate

$$\Big|\sum_{j=1}^{n} \widetilde{M}_{0jk}(\rho)T_{0jl}(\rho)\Big| \leq \sum_{j=1}^{n} \gamma_4|\rho|^{p_0-m_j} \cdot \gamma_1|\rho|^{m_j} \leq \gamma_5|\rho|^{p_0} \tag{6.36}$$

for $\rho \in S_0$ with $|\rho| > R_0$ and for $k, l = 0, 1, \ldots, n-1$.

Finally, take any point $\lambda = \rho^n$ in $\mathbb{C}$ with $\rho = a + ib \in S_0$, with $|\rho| > R_0$, and with $\Delta_0(\rho) \neq 0$. Clearly $\rho \in T_0$ with $|\rho| > R_0$, $\rho \in G_0$, and $b \geq 0$, and λ belongs to the resolvent set $\rho(L)$ with the Green's function $G(t, s; \lambda)$ given by (6.26). Applying the estimates (6.20), (6.36), (6.4), and (6.32) to (6.26), we see that

$$|G(t, s; \lambda)| \leq \frac{2}{|\rho|^{n-1}} + \frac{1}{n|\rho|^{n-1}|\Delta_0(\rho)|} \sum_{k,l=0}^{n-1} \gamma_5|\rho|^{p_0} \cdot 2 \cdot 2,$$

or

$$|G(t, s; \lambda)| \leq \frac{2}{|\rho|^{n-1}} + \frac{\gamma|\rho|^{p_0}}{n|\rho|^{n-1}|\Delta_0(\rho)|} \tag{6.37}$$

for $t \neq s$ in $[0, 1]$, where (6.37) is valid for $\lambda = \rho^n$ in $\mathbb{C}$ with $\rho \in S_0$ and $|\rho| > R_0$ and with $\Delta_0(\rho) \neq 0$. Cf. equation (9.8) in Chapter 4 of [**34**]. This is our principal result for the growth rate of the Green's function $G(t, s; \lambda)$ relative to the sector S_0 when $\lambda = \rho^n$ and $\rho \in S_0$.

We can derive alternate representations of the resolvent $R_\lambda(L)$ and the Green's function $G(t, s; \lambda)$ that are in terms of the characteristic determinant $\Delta_1(\rho)$ and the sectors S_1, T_1. The treatment for Δ_1 and S_1, T_1 is similar to the treatment for Δ_0 and S_0, T_0. For $\rho \in T_1$ with $|\rho| > R_0$, let $v_{10}(\cdot, \rho), v_{11}(\cdot, \rho), \ldots, v_{1\,n-1}(\cdot, \rho)$ be the basis for the solution space of the differential equation (2.1) determined in Theorem 4.4, and then form the modified solutions of the differential equation (2.1):

$$\begin{aligned}
u_{10}(t, \rho) &= e^{-i\rho\omega_0} v_{10}(t, \rho), \\
u_{1k}(t, \rho) &= v_{1k}(t, \rho), && k = 1, \ldots, \nu - 1, \\
u_{1\nu}(t, \rho) &= v_{1\nu}(t, \rho), \\
u_{1k}(t, \rho) &= e^{-i\rho\omega_k} v_{1k}(t, \rho), && k = \nu + 1, \ldots, n - 1,
\end{aligned}$$

for $0 \leq t \leq 1$. These functions also form a basis for the solution space of the differential equation (2.1), and from equation (5.37) we have the bounds

$$|u_{1k}(t, \rho)| \leq 2, \qquad 0 \leq t \leq 1, \quad k = 0, 1, \ldots, n-1, \tag{6.38}$$

for $\rho = a + ib$ in the sector S_1 (where $b \leq 0$) with $|\rho| > R_0$.

For $\lambda = \rho^n$ in $\mathbb{C}$ with $\rho \in T_1$ satisfying $|\rho| > R_0$, we again compute the Green's function $g(t,s;\lambda)$ of the differential operator $\lambda I - L_0$, but now we use the basis $v_{1k}(\,\cdot\,,\rho)$, $k = 0,1,\ldots,n-1$:

$$(6.39)\qquad \begin{aligned} g(t,s;\lambda) &= \sum_{k=0}^{n-1} v_{1k}(t,\rho)\eta_{1k}(s,\rho), && 0 \le s < t \le 1, \\ g(t,s;\lambda) &= 0, && 0 \le t < s \le 1, \end{aligned}$$

where the functions $\eta_{1k}(\,\cdot\,,\rho)$, $k = 0,1,\ldots,n-1$, belong to $H^n[0,1]$ and are determined by the linear system

$$(6.40)\qquad \sum_{k=0}^{n-1} v_{1k}^{(\alpha)}(s,\rho)\eta_{1k}(s,\rho) = -\delta_{\alpha\,n-1} i^n, \qquad \alpha = 0,1,\ldots,n-1,$$

for $0 \le s \le 1$. In solving for the functions $\eta_{1k}(\,\cdot\,,\rho)$, $k = 0,1,\ldots,n-1$, we use the same method as used above for the sector T_0.

Next, recall that $\omega_n = \omega_0 = 1$ and $\omega_\nu = -1$. To simplify the discussion, we set

$$v_{1n}(t,\rho) := v_{10}(t,\rho), \qquad \eta_{1n}(s,\rho) := \eta_{10}(s,\rho)$$

for $0 \le t, s \le 1$ and for $\rho \in T_1$ with $|\rho| > R_0$. With this change of notation, we can then rewrite (6.39) in the form

$$(6.41)\qquad \begin{aligned} g(t,s;\lambda) &= \sum_{k=1}^{n} v_{1k}(t,\rho)\eta_{1k}(s,\rho), && 0 \le s < t \le 1, \\ g(t,s;\lambda) &= 0, && 0 \le t < s \le 1, \end{aligned}$$

for $\lambda = \rho^n$ in $\mathbb{C}$ and $\rho \in T_1$ with $|\rho| > R_0$. The representation (6.41) can be expressed in the alternate form

$$g(t,s;\lambda) = k_1(t,s;\rho) + \ell_1(t,s;\rho)$$

for $t \ne s$ in $[0,1]$ and for $\lambda = \rho^n$ in $\mathbb{C}$ and $\rho \in T_1$ with $|\rho| > R_0$, where

$$(6.42)\qquad \begin{aligned} k_1(t,s;\rho) &:= \sum_{k=1}^{\nu} v_{1k}(t,\rho)\eta_{1k}(s,\rho), && 0 \le s < t \le 1, \\ k_1(t,s;\rho) &:= -\sum_{k=\nu+1}^{n} v_{1k}(t,\rho)\eta_{1k}(s,\rho), && 0 \le t < s \le 1, \end{aligned}$$

and

$$\ell_1(t,s;\rho) := \sum_{k=\nu+1}^{n} v_{1k}(t,\rho)\eta_{1k}(s,\rho), \qquad 0 \le t, s \le 1.$$

For each $\rho \in T_1$ with $|\rho| > R_0$, let $K_{1\rho}$ be the integral operator on $L^2[0,1]$ defined by

$$K_{1\rho}u(t) := \int_0^1 k_1(t,s;\rho)u(s)\,ds, \qquad 0 \le t \le 1,$$

for $u \in L^2[0,1]$. If $u \in L^2[0,1]$ and $v = K_{1\rho}u$, then it follows that v belongs to $H^n[0,1]$, $(\rho^n I - \ell)v = u$, and

$$\begin{aligned} v^{(j)}(t) &= \int_0^1 \frac{\partial^j k_1}{\partial t^j}(t,s;\rho)u(s)\,ds \\ &= \sum_{k=1}^{\nu} v_{1k}^{(j)}(t,\rho)\int_0^t \eta_{1k}(s,\rho)u(s)\,ds - \sum_{k=\nu+1}^{n} v_{1k}^{(j)}(t,\rho)\int_t^1 \eta_{1k}(s,\rho)u(s)\,ds \end{aligned} \tag{6.43}$$

for $0 \le t \le 1$ and for $j = 0, 1, \ldots, n-1$. The kernel $k_1(t,s;\rho)$ satisfies the growth rate

$$|k_1(t,s;\rho)| \le \frac{2}{|\rho|^{n-1}} \tag{6.44}$$

for $t \neq s$ in $[0,1]$ and for $\rho \in S_1$ with $|\rho| > R_0$.

Finally, fix any point $\lambda = \rho^n$ in $\mathbb{C}$ with $\rho \in G_1$, and assume that $\Delta_1(\rho) \neq 0$, so the point λ belongs to the resolvent set $\rho(L)$. We are going to establish representations of the resolvent $R_\lambda(L)$ and the associated Green's function $G(t,s;\lambda)$. Indeed, take any $u \in L^2[0,1]$, and set

$$v = K_{1\rho}u \qquad \text{and} \qquad w = R_\lambda(L)u.$$

Clearly the functions v and w belong to $H^n[0,1]$, and $(\lambda I - \ell)v = u = (\lambda I - \ell)w$. Thus, there exist constants $c_0, c_1, \ldots, c_{n-1}$ (depending on ρ) such that

$$R_\lambda(L)u(t) = w(t) = v(t) + \sum_{k=0}^{n-1} c_k u_{1k}(t,\rho), \qquad 0 \le t \le 1.$$

Here the functions $u_{1k}(\,\cdot\,,\rho)$, $k = 0, 1, \ldots, n-1$, are the basis functions formed above — the characteristic determinant $\Delta_1(\rho)$ is defined in terms of them.

Applying the boundary value B_i to the last equation, we obtain the linear system

$$\sum_{k=0}^{n-1} M_{1ik}(\rho)c_k = -B_i(v), \qquad i = 1, \ldots, n, \tag{6.45}$$

for the constants $c_0, c_1, \ldots, c_{n-1}$, where $M_{1ik}(\rho) = B_i(u_{1k}(\,\cdot\,,\rho))$ for $i = 1, \ldots, n$, $k = 0, 1, \ldots, n-1$. Note that $\det(M_{1ik}(\rho)) = \Delta_1(\rho) \neq 0$. Now fix an index i with $1 \le i \le n$, and consider the quantity $B_i(v) = B_i(K_{1\rho}u)$. From equation (6.43) we have

$$\begin{aligned} B_i(v) &= -\sum_{j=0}^{m_i} \alpha_{ij} \sum_{k=\nu+1}^{n} v_{1k}^{(j)}(0,\rho)\int_0^1 \eta_{1k}(s,\rho)u(s)\,ds \\ &\quad + \sum_{j=0}^{m_i} \beta_{ij} \sum_{k=1}^{\nu} v_{1k}^{(j)}(1,\rho)\int_0^1 \eta_{1k}(s,\rho)u(s)\,ds \\ &= \frac{1}{n\rho^{n-1}} \sum_{k=1}^{n} \mathcal{T}_{1ik}(\rho)\int_0^1 U_{1k}(s,\rho)u(s)\,ds \end{aligned} \tag{6.46}$$

for $i = 1, \ldots, n$, where the functions $\mathcal{T}_{1ik}(\rho)$ are analytic functions of ρ on G_1, and for fixed ρ in T_1 with $|\rho| > R_0$ the functions $U_{1k}(\,\cdot\,,\rho)$ belong to $H^n[0,1]$.

In terms of the matrix $(M_{1ik}(\rho))$, let $\widetilde{M}_{1ik}(\rho)$ denote the cofactor of the entry $M_{1ik}(\rho)$. Clearly these cofactors are analytic functions in the ρ variable on the open set G_1, and using these cofactors, we proceed to solve for the constants $c_0, c_1, \ldots, c_{n-1}$ in the linear system (6.45):

$$c_k = \frac{-1}{\Delta_1(\rho)} \sum_{j=1}^{n} \widetilde{M}_{1jk}(\rho) B_j(v),$$

or

$$c_k = \frac{-1}{n\rho^{n-1}\Delta_1(\rho)} \sum_{l=1}^{n} \sum_{j=1}^{n} \widetilde{M}_{1jk}(\rho) \mathcal{T}_{1jl}(\rho) \int_0^1 U_{1l}(s,\rho) u(s)\, ds \tag{6.47}$$

for $k = 0, 1, \ldots, n-1$.

Combining the above results, we conclude that

$$\begin{aligned} R_\lambda(L)u(t) &= K_{1\rho}u(t) \\ &\quad - \frac{1}{n\rho^{n-1}\Delta_1(\rho)} \sum_{k=0}^{n-1} \sum_{j,l=1}^{n} \widetilde{M}_{1jk}(\rho) \mathcal{T}_{1jl}(\rho) u_{1k}(t,\rho) \int_0^1 U_{1l}(s,\rho) u(s)\, ds \end{aligned} \tag{6.48}$$

for $0 \le t \le 1$ and for $u \in L^2[0,1]$, where (6.48) is valid for $\lambda = \rho^n$ in $\mathbb{C}$ with $\rho \in G_1$ and with $\Delta_1(\rho) \neq 0$. Also, from (6.48) the associated Green's function is given by

$$\begin{aligned} G(t,s;\lambda) &= k_1(t,s;\rho) \\ &\quad - \frac{1}{n\rho^{n-1}\Delta_1(\rho)} \sum_{k=0}^{n-1} \sum_{j,l=1}^{n} \widetilde{M}_{1jk}(\rho) \mathcal{T}_{1jl}(\rho) u_{1k}(t,\rho) U_{1l}(s,\rho) \end{aligned} \tag{6.49}$$

for $t \neq s$ in $[0,1]$, where (6.49) is valid for $\lambda = \rho^n$ in $\mathbb{C}$ with $\rho \in G_1$ and with $\Delta_1(\rho) \neq 0$.

The functions appearing in equations (6.48) and (6.49) satisfy the following bounds and growth rates. For the basis functions $u_{1k}(t,\rho)$ a bound is given in equation (6.38). The kernel $k_1(t,s;\rho)$ satisfies the growth rate given in equation (6.44). The functions $\mathcal{T}_{1ik}(\rho)$, $U_{1k}(s,\rho)$, and $\widetilde{M}_{1ik}(\rho)$ satisfy the bounds and growth rates

$$|\mathcal{T}_{1ik}(\rho)| \le \gamma_1 |\rho|^{m_i} \tag{6.50}$$

for $\rho \in T_1$ with $|\rho| > R_0$ and for $i = 1, \ldots, n$, $k = 1, \ldots, n$;

$$|U_{1k}(s,\rho)| \le 2, \qquad 0 \le s \le 1, \quad k = 1, \ldots, n, \tag{6.51}$$

for $\rho = a + ib$ belonging to the sector S_1 (where $b \le 0$) with $|\rho| > R_0$; and

$$|\widetilde{M}_{1ik}(\rho)| \le \gamma_2 |\rho|^{p_0 - m_i} \tag{6.52}$$

for $\rho \in S_1$ with $|\rho| > R_0$ and for $i = 1, \ldots, n$, $k = 0, 1, \ldots, n-1$. Thus, for the Green's function we obtain the estimate

$$|G(t,s;\lambda)| \le \frac{2}{|\rho|^{n-1}} + \frac{\gamma |\rho|^{p_0}}{n|\rho|^{n-1}|\Delta_1(\rho)|} \tag{6.53}$$

for $t \neq s$ in $[0,1]$, where (6.53) is valid for $\lambda = \rho^n$ in $\mathbb{C}$ with $\rho \in S_1$ and $|\rho| > R_0$ and with $\Delta_1(\rho) \neq 0$. Cf. equation (6.37) and equation (9.8) in Chapter 4 of **[34]**. In Chapter 9 we will use the estimates (6.37) and (6.53) of the Green's function $G(t,s;\lambda)$ to establish the completeness of the generalized eigenfunctions of the differential operator L.

6.2. The Green's Function for n Odd

Assume that n is odd: $n = 2\nu - 1 \geq 3$. For the odd order case the sectors S_0 and S_1 are given by

$$S_0\colon \text{ all } \rho = |\rho|e^{i\theta} \in \mathbb{C} \text{ with } -\frac{\pi}{2n} \leq \theta \leq \frac{\pi}{2n},$$
$$S_1\colon \text{ all } \rho = |\rho|e^{i\theta} \in \mathbb{C} \text{ with } \pi - \frac{\pi}{2n} \leq \theta \leq \pi + \frac{\pi}{2n},$$

with T_0 and T_1 the corresponding translated sectors. In the last part of this chapter we again construct representations of the resolvent and the Green's function, and then use these representation to derive their growth rates for ρ belonging to the sectors S_0 and S_1. The treatment closely follows the development in the first part for the case n even. Consequently, we will only indicate the main features of the theory for the odd order case.

For each $\rho \in T_0$ with $|\rho| > R_0$, let $v_{00}(\cdot,\rho), v_{01}(\cdot,\rho), \ldots, v_{0\,n-1}(\cdot,\rho)$ be the basis for the solution space of the differential equation (2.1) determined in Theorem 4.6, and then form the modified solutions of the differential equation (2.1):

$$u_{0k}(t,\rho) = v_{0k}(t,\rho), \qquad k = 0, 1, \ldots, \nu - 1,$$
$$u_{0k}(t,\rho) = e^{-i\rho\omega_k} v_{0k}(t,\rho), \qquad k = \nu, \ldots, n-1,$$

for $0 \leq t \leq 1$ and for $\rho \in T_0$ with $|\rho| > R_0$. These functions also form a basis for the solution space of the differential equation (2.1) for $\rho \in T_0$ with $|\rho| > R_0$, and from equation (5.45) we have the bounds

$$|u_{0k}(t,\rho)| \leq 2, \qquad 0 \leq t \leq 1, \quad k = 0, 1, \ldots, n-1, \tag{6.54}$$

for $\rho = a + ib \in S_0$ with $|\rho| > R_0$ and $b \geq 0$. We must form another set of modified solutions in order to treat the case of $\rho = a + ib \in S_0$ with $|\rho| > R_0$ and $b \leq 0$.

Indeed, set $v_{0n}(t,\rho) := v_{00}(t,\rho)$ for $0 \leq t \leq 1$ and for $\rho \in T_0$ with $|\rho| > R_0$, and then set

$$u_{0n}(t,\rho) := e^{-i\rho\omega_0} u_{00}(t,\rho) = e^{-i\rho\omega_n} v_{0n}(t,\rho)$$

for $0 \leq t \leq 1$ and for $\rho \in T_0$ with $|\rho| > R_0$. Clearly the functions $u_{0k}(\cdot,\rho)$, $k = 1, \ldots, n$, form a basis for the solution space of the differential equation (2.1), and they satisfy the bounds

$$|u_{0k}(t,\rho)| \leq 2, \qquad 0 \leq t \leq 1, \quad k = 1, \ldots, n, \tag{6.55}$$

for $\rho = a + ib \in S_0$ with $|\rho| > R_0$ and $b \leq 0$. In terms of this new basis, we can form the new characteristic determinant

$$\Delta_0^*(\rho) := \det \begin{pmatrix} B_1(u_{01}(\cdot,\rho)) & \cdots & B_1(u_{0n}(\cdot,\rho)) \\ \vdots & & \vdots \\ B_n(u_{01}(\cdot,\rho)) & \cdots & B_n(u_{0n}(\cdot,\rho)) \end{pmatrix} = e^{-i\rho}\Delta_0(\rho)$$

for $\rho \in G_0$.

Next, as earlier in this chapter let L_0 be the nth order differential operator in $L^2[0,1]$ defined by

$$\mathcal{D}(L_0) = \{u \in H^n[0,1] \mid u^{(n-i)}(0) = 0,\ i = 1, \ldots, n\}, \qquad L_0 u = \ell u.$$

Then the resolvent set $\rho(L_0)$ is equal to $\mathbb{C}$, and the Green's function $g(t,s;\lambda)$ of the differential operator $\lambda I - L_0$ is given by

$$(6.56)\qquad \begin{aligned} g(t,s;\lambda) &= \sum_{k=0}^{n-1} v_{0k}(t,\rho)\eta_{0k}(s,\rho), && 0 \le s < t \le 1,\\ g(t,s;\lambda) &= 0, && 0 \le t < s \le 1, \end{aligned}$$

for $\lambda = \rho^n$ in $\mathbb{C}$ and $\rho \in T_0$ with $|\rho| > R_0$, where the functions $\eta_{0k}(\,\cdot\,,\rho)$ belong to $H^n[0,1]$ and are determined by the linear system

$$(6.57)\qquad \sum_{k=0}^{n-1} v_{0k}^{(\alpha)}(s,\rho)\eta_{0k}(s,\rho) = -\delta_{\alpha\, n-1} i^n, \qquad \alpha = 0,1,\dots,n-1,$$

for $0 \le s \le 1$. The functions $\eta_{0k}(\,\cdot\,,\rho)$, $k = 0,1,\dots,n-1$, are determined exactly as in the case n even.

We can rewrite (6.56) in the form

$$g(t,s;\lambda) = k_0(t,s;\rho) + \ell_0(t,s;\rho)$$

for $t \ne s$ in $[0,1]$ and for $\lambda = \rho^n$ in $\mathbb{C}$ and $\rho \in T_0$ with $|\rho| > R_0$, where

$$(6.58)\qquad \begin{aligned} k_0(t,s;\rho) &:= \sum_{k=0}^{\nu-1} v_{0k}(t,\rho)\eta_{0k}(s,\rho), && 0 \le s < t \le 1,\\ k_0(t,s;\rho) &:= -\sum_{k=\nu}^{n-1} v_{0k}(t,\rho)\eta_{0k}(s,\rho), && 0 \le t < s \le 1, \end{aligned}$$

and

$$\ell_0(t,s;\rho) := \sum_{k=\nu}^{n-1} v_{0k}(t,\rho)\eta_{0k}(s,\rho), \qquad 0 \le t,s \le 1.$$

The Green's function $g(t,s;\lambda)$ is the kernel of the integral operator $R_\lambda(L_0)$, which assigns to each u in $L^2[0,1]$ the unique solution $z \in \mathcal{D}(L_0)$ of the initial value problem $(\lambda I - L_0)z = u$. On the other hand, the function $\ell_0(t,s;\rho)$ is the kernel of an integral operator which maps $L^2[0,1]$ into the solution space of the differential equation $(\rho^n I - \ell)u = 0$.

For each $\rho \in T_0$ with $|\rho| > R_0$, let $K_{0\rho}$ be the integral operator on $L^2[0,1]$ defined by

$$K_{0\rho}u(t) := \int_0^1 k_0(t,s;\rho)u(s)\,ds, \qquad 0 \le t \le 1,$$

for $u \in L^2[0,1]$. From the above remarks it follows that if $u \in L^2[0,1]$ and $v = K_{0\rho}u$, then v belongs to $H^n[0,1]$, $(\rho^n I - \ell)v = u$, and

$$(6.59)\qquad \begin{aligned} v^{(j)}(t) &= \int_0^1 \frac{\partial^j k_0}{\partial t^j}(t,s;\rho)u(s)\,ds\\ &= \sum_{k=0}^{\nu-1} v_{0k}^{(j)}(t,\rho)\int_0^t \eta_{0k}(s,\rho)u(s)\,ds - \sum_{k=\nu}^{n-1} v_{0k}^{(j)}(t,\rho)\int_t^1 \eta_{0k}(s,\rho)u(s)\,ds \end{aligned}$$

for $0 \le t \le 1$ and for $j = 0,1,\dots,n-1$. The kernel $k_0(t,s;\rho)$ satisfies the growth rate

$$(6.60)\qquad |k_0(t,s;\rho)| \le \frac{4}{|\rho|^{n-1}}$$

for $t \neq s$ in $[0,1]$ and for $\rho = a + ib \in S_0$ with $|\rho| > R_0$ and $b \geq 0$.

Now fix any point $\lambda = \rho^n$ in $\mathbb{C}$ with $\rho \in G_0$ and with $\Delta_0(\rho) \neq 0$, so the point λ belongs to the resolvent set $\rho(L)$. Take any function $u \in L^2[0,1]$, and set

$$v = K_{0\rho}u \qquad \text{and} \qquad w = R_\lambda(L)u.$$

Then v and w belong to $H^n[0,1]$, and $(\lambda I - \ell)v = u = (\lambda I - \ell)w$. Thus, there exist constants $c_0, c_1, \dots, c_{n-1}$ (depending on ρ) such that

$$R_\lambda(L)u(t) = w(t) = v(t) + \sum_{k=0}^{n-1} c_k u_{0k}(t,\rho), \qquad 0 \leq t \leq 1.$$

The functions $u_{0k}(\,\cdot\,,\rho)$, $k = 0,1,\dots,n-1$, are the modified solutions of the differential equation (2.1) introduced earlier; the characteristic determinant $\Delta_0(\rho)$ is defined in terms of them. Applying the boundary value B_i to the last equation, we obtain the linear system

$$\sum_{k=0}^{n-1} M_{0ik}(\rho)c_k = -B_i(v), \qquad i = 1,\dots,n, \tag{6.61}$$

for the constants $c_0, c_1, \dots, c_{n-1}$, where $M_{0ik}(\rho) = B_i(u_{0k}(\,\cdot\,,\rho))$ for $i = 1,\dots,n$, $k = 0,1,\dots,n-1$, and where

$$\begin{aligned} B_i(v) &= -\sum_{j=0}^{m_i} \alpha_{ij} \sum_{k=\nu}^{n-1} v_{0k}^{(j)}(0,\rho) \int_0^1 \eta_{0k}(s,\rho)u(s)\,ds \\ &\quad + \sum_{j=0}^{m_i} \beta_{ij} \sum_{k=0}^{\nu-1} v_{0k}^{(j)}(1,\rho) \int_0^1 \eta_{0k}(s,\rho)u(s)\,ds \\ &= \frac{1}{n\rho^{n-1}} \sum_{k=0}^{n-1} \mathcal{T}_{0ik}(\rho) \int_0^1 U_{0k}(s,\rho)u(s)\,ds \end{aligned} \tag{6.62}$$

for $i = 1,\dots,n$. In the last equation the functions $\mathcal{T}_{0ik}(\rho)$ are analytic functions of ρ on G_0, and for fixed ρ in T_0 with $|\rho| > R_0$ the functions $U_{0k}(\,\cdot\,,\rho)$ belong to $H^n[0,1]$.

In terms of the matrix $(M_{0ik}(\rho))$, let $\widetilde{M}_{0ik}(\rho)$ denote the cofactor of the entry $M_{0ik}(\rho)$. Clearly these cofactors are analytic functions in the ρ variable on the open set G_0, and in terms of them we can solve for the constants $c_0, c_1, \dots, c_{n-1}$ that appear in the linear system (6.61):

$$c_k = \frac{-1}{\Delta_0(\rho)} \sum_{j=1}^{n} \widetilde{M}_{0jk}(\rho)B_j(v),$$

or

$$c_k = \frac{-1}{n\rho^{n-1}\Delta_0(\rho)} \sum_{l=0}^{n-1}\sum_{j=1}^{n} \widetilde{M}_{0jk}(\rho)\mathcal{T}_{0jl}(\rho) \int_0^1 U_{0l}(s,\rho)u(s)\,ds \tag{6.63}$$

for $k = 0,1,\dots,n-1$. Combining the above results, we conclude that

$$\begin{aligned} R_\lambda(L)u(t) &= K_{0\rho}u(t) \\ &\quad - \frac{1}{n\rho^{n-1}\Delta_0(\rho)} \sum_{k,l=0}^{n-1}\sum_{j=1}^{n} \widetilde{M}_{0jk}(\rho)\mathcal{T}_{0jl}(\rho)u_{0k}(t,\rho) \int_0^1 U_{0l}(s,\rho)u(s)\,ds \end{aligned} \tag{6.64}$$

for $0 \le t \le 1$ and for $u \in L^2[0,1]$, where (6.64) is valid for $\lambda = \rho^n$ in $\mathbb{C}$ with $\rho \in G_0$ and with $\Delta_0(\rho) \neq 0$. The associated Green's function is then given by

$$G(t,s;\lambda) = k_0(t,s;\rho) - \frac{1}{n\rho^{n-1}\Delta_0(\rho)} \sum_{k,l=0}^{n-1} \sum_{j=1}^{n} \widetilde{M}_{0jk}(\rho) \mathcal{T}_{0jl}(\rho) u_{0k}(t,\rho) U_{0l}(s,\rho) \tag{6.65}$$

for $t \neq s$ in $[0,1]$, where (6.65) is valid for $\lambda = \rho^n$ in $\mathbb{C}$ with $\rho \in G_0$ and with $\Delta_0(\rho) \neq 0$.

Bounds and growth rates for the various functions appearing in equations (6.64) and (6.65) are given as follows. For the basis functions $u_{0k}(t,\rho)$, $k = 0, 1, \dots, n-1$, we have the bounds given previously in equation (6.54). For the kernel $k_0(t,s;\rho)$ the growth rate is determined by equation (6.60). Note the restriction $\operatorname{Im}\rho \ge 0$ that applies to these estimates. The functions $\mathcal{T}_{0ik}(\rho)$, $U_{0k}(s,\rho)$, and $\widetilde{M}_{0ik}(\rho)$ satisfy the bounds and growth rates:

$$|\mathcal{T}_{0ik}(\rho)| \le \gamma_1 |\rho|^{m_i} \tag{6.66}$$

for $\rho \in T_0$ with $|\rho| > R_0$ and for $i = 1, \dots, n$, $k = 0, 1, \dots, n-1$;

$$|U_{0k}(s,\rho)| \le 2, \qquad 0 \le s \le 1, \quad k = 0, 1, \dots, n-1, \tag{6.67}$$

for $\rho = a + ib \in S_0$ with $|\rho| > R_0$ and $b \ge 0$; and

$$|\widetilde{M}_{0ik}(\rho)| \le \gamma_2 |\rho|^{p_0 - m_i} \tag{6.68}$$

for $\rho = a + ib \in S_0$ with $|\rho| > R_0$ and $b \ge 0$ and for $i = 1, \dots, n$, $k = 0, 1, \dots, n-1$. Thus, for the Green's function we obtain the estimate

$$|G(t,s;\lambda)| \le \frac{4}{|\rho|^{n-1}} + \frac{\gamma|\rho|^{p_0}}{n|\rho|^{n-1}|\Delta_0(\rho)|} \tag{6.69}$$

for $t \neq s$ in $[0,1]$, where (6.69) is valid for $\lambda = \rho^n$ in $\mathbb{C}$ with $\rho = a + ib \in S_0$, with $|\rho| > R_0$ and $b \ge 0$, and with $\Delta_0(\rho) \neq 0$.

The representation (6.65) of the Green's function $G(t,s;\lambda)$ is still valid for $\lambda = \rho^n$ in $\mathbb{C}$ with $\rho = a + ib \in S_0$ and with $|\rho| > R_0$ and $b \le 0$, but in this form it is difficult to obtain the necessary bounds and growth rates. To remedy this situation, we make simple modifications in the above work: we first alter the integral operator $K_{0\rho}$, and then replace the basis $u_{0k}(\cdot, \rho)$, $k = 0, 1, \dots, n-1$, with the basis $u_{0k}(\cdot, \rho)$, $k = 1, \dots, n$.

Recall that $\omega_n = \omega_0 = 1$, and earlier we introduced the notations $v_{0n}(t,\rho) := v_{00}(t,\rho)$ and $u_{0n}(t,\rho) := e^{-i\rho\omega_0} u_{00}(t,\rho)$ for $0 \le t \le 1$ and for $\rho \in T_0$ with $|\rho| > R_0$. Now set $\eta_{0n}(s,\rho) := \eta_{00}(s,\rho)$ for $0 \le s \le 1$ and for $\rho \in T_0$ with $|\rho| > R_0$. With this change in notation, the representation (6.56) of the Green's function $g(t,s;\lambda)$ can be rewritten as

$$\begin{aligned} g(t,s;\lambda) &= \sum_{k=1}^{n} v_{0k}(t,\rho)\eta_{0k}(s,\rho), && 0 \le s < t \le 1, \\ g(t,s;\lambda) &= 0, && 0 \le t < s \le 1, \end{aligned} \tag{6.70}$$

for $\lambda = \rho^n$ in $\mathbb{C}$ and $\rho \in T_0$ with $|\rho| > R_0$. We can then rewrite (6.70) in the alternate form

$$g(t,s;\lambda) = k_0^*(t,s;\rho) + \ell_0^*(t,s;\rho)$$

for $t \neq s$ in $[0,1]$ and for $\lambda = \rho^n$ in $\mathbb{C}$ and $\rho \in T_0$ with $|\rho| > R_0$, where

$$k_0^*(t,s;\rho) := \sum_{k=1}^{\nu-1} v_{0k}(t,\rho)\eta_{0k}(s,\rho), \qquad 0 \le s < t \le 1,$$

(6.71)

$$k_0^*(t,s;\rho) := -\sum_{k=\nu}^{n} v_{0k}(t,\rho)\eta_{0k}(s,\rho), \qquad 0 \le t < s \le 1,$$

and

$$\ell_0^*(t,s;\rho) := \sum_{k=\nu}^{n} v_{0k}(t,\rho)\eta_{0k}(s,\rho), \qquad 0 \le t,s \le 1.$$

For $\rho \in T_0$ with $|\rho| > R_0$, let $K_{0\rho}^*$ be the integral operator on $L^2[0,1]$ defined by

$$K_{0\rho}^* u(t) := \int_0^1 k_0^*(t,s;\rho)u(s)\,ds, \qquad 0 \le t \le 1,$$

for $u \in L^2[0,1]$. If $u \in L^2[0,1]$ and $v = K_{0\rho}^* u$, then it follows that v belongs to $H^n[0,1]$, $(\rho^n I - \ell)v = u$, and the derivatives of v are given by

(6.72)

$$\begin{aligned} v^{(j)}(t) &= \int_0^1 \frac{\partial^j k_0^*}{\partial t^j}(t,s;\rho)u(s)\,ds \\ &= \sum_{k=1}^{\nu-1} v_{0k}^{(j)}(t,\rho)\int_0^t \eta_{0k}(s,\rho)u(s)\,ds - \sum_{k=\nu}^{n} v_{0k}^{(j)}(t,\rho)\int_t^1 \eta_{0k}(s,\rho)u(s)\,ds \end{aligned}$$

for $0 \le t \le 1$ and for $j = 0,1,\ldots,n-1$. The kernel $k_0^*(t,s;\rho)$ satisfies the growth rate

$$|k_0^*(t,s;\rho)| \le \frac{4}{|\rho|^{n-1}} \tag{6.73}$$

for $t \neq s$ in $[0,1]$ and for $\rho \in S_0$ with $|\rho| > R_0$ and $b \le 0$.

Using the integral operator $K_{0\rho}^*$, the alternate basis $u_{0k}(\,\cdot\,,\rho)$, $k = 1,\ldots,n$, and the alternate characteristic determinant

$$\Delta_0^*(\rho) = \det\begin{pmatrix} B_1(u_{01}(\,\cdot\,,\rho)) & \cdots & B_1(u_{0n}(\,\cdot\,,\rho)) \\ \vdots & & \vdots \\ B_n(u_{01}(\,\cdot\,,\rho)) & \cdots & B_n(u_{0n}(\,\cdot\,,\rho)) \end{pmatrix} = e^{-i\rho}\Delta_0(\rho), \qquad \rho \in G_0,$$

we can establish alternate representations for the resolvent $R_\lambda(L)$ and the associated Green's function $G(t,s;\lambda)$. Fix any point $\lambda = \rho^n$ in $\mathbb{C}$ with $\rho \in G_0$, and assume that $\Delta_0(\rho) \neq 0$. Clearly $\Delta_0^*(\rho) \neq 0$, and the point λ belongs to the resolvent set $\rho(L)$. Take any function $u \in L^2[0,1]$, and set $v = K_{0\rho}^* u$ and $w = R_\lambda(L)u$. Then the functions v and w belong to $H^n[0,1]$, $(\lambda I - \ell)v = u = (\lambda I - \ell)w$, and there exist constants $c_1,\ldots,c_n$ such that

$$R_\lambda(L)u(t) = w(t) = v(t) + \sum_{k=1}^{n} c_k u_{0k}(t,\rho), \qquad 0 \le t \le 1.$$

The constants $c_1,\ldots,c_n$ are determined by the linear system

$$\sum_{k=1}^{n} M_{0ik}^*(\rho)c_k = -B_i(v), \qquad i = 1,\ldots,n, \tag{6.74}$$

where $M^*_{0ik}(\rho) = B_i(u_{0k}(\,\cdot\,,\rho))$ for $i = 1,\dots,n$, $k = 1,\dots,n$, and where

$$B_i(v) = \frac{1}{n\rho^{n-1}} \sum_{k=1}^{n} \mathcal{T}^*_{0ik}(\rho) \int_0^1 U^*_{0k}(s,\rho)u(s)\,ds \tag{6.75}$$

for $i = 1,\dots,n$. In the last equation the functions $\mathcal{T}^*_{0ik}(\rho)$ are analytic functions of ρ on the open set G_0, and for fixed ρ in T_0 with $|\rho| > R_0$ the functions $U^*_{0k}(\,\cdot\,,\rho)$ belong to $H^n[0,1]$.

For the matrix $(M^*_{0ik}(\rho))$, let $\widetilde{M}^*_{0ik}(\rho)$ denote the cofactor of the entry $M^*_{0ik}(\rho)$. These cofactors are analytic functions in the ρ variable on the open set G_0. They appear naturally when solving for $c_1,\dots,c_n$ in the linear system (6.74):

$$c_k = \frac{-1}{\Delta^*_0(\rho)} \sum_{j=1}^{n} \widetilde{M}^*_{0jk}(\rho)B_j(v),$$

or

$$c_k = \frac{-1}{n\rho^{n-1}\Delta^*_0(\rho)} \sum_{l=1}^{n}\sum_{j=1}^{n} \widetilde{M}^*_{0jk}(\rho)\mathcal{T}^*_{0jl}(\rho) \int_0^1 U^*_{0l}(s,\rho)u(s)\,ds \tag{6.76}$$

for $k = 1,\dots,n$. From these results we conclude that

$$\begin{aligned} R_\lambda(L)u(t) &= K^*_{0\rho}u(t) \\ &\quad - \frac{1}{n\rho^{n-1}\Delta^*_0(\rho)} \sum_{k,l=1}^{n}\sum_{j=1}^{n} \widetilde{M}^*_{0jk}(\rho)\mathcal{T}^*_{0jl}(\rho)u_{0k}(t,\rho) \int_0^1 U^*_{0l}(s,\rho)u(s)\,ds \end{aligned} \tag{6.77}$$

for $0 \le t \le 1$ and for $u \in L^2[0,1]$, where (6.77) is valid for $\lambda = \rho^n$ in $\mathbb{C}$ with $\rho \in G_0$ and with $\Delta^*_0(\rho) \neq 0$ (equivalently $\Delta_0(\rho) \neq 0$). From (6.77) the associated Green's function is given by

$$\begin{aligned} G(t,s;\lambda) &= k^*_0(t,s;\rho) \\ &\quad - \frac{1}{n\rho^{n-1}\Delta^*_0(\rho)} \sum_{k,l=1}^{n}\sum_{j=1}^{n} \widetilde{M}^*_{0jk}(\rho)\mathcal{T}^*_{0jl}(\rho)u_{0k}(t,\rho)U^*_{0l}(s,\rho) \end{aligned} \tag{6.78}$$

for $t \neq s$ in $[0,1]$, valid for $\lambda = \rho^n$ in $\mathbb{C}$ with $\rho \in G_0$ and with $\Delta^*_0(\rho) \neq 0$.

Finally, the functions appearing in (6.77) and (6.78) satisfy the following bounds and growth rates. For the basis functions $u_{0k}(t,\rho)$, $k = 1,\dots,n$, we have already established bounds in equation (6.55). For the kernel $k^*_0(t,s;\rho)$ the growth rate is given in equation (6.73). Note the restriction $\operatorname{Im}\rho \le 0$ that applies to these estimates. For the functions $\mathcal{T}^*_{0ik}(\rho)$, $U^*_{0k}(s,\rho)$, and $\widetilde{M}^*_{0ik}(\rho)$ we have the following bounds and growth rates:

$$|\mathcal{T}^*_{0ik}(\rho)| \le \gamma_1|\rho|^{m_i} \tag{6.79}$$

for $\rho \in T_0$ with $|\rho| > R_0$ and for $i = 1,\dots,n$, $k = 1,\dots,n$;

$$|U^*_{0k}(s,\rho)| \le 2, \qquad 0 \le s \le 1, \quad k = 1,\dots,n, \tag{6.80}$$

for $\rho = a + ib \in S_0$ with $|\rho| > R_0$ and $b \le 0$; and

$$|\widetilde{M}^*_{0ik}(\rho)| \le \gamma_2|\rho|^{p_0 - m_i} \tag{6.81}$$

for $\rho = a + ib \in S_0$ with $|\rho| > R_0$ and $b \leq 0$ and for $i = 1, \ldots, n$, $k = 1, \ldots, n$. We conclude that the Green's function satisfies the estimate

$$|G(t,s;\lambda)| \leq \frac{4}{|\rho|^{n-1}} + \frac{\gamma|\rho|^{p_0}}{n|\rho|^{n-1}|\Delta_0^*(\rho)|},$$

or

$$|G(t,s;\lambda)| \leq \frac{4}{|\rho|^{n-1}} + \frac{\gamma|\rho|^{p_0}}{n|\rho|^{n-1}e^b|\Delta_0(\rho)|} \tag{6.82}$$

for $t \neq s$ in $[0,1]$, where (6.82) is valid for $\lambda = \rho^n$ in $\mathbb{C}$ with $\rho = a + ib \in S_0$, with $|\rho| > R_0$ and $b \leq 0$, and with $\Delta_0(\rho) \neq 0$. Compare this growth rate to the growth rate given previously in equation (6.69). Note the extra factor e^b that appears in the denominator of (6.82).

To develop the growth rate of the Green's function $G(t,s;\lambda)$ for $\lambda = \rho^n$ with ρ belonging to the sector S_1, we simply recopy the above material, and then make the few changes needed for the new material. Let us proceed with these extensions of the theory.

For each $\rho \in T_1$ with $|\rho| > R_0$, let $v_{10}(\cdot,\rho), v_{11}(\cdot,\rho), \ldots, v_{1\,n-1}(\cdot,\rho)$ be the basis for the solution space of the differential equation (2.1) determined in Theorem 4.7, and then form the modified solutions of the differential equation (2.1):

$$\begin{aligned} u_{1k}(t,\rho) &:= e^{-i\rho\omega_k} v_{1k}(t,\rho), && k = 0, 1, \ldots, \nu - 1, \\ u_{1k}(t,\rho) &:= v_{1k}(t,\rho), && k = \nu, \ldots, n-1, \end{aligned}$$

for $0 \leq t \leq 1$. These functions also form a basis for the solution space of the differential equation (2.1), and from equation (5.52) we have the bounds

$$|u_{1k}(t,\rho)| \leq 2, \qquad 0 \leq t \leq 1, \quad k = 0, 1, \ldots, n-1, \tag{6.83}$$

for $\rho = a + ib \in S_1$ with $|\rho| > R_0$ and $b \leq 0$. We must form another set of modified solutions in order to treat the case of $\rho = a + ib \in S_1$ with $|\rho| > R_0$ and $b \geq 0$.

Indeed, set $v_{1n}(t,\rho) := v_{10}(t,\rho)$ for $0 \leq t \leq 1$ and for $\rho \in T_1$ with $|\rho| > R_0$, and then set

$$u_{1n}(t,\rho) := e^{i\rho\omega_0} u_{10}(t,\rho) = v_{1n}(t,\rho)$$

for $0 \leq t \leq 1$ and for $\rho \in T_1$ with $|\rho| > R_0$. Clearly the functions $u_{1k}(\cdot,\rho)$, $k = 1, \ldots, n$, form a basis for the solution space of the differential equation (2.1), with

$$|u_{1n}(t,\rho)| \leq 2e^{-bt} \tag{6.84}$$

for $0 \leq t \leq 1$ and for $\rho = a + ib \in T_1$ with $|\rho| > R_0$, and hence, for $\rho = a + ib \in S_1$ with $|\rho| > R_0$ and $b \geq 0$, we have

$$|u_{1k}(t,\rho)| \leq 2, \qquad 0 \leq t \leq 1, \quad k = 1, \ldots, n. \tag{6.85}$$

In terms of this new basis, we can form the new characteristic determinant

$$\Delta_1^*(\rho) := \det \begin{pmatrix} B_1(u_{11}(\cdot,\rho)) & \cdots & B_1(u_{1n}(\cdot,\rho)) \\ \vdots & & \vdots \\ B_n(u_{11}(\cdot,\rho)) & \cdots & B_n(u_{1n}(\cdot,\rho)) \end{pmatrix} = e^{i\rho}\Delta_1(\rho)$$

for $\rho \in G_1$.

For $\lambda = \rho^n$ in $\mathbb{C}$ with $\rho \in T_1$ satisfying $|\rho| > R_0$, we again compute the Green's function $g(t, s; \lambda)$ of the differential operator $\lambda I - L_0$, but now we use the basis $v_{1k}(\,\cdot\,, \rho)$, $k = 0, 1, \ldots, n-1$:

$$(6.86)\qquad \begin{aligned} g(t, s; \lambda) &= \sum_{k=0}^{n-1} v_{1k}(t, \rho)\eta_{1k}(s, \rho), && 0 \le s < t \le 1, \\ g(t, s; \lambda) &= 0, && 0 \le t < s \le 1, \end{aligned}$$

where the functions $\eta_{1k}(\,\cdot\,, \rho)$, $k = 0, 1, \ldots, n-1$, belong to $H^n[0, 1]$ and are determined by the linear system

$$(6.87)\qquad \sum_{k=0}^{n-1} v_{1k}^{(\alpha)}(s, \rho)\eta_{1k}(s, \rho) = -\delta_{\alpha\, n-1} i^n, \qquad \alpha = 0, 1, \ldots, n-1,$$

for $0 \le s \le 1$. The functions $\eta_{1k}(\,\cdot\,, \rho)$, $k = 0, 1, \ldots, n-1$, are calculated as in the previous cases.

We can rewrite (6.86) in the form

$$g(t, s; \lambda) = k_1(t, s; \rho) + \ell_1(t, s; \rho)$$

for $t \ne s$ in $[0, 1]$ and for $\lambda = \rho^n$ in $\mathbb{C}$ and $\rho \in T_1$ with $|\rho| > R_0$, where

$$(6.88)\qquad \begin{aligned} k_1(t, s; \rho) &:= \sum_{k=\nu}^{n-1} v_{1k}(t, \rho)\eta_{1k}(s, \rho), && 0 \le s < t \le 1, \\ k_1(t, s; \rho) &:= -\sum_{k=0}^{\nu-1} v_{1k}(t, \rho)\eta_{1k}(s, \rho), && 0 \le t < s \le 1, \end{aligned}$$

and

$$\ell_1(t, s; \rho) := \sum_{k=0}^{\nu-1} v_{1k}(t, \rho)\eta_{1k}(s, \rho), \qquad 0 \le t, s \le 1.$$

For each $\rho \in T_1$ with $|\rho| > R_0$, let $K_{1\rho}$ be the integral operator on $L^2[0, 1]$ defined by

$$K_{1\rho}u(t) := \int_0^1 k_1(t, s; \rho)u(s)\, ds, \qquad 0 \le t \le 1,$$

for $u \in L^2[0, 1]$. If $u \in L^2[0, 1]$ and $v = K_{1\rho}u$, then it follows that v belongs to $H^n[0, 1]$, $(\rho^n I - \ell)v = u$, and

$$(6.89)\qquad \begin{aligned} v^{(j)}(t) &= \int_0^1 \frac{\partial^j k_1}{\partial t^j}(t, s; \rho)u(s)\, ds \\ &= \sum_{k=\nu}^{n-1} v_{1k}^{(j)}(t, \rho)\int_0^t \eta_{1k}(s, \rho)u(s)\, ds - \sum_{k=0}^{\nu-1} v_{1k}^{(j)}(t, \rho)\int_t^1 \eta_{1k}(s, \rho)u(s)\, ds \end{aligned}$$

for $0 \le t \le 1$ and for $j = 0, 1, \ldots, n-1$. The kernel $k_1(t, s; \rho)$ satisfies the growth rate

$$(6.90)\qquad |k_1(t, s; \rho)| \le \frac{4}{|\rho|^{n-1}}$$

for $t \ne s$ in $[0, 1]$ and for $\rho = a + ib \in S_1$ with $|\rho| > R_0$ and $b \le 0$.

Fix any point $\lambda = \rho^n$ in $\mathbb{C}$ with $\rho \in G_1$, and assume that $\Delta_1(\rho) \neq 0$, so the point λ belongs to the resolvent set $\rho(L)$. Take any function $u \in L^2[0,1]$, and set

$$v = K_{1\rho}u \qquad \text{and} \qquad w = R_\lambda(L)u.$$

Then v and w belong to $H^n[0,1]$, and $(\lambda I - \ell)v = u = (\lambda I - \ell)w$. Thus, there exist constants $c_0, c_1, \ldots, c_{n-1}$ (depending on ρ) such that

$$R_\lambda(L)u(t) = w(t) = v(t) + \sum_{k=0}^{n-1} c_k u_{1k}(t,\rho), \qquad 0 \le t \le 1.$$

The constants $c_0, c_1, \ldots, c_{n-1}$ are determined by the linear system

$$\sum_{k=0}^{n-1} M_{1ik}(\rho) c_k = -B_i(v), \qquad i = 1, \ldots, n, \tag{6.91}$$

where $M_{1ik}(\rho) = B_i(u_{1k}(\,\cdot\,,\rho))$ for $i = 1, \ldots, n$, $k = 0, 1, \ldots, n-1$, and where

$$B_i(v) = \frac{1}{n\rho^{n-1}} \sum_{k=0}^{n-1} \mathcal{T}_{1ik}(\rho) \int_0^1 U_{1k}(s,\rho)u(s)\,ds \tag{6.92}$$

for $i = 1, \ldots, n$. In the last equation the functions $\mathcal{T}_{1ik}(\rho)$ are analytic functions of ρ on G_1, and for fixed ρ in T_1 with $|\rho| > R_0$ the functions $U_{1k}(\,\cdot\,,\rho)$ belong to $H^n[0,1]$.

In terms of the matrix $(M_{1ik}(\rho))$, let $\widetilde{M}_{1ik}(\rho)$ denote the cofactor of the entry $M_{1ik}(\rho)$. Clearly the cofactors $\widetilde{M}_{1ik}(\rho)$, $i = 1, \ldots, n$, $k = 0, 1, \ldots, n-1$, are analytic functions in the ρ variable on the open set G_1. These cofactors arise naturally when we solve for the constants $c_0, c_1, \ldots, c_{n-1}$ in the linear system (6.91):

$$c_k = \frac{-1}{\Delta_1(\rho)} \sum_{j=1}^{n} \widetilde{M}_{1jk}(\rho) B_j(v),$$

or

$$c_k = \frac{-1}{n\rho^{n-1}\Delta_1(\rho)} \sum_{l=0}^{n-1} \sum_{j=1}^{n} \widetilde{M}_{1jk}(\rho)\mathcal{T}_{1jl}(\rho) \int_0^1 U_{1l}(s,\rho)u(s)\,ds \tag{6.93}$$

for $k = 0, 1, \ldots, n-1$. Combining these results, we conclude that

$$\begin{aligned} R_\lambda(L)u(t) &= K_{1\rho}u(t) \\ &\quad - \frac{1}{n\rho^{n-1}\Delta_1(\rho)} \sum_{k,l=0}^{n-1} \sum_{j=1}^{n} \widetilde{M}_{1jk}(\rho)\mathcal{T}_{1jl}(\rho)u_{1k}(t,\rho) \int_0^1 U_{1l}(s,\rho)u(s)\,ds \end{aligned} \tag{6.94}$$

for $0 \le t \le 1$ and for $u \in L^2[0,1]$, where (6.94) is valid for $\lambda = \rho^n$ in $\mathbb{C}$ with $\rho \in G_1$ and with $\Delta_1(\rho) \neq 0$. The associated Green's function is then given by

$$\begin{aligned} G(t,s;\lambda) &= k_1(t,s;\rho) \\ &\quad - \frac{1}{n\rho^{n-1}\Delta_1(\rho)} \sum_{k,l=0}^{n-1} \sum_{j=1}^{n} \widetilde{M}_{1jk}(\rho)\mathcal{T}_{1jl}(\rho)u_{1k}(t,\rho)U_{1l}(s,\rho) \end{aligned} \tag{6.95}$$

for $t \neq s$ in $[0,1]$, where (6.95) is valid for $\lambda = \rho^n$ in $\mathbb{C}$ with $\rho \in G_1$ and with $\Delta_1(\rho) \neq 0$.

Bounds and growth rates for the functions appearing in (6.94) and (6.95) are as follows. For the basis functions $u_{1k}(t,\rho)$, $k = 0, 1, \ldots, n-1$, bounds are given in equation (6.83). For the kernel $k_1(t,s;\rho)$ the required growth rate is given in equation (6.90). Note the restriction $\operatorname{Im}\rho \le 0$ that applies to these estimates. For the functions $\mathcal{T}_{1ik}(\rho)$, $U_{1k}(s,\rho)$, and $\widetilde{M}_{1ik}(\rho)$, we have the estimates

$$|\mathcal{T}_{1ik}(\rho)| \le \gamma_1 |\rho|^{m_i} \tag{6.96}$$

for $\rho \in T_1$ with $|\rho| > R_0$ and for $i = 1,\dots,n$, $k = 0,1,\dots,n-1$;

$$|U_{1k}(s,\rho)| \le 2, \qquad 0 \le s \le 1, \quad k = 0,1,\dots,n-1, \tag{6.97}$$

for $\rho = a + ib \in S_1$ with $|\rho| > R_0$ and $b \le 0$; and

$$|\widetilde{M}_{1ik}(\rho)| \le \gamma_2 |\rho|^{p_0 - m_i} \tag{6.98}$$

for $\rho = a+ib \in S_1$ with $|\rho| > R_0$ and $b \le 0$ and for $i = 1,\dots,n$, $k = 0,1,\dots,n-1$. Therefore, for the Green's function we obtain the estimate

$$|G(t,s;\lambda)| \le \frac{4}{|\rho|^{n-1}} + \frac{\gamma|\rho|^{p_0}}{n|\rho|^{n-1}|\Delta_1(\rho)|} \tag{6.99}$$

for $t \ne s$ in $[0,1]$, where (6.99) is valid for $\lambda = \rho^n$ in $\mathbb{C}$ with $\rho = a + ib \in S_1$, with $|\rho| > R_0$ and $b \le 0$, and with $\Delta_1(\rho) \ne 0$.

To derive analogous estimates for $\lambda = \rho^n$ in $\mathbb{C}$ with $\rho = a + ib \in S_1$ and with $|\rho| > R_0$ and $b \ge 0$, we make simple modifications in the above work. Recall that $\omega_n = \omega_0 = 1$, and earlier we introduced the notations $v_{1n}(t,\rho) := v_{10}(t,\rho)$ and $u_{1n}(t,\rho) := e^{i\rho\omega_0} u_{10}(t,\rho)$ for $0 \le t \le 1$ and for $\rho \in T_1$ with $|\rho| > R_0$. Now set $\eta_{1n}(s,\rho) := \eta_{10}(s,\rho)$ for $0 \le s \le 1$ and for $\rho \in T_1$ with $|\rho| > R_0$. With this change in notation, we can rewrite the representation (6.86) of the Green's function $g(t,s;\lambda)$ as

$$\begin{aligned} g(t,s;\lambda) &= \sum_{k=1}^{n} v_{1k}(t,\rho)\eta_{1k}(s,\rho), && 0 \le s < t \le 1, \\ g(t,s;\lambda) &= 0, && 0 \le t < s \le 1, \end{aligned} \tag{6.100}$$

for $\lambda = \rho^n$ in $\mathbb{C}$ and $\rho \in T_1$ with $|\rho| > R_0$. We then rewrite (6.100) in the alternate form

$$g(t,s;\lambda) = k_1^*(t,s;\rho) + \ell_1^*(t,s;\rho)$$

for $t \ne s$ in $[0,1]$ and for $\lambda = \rho^n$ in $\mathbb{C}$ and $\rho \in T_1$ with $|\rho| > R_0$, where

$$\begin{aligned} k_1^*(t,s;\rho) &:= \sum_{k=\nu}^{n} v_{1k}(t,\rho)\eta_{1k}(s,\rho), && 0 \le s < t \le 1, \\ k_1^*(t,s;\rho) &:= -\sum_{k=1}^{\nu-1} v_{1k}(t,\rho)\eta_{1k}(s,\rho), && 0 \le t < s \le 1, \end{aligned} \tag{6.101}$$

and

$$\ell_1^*(t,s;\rho) := \sum_{k=1}^{\nu-1} v_{1k}(t,\rho)\eta_{1k}(s,\rho), \qquad 0 \le t,s \le 1.$$

For $\rho \in T_1$ with $|\rho| > R_0$, let $K_{1\rho}^*$ be the integral operator on $L^2[0,1]$ defined by

$$K_{1\rho}^* u(t) := \int_0^1 k_1^*(t,s;\rho)u(s)\,ds, \qquad 0 \le t \le 1,$$

for $u \in L^2[0,1]$. If $u \in L^2[0,1]$ and $v = K^*_{1\rho}u$, then it follows that v belongs to $H^n[0,1]$, $(\rho^n I - \ell)v = u$, and the derivatives of v are given by

$$
\begin{aligned}
v^{(j)}(t) &= \int_0^1 \frac{\partial^j k_1^*}{\partial t^j}(t,s;\rho)u(s)\,ds \\
&= \sum_{k=\nu}^{n} v_{1k}^{(j)}(t,\rho)\int_0^t \eta_{1k}(s,\rho)u(s)\,ds - \sum_{k=1}^{\nu-1} v_{1k}^{(j)}(t,\rho)\int_t^1 \eta_{1k}(s,\rho)u(s)\,ds
\end{aligned} \tag{6.102}
$$

for $0 \le t \le 1$ and for $j = 0, 1, \ldots, n-1$. The kernel $k_1^*(t,s;\rho)$ satisfies the growth rate

$$
|k_1^*(t,s;\rho)| \le \frac{4}{|\rho|^{n-1}} \tag{6.103}
$$

for $t \neq s$ in $[0,1]$ and for $\rho = a + ib \in S_1$ with $|\rho| > R_0$ and $b \ge 0$.

Using the integral operator $K^*_{1\rho}$, the alternate basis $u_{1k}(\,\cdot\,,\rho)$, $k = 1, \ldots, n$, and the alternate characteristic determinant

$$
\Delta_1^*(\rho) = \det\begin{pmatrix} B_1(u_{11}(\,\cdot\,,\rho)) & \cdots & B_1(u_{1n}(\,\cdot\,,\rho)) \\ \vdots & & \vdots \\ B_n(u_{11}(\,\cdot\,,\rho)) & \cdots & B_n(u_{1n}(\,\cdot\,,\rho)) \end{pmatrix} = e^{i\rho}\Delta_1(\rho), \qquad \rho \in G_1,
$$

we now proceed to establish alternate representations for the resolvent $R_\lambda(L)$ and the associated Green's function $G(t,s;\lambda)$. Fix any point $\lambda = \rho^n$ in $\mathbb{C}$ with $\rho \in G_1$, and assume that $\Delta_1(\rho) \neq 0$. Clearly $\Delta_1^*(\rho) \neq 0$, and the point λ belongs to the resolvent set $\rho(L)$. Take any function $u \in L^2[0,1]$, and set

$$
v = K^*_{1\rho}u \qquad \text{and} \qquad w = R_\lambda(L)u.
$$

The functions v and w belong to $H^n[0,1]$, $(\lambda I - \ell)v = u = (\lambda I - \ell)w$, and there exist constants $c_1, \ldots, c_n$ such that

$$
R_\lambda(L)u(t) = w(t) = v(t) + \sum_{k=1}^{n} c_k u_{1k}(t,\rho), \qquad 0 \le t \le 1.
$$

The constants $c_1, \ldots, c_n$ are determined by the linear system

$$
\sum_{k=1}^{n} M^*_{1ik}(\rho)c_k = -B_i(v), \qquad i = 1, \ldots, n, \tag{6.104}
$$

where $M^*_{1ik}(\rho) = B_i(u_{1k}(\,\cdot\,,\rho))$ for $i = 1, \ldots, n$, $k = 1, \ldots, n$, and where

$$
B_i(v) = \frac{1}{n\rho^{n-1}} \sum_{k=1}^{n} T^*_{1ik}(\rho) \int_0^1 U^*_{1k}(s,\rho)u(s)\,ds \tag{6.105}
$$

for $i = 1, \ldots, n$. In equation (6.105) the functions $T^*_{1ik}(\rho)$ are analytic functions of ρ on the open setG_1, and for fixed ρ in T_1 with $|\rho| > R_0$ the functions $U^*_{1k}(\,\cdot\,,\rho)$ belong to $H^n[0,1]$.

In terms of the matrix $(M^*_{1ik}(\rho))$, let $\widetilde{M}^*_{1ik}(\rho)$ denote the cofactor of the entry $M^*_{1ik}(\rho)$. These cofactors are analytic functions in the ρ variable on the open set G_1; they appear naturally when we solve for $c_1, \ldots, c_n$ in the linear system (6.104):

$$
c_k = \frac{-1}{\Delta_1^*(\rho)} \sum_{j=1}^{n} \widetilde{M}^*_{1jk}(\rho)B_j(v),
$$

or

$$c_k = \frac{-1}{n\rho^{n-1}\Delta_1^*(\rho)} \sum_{l=1}^{n} \sum_{j=1}^{n} \widetilde{M}_{1jk}^*(\rho)\mathcal{T}_{1jl}^*(\rho)\int_0^1 U_{1l}^*(s,\rho)u(s)\,ds \tag{6.106}$$

for $k = 1, \ldots, n$. Combining these results, we conclude that

$$\begin{aligned} R_\lambda(L)u(t) &= K_{1\rho}^* u(t) \\ &- \frac{1}{n\rho^{n-1}\Delta_1^*(\rho)} \sum_{k,l=1}^{n} \sum_{j=1}^{n} \widetilde{M}_{1jk}^*(\rho)\mathcal{T}_{1jl}^*(\rho)u_{1k}(t,\rho)\int_0^1 U_{1l}^*(s,\rho)u(s)\,ds \end{aligned} \tag{6.107}$$

for $0 \le t \le 1$ and for $u \in L^2[0,1]$, where (6.107) is valid for $\lambda = \rho^n$ in $\mathbb{C}$ with $\rho \in G_1$ and with $\Delta_1^*(\rho) \neq 0$ (equivalently $\Delta_1(\rho) \neq 0$). Also, from (6.107) the associated Green's function is given by

$$\begin{aligned} G(t,s;\lambda) &= k_1^*(t,s;\rho) \\ &- \frac{1}{n\rho^{n-1}\Delta_1^*(\rho)} \sum_{k,l=1}^{n} \sum_{j=1}^{n} \widetilde{M}_{1jk}^*(\rho)\mathcal{T}_{1jl}^*(\rho)u_{1k}(t,\rho)U_{1l}^*(s,\rho) \end{aligned} \tag{6.108}$$

for $t \neq s$ in $[0,1]$, valid for $\lambda = \rho^n$ in $\mathbb{C}$ with $\rho \in G_1$ and with $\Delta_1^*(\rho) \neq 0$.

Bounds and growth rates for the functions appearing in (6.107) and (6.108) are as follows. For the basis functions $u_{1k}(t,\rho)$, $k = 1, \ldots, n$, we have the bounds given in equation (6.85). For the kernel $k_1^*(t,s;\rho)$ the growth rate is given in equation (6.103). Note the restriction $\operatorname{Im}\rho \ge 0$ that applies to these inequalities. For the functions $\mathcal{T}_{1ik}^*(\rho)$, $U_{1k}^*(s,\rho)$, and $\widetilde{M}_{1ik}^*(\rho)$, we obtain the following bounds and growth rates:

$$|\mathcal{T}_{1ik}^*(\rho)| \le \gamma_1 |\rho|^{m_i} \tag{6.109}$$

for $\rho \in T_1$ with $|\rho| > R_0$ and for $i = 1, \ldots, n$, $k = 1, \ldots, n$;

$$|U_{1k}^*(s,\rho)| \le 2, \qquad 0 \le s \le 1, \quad k = 1, \ldots, n, \tag{6.110}$$

for $\rho = a + ib \in S_1$ with $|\rho| > R_0$ and $b \ge 0$; and

$$|\widetilde{M}_{1ik}^*(\rho)| \le \gamma_2 |\rho|^{p_0 - m_i} \tag{6.111}$$

$\rho = a + ib \in S_1$ with $|\rho| > R_0$ and $b \ge 0$ and for $i = 1, \ldots, n$, $k = 1, \ldots, n$. We conclude that the Green's function satisfies the estimate

$$|G(t,s;\lambda)| \le \frac{4}{|\rho|^{n-1}} + \frac{\gamma|\rho|^{p_0}}{n|\rho|^{n-1}|\Delta_1^*(\rho)|},$$

or

$$|G(t,s;\lambda)| \le \frac{4}{|\rho|^{n-1}} + \frac{\gamma|\rho|^{p_0}}{n|\rho|^{n-1}e^{-b}|\Delta_1(\rho)|} \tag{6.112}$$

for $t \neq s$ in $[0,1]$, where equation (6.112) is valid for $\lambda = \rho^n \in \mathbb{C}$ with $\rho = a + ib \in S_1$, with $|\rho| > R_0$ and $b \ge 0$, and with $\Delta_1(\rho) \neq 0$. Compare this growth rate to the growth rate given in equation (6.99). Note the additional factor e^{-b} that appears in the denominator of (6.112). In Chapter 9 we will use the estimates (6.69), (6.82) and (6.99), (6.112) of the Green's function $G(t,s;\lambda)$ to establish the completeness of the generalized eigenfunctions of the differential operator L.

CHAPTER 7

The Eigenvalues for n Even

In this chapter we compute the eigenvalues of the differential operator L for the case n even. We assume the hypotheses of Chapters 3–5: (i) $n = 2\nu \geq 2$; (ii) the differential operator L is either regular or simply irregular; (iii) the integers p and q have been determined with $-\infty < p \leq q \leq p_0$ and with $a_p \neq 0$, $c_p \neq 0$, and $a_\kappa = c_\kappa = 0$ for $\kappa = p+1, \dots, p_0$ and $b_\kappa = 0$ for $\kappa = q+1, \dots, p_0$; (iv) the translated sectors T_0 and T_1 have been chosen with condition (3.31) being satisfied for the case $p = q$; (v) the integer m has been fixed with $m > n$, $m > p_0$, and $-(m - p_0 - 1) \leq p \leq p_0$; and (vi) the functions π_i, $i = 0, 1, 2$, have been determined as per Chapter 3 or equation (5.29), viz.

$$\pi_2(\rho) = \sum_{\kappa=-(m-p_0-1)}^{p} a_\kappa \rho^\kappa, \quad \pi_1(\rho) = \sum_{\kappa=-(m-p_0-1)}^{q} b_\kappa \rho^\kappa, \quad \pi_0(\rho) = \sum_{\kappa=-(m-p_0-1)}^{p} c_\kappa \rho^\kappa \tag{7.1}$$

for $\rho \neq 0$ in $\mathbb{C}$.

Let Δ_0 and Δ_1 be the characteristic determinants of L determined in Theorems 5.1 and 5.2, respectively. Δ_0 is analytic on the open set $G_0 = \{\rho \in \operatorname{Int} T_0 \mid |\rho| > R_0\}$ and has the representation

$$\begin{aligned} \Delta_0(\rho) &= \pi_2(\rho)e^{2i\rho} + \pi_1(\rho)e^{i\rho} + \pi_0(\rho) \\ &\quad + \Phi_{02}(\rho)e^{2i\rho} + \Phi_{01}(\rho)e^{i\rho} + \Phi_{00}(\rho) \end{aligned} \tag{7.2}$$

for $\rho \in G_0$, where the functions Φ_{0i}, $i = 0, 1, 2$, are analytic on G_0 with

$$|\Phi_{0i}(\rho)| \leq \gamma |\rho|^{-(m-p_0)} \tag{7.3}$$

for $\rho \in G_0$ and for $i = 0, 1, 2$. Similarly, the characteristic determinant Δ_1 is analytic on the open set $G_1 = \{\rho \in \operatorname{Int} T_1 \mid |\rho| > R_0\}$ and has the representation

$$\begin{aligned} \Delta_1(\rho) &= \pi_2(\rho) + \pi_1(\rho)e^{-i\rho} + \pi_0(\rho)e^{-2i\rho} \\ &\quad + \Phi_{12}(\rho) + \Phi_{11}(\rho)e^{-i\rho} + \Phi_{10}(\rho)e^{-2i\rho} \end{aligned} \tag{7.4}$$

for $\rho \in G_1$, where the functions Φ_{1i}, $i = 0, 1, 2$, are analytic on G_1 with

$$|\Phi_{1i}(\rho)| \leq \gamma |\rho|^{-(m-p_0)} \tag{7.5}$$

for $\rho \in G_1$ and for $i = 0, 1, 2$.

To determine the eigenvalues of L, we calculate the zeros of the characteristic determinants Δ_0 and Δ_1. The analysis divides quite naturally into several cases determined by the relative size of the integers p and q.

Assume that $p = q$. This case divides below into Case 1 and Case 2. Case 1 corresponds to simple eigenvalues, while Case 2 allows the possibility of multiple eigenvalues. These two cases include all the cases in which L is regular, as well as many cases in which L is simply irregular. Consider first the characteristic determinant Δ_0 on the open set G_0. Let

$$f_0(\rho) := a_p e^{2i\rho} + b_p e^{i\rho} + c_p$$

for $\rho \in \mathbb{C}$, an entire function, and let

$$g_0(\rho) := \sum_{\kappa=-(m-p_0-1)}^{p-1} \frac{a_\kappa}{\rho^{p-\kappa}} e^{2i\rho} + \sum_{\kappa=-(m-p_0-1)}^{p-1} \frac{b_\kappa}{\rho^{p-\kappa}} e^{i\rho} + \sum_{\kappa=-(m-p_0-1)}^{p-1} \frac{c_\kappa}{\rho^{p-\kappa}}$$
$$+ \frac{1}{\rho^p} \left[\Phi_{02}(\rho)e^{2i\rho} + \Phi_{01}(\rho)e^{i\rho} + \Phi_{00}(\rho)\right]$$

for $\rho \in G_0$, where the function g_0 is analytic on the open set G_0. From (7.2) we obtain the representation

$$\Delta_0(\rho) = \rho^p[f_0(\rho) + g_0(\rho)] \quad \text{for } \rho \in G_0. \tag{7.6}$$

Next, consider the characteristic determinant Δ_1 on the open set G_1. Here we let

$$f_1(\rho) := a_p + b_p e^{-i\rho} + c_p e^{-2i\rho} = e^{-2i\rho} f_0(\rho)$$

for $\rho \in \mathbb{C}$, an entire function, and let

$$g_1(\rho) := \sum_{\kappa=-(m-p_0-1)}^{p-1} \frac{a_\kappa}{\rho^{p-\kappa}} + \sum_{\kappa=-(m-p_0-1)}^{p-1} \frac{b_\kappa}{\rho^{p-\kappa}} e^{-i\rho} + \sum_{\kappa=-(m-p_0-1)}^{p-1} \frac{c_\kappa}{\rho^{p-\kappa}} e^{-2i\rho}$$
$$+ \frac{1}{\rho^p} \left[\Phi_{12}(\rho) + \Phi_{11}(\rho)e^{-i\rho} + \Phi_{10}(\rho)e^{-2i\rho}\right]$$

for $\rho \in G_1$, where the function g_1 is analytic on the open set G_1. From (7.4) we have

$$\Delta_1(\rho) = \rho^p[f_1(\rho) + g_1(\rho)] \quad \text{for } \rho \in G_1. \tag{7.7}$$

Recall that the constant $d > 0$ has been chosen in Chapter 3 to satisfy condition (3.31):

$$|a_p|e^{-2d} + |b_p|e^{-d} + |c_p|e^{-2d} \leq \frac{1}{4}|a_p| = \frac{1}{4}|c_p|, \tag{7.8}$$

and that the translated sectors T_0, T_1 have been selected such that the sectors S_0, S_1 lie in the interiors of T_0, T_1, respectively, and such that the horizontal strip

$$\Gamma = \{\rho = a + ib \in \mathbb{C} \mid a \geq -\pi \text{ and } |b| \leq d\}$$

lies in the interiors of both T_0 and T_1.

Let us examine the functions f_0 and g_0 which make up the characteristic determinant Δ_0. First, if $\rho = a+ib \in \mathbb{C}$ with $b \geq d$, then $|e^{i\rho}| = e^{-b} \leq e^{-d}$, $|e^{2i\rho}| \leq e^{-2d}$, and by inequality (7.8)

$$|f_0(\rho)| \geq |c_p| - |a_p||e^{2i\rho}| - |b_p||e^{i\rho}|$$
$$\geq |c_p| - |a_p|e^{-2d} - |b_p|e^{-d} \geq |c_p| - \frac{1}{4}|c_p|.$$

Thus,

$$|f_0(\rho)| \geq \frac{3}{4}|c_p| = \frac{3}{4}|a_p| \quad \text{for } \rho = a + ib \in \mathbb{C} \text{ with } b \geq d.$$

Second, take any $\rho = a + ib \in G_0$ with $b \geq -d$. Clearly $|e^{i\rho}| = e^{-b} \leq e^d$ and $|e^{2i\rho}| \leq e^{2d}$. Also, since $-(m - p_0 - 1) \leq p \leq p_0$, we have $m - p_0 + p \geq 1$, and hence, by (7.3)

$$|g_0(\rho)| \leq \frac{\gamma_1}{|\rho|} + \frac{1}{|\rho|^p} \cdot \frac{\gamma_2}{|\rho|^{m-p_0}}$$
$$= \frac{\gamma_1}{|\rho|} + \frac{\gamma_2}{|\rho|^{m-p_0+p}} \leq \frac{\gamma_1 + \gamma_2}{|\rho|}.$$

Therefore,

$$|g_0(\rho)| \leq \frac{\gamma_0}{|\rho|} \quad \text{for } \rho = a + ib \in G_0 \text{ with } b \geq -d.$$

Third, if $\rho = a + ib \in \mathbb{C}$ with $b \leq -d$, then $|e^{-i\rho}| = e^b \leq e^{-d}$, $|e^{-2i\rho}| \leq e^{-2d}$, and by inequality (7.8) again

$$|f_1(\rho)| = |a_p + b_p e^{-i\rho} + c_p e^{-2i\rho}|$$
$$\geq |a_p| - |b_p| e^{-d} - |c_p| e^{-2d} \geq |a_p| - \frac{1}{4}|a_p|.$$

Hence,

$$|f_1(\rho)| \geq \frac{3}{4}|a_p| \quad \text{for } \rho = a + ib \in \mathbb{C} \text{ with } b \leq -d.$$

Fourth, consider the function g_1 appearing in the representation (7.7) for $\Delta_1(\rho)$. Take any point $\rho = a + ib \in G_1$ with $b \leq d$. Then $|e^{-i\rho}| = e^b \leq e^d$, $|e^{-2i\rho}| \leq e^{2d}$, and by (7.5)

$$|g_1(\rho)| \leq \frac{\gamma_3}{|\rho|} + \frac{1}{|\rho|^p} \cdot \frac{\gamma_4}{|\rho|^{m-p_0}} \leq \frac{\gamma_3 + \gamma_4}{|\rho|},$$

and hence,

$$|g_1(\rho)| \leq \frac{\gamma_0}{|\rho|} \quad \text{for } \rho = a + ib \in G_1 \text{ with } b \leq d.$$

In terms of the constant γ_0 appearing in the estimates for g_0 and g_1, choose a constant $r_1 > R_0$ such that

$$\frac{\gamma_0}{|\rho|} \leq \frac{1}{4}|a_p| \quad \text{for all } \rho \in \mathbb{C} \text{ with } |\rho| \geq r_1.$$

It follows from the above that if $\rho = a + ib \in G_0$ with $|\rho| \geq r_1$ and $b \geq d$, then

$$\begin{aligned} |\Delta_0(\rho)| &\geq |\rho|^p\{|f_0(\rho)| - |g_0(\rho)|\} \\ &\geq |\rho|^p\left\{\frac{3}{4}|a_p| - \frac{1}{4}|a_p|\right\} = \frac{1}{2}|a_p||\rho|^p > 0. \end{aligned} \tag{7.9}$$

On the other hand, if $\rho = a + ib \in G_1$ with $|\rho| \geq r_1$ and $b \leq -d$, then

$$\begin{aligned} |\Delta_1(\rho)| &\geq |\rho|^p\{|f_1(\rho)| - |g_1(\rho)|\} \\ &\geq |\rho|^p\left\{\frac{3}{4}|a_p| - \frac{1}{4}|a_p|\right\} = \frac{1}{2}|a_p||\rho|^p > 0. \end{aligned} \tag{7.10}$$

The estimates (7.9) and (7.10) are our initial growth rates for the characteristic determinants Δ_0 and Δ_1 relative to the open sets G_0 and G_1, respectively. They will play a crucial role in Chapter 9 where we show that the generalized eigenfunctions of the differential operator L are complete in the Hilbert space $L^2[0,1]$. As an immediate application of (7.9) and (7.10), we have the following theorem which establishes *apriori estimates* for the eigenvalues of L relative to the sets G_0 and G_1.

THEOREM 7.1. *Assume that* $p = q$. *Let* $\lambda = \rho^n \in \mathbb{C}$ *with* $\rho = a + ib \in G_0$ *and with* $|\rho| \geq r_1$.

(a) *If* $b \geq d$, *then* $\Delta_0(\rho) \neq 0$ *and* $\lambda \in \rho(L)$.

(b) *If* λ *is an eigenvalue of* L, *then* $\Delta_0(\rho) = 0$ *and* $b < d$.

In addition, let $\lambda = \rho^n \in \mathbb{C}$ *with* $\rho = a + ib \in G_1$ *and with* $|\rho| \geq r_1$.

(c) *If* $b \leq -d$, *then* $\Delta_1(\rho) \neq 0$ *and* $\lambda \in \rho(L)$.

(d) *If* λ *is an eigenvalue of* L, *then* $\Delta_1(\rho) = 0$ *and* $b > -d$.

It follows from the theorem that the resolvent set $\rho(L)$ is nonempty. Thus, the differential operator L is a Fredholm operator with Fredholm set $\Phi(L) = \mathbb{C}$ and with resolvent set $\rho(L) \neq \emptyset$: this implies that the spectrum $\sigma(L)$ is a countable set having no limit points in $\mathbb{C}$. See [**34**, p. 58 or p.60].

We next focus our search for the zeros of Δ_0 on the horizontal strip Γ. Let ξ_0 and η_0 be the roots of the quadratic polynomial $Q(z) := a_p z^2 + b_p z + c_p$, so

$$Q(z) = a_p[z - \xi_0][z - \eta_0]$$

with $a_p \xi_0 \eta_0 = c_p$, $\xi_0 \neq 0$, $\eta_0 \neq 0$ and with $\eta_0 = -1/(\omega_p \xi_0)$ by (3.30). Then the function f_0 can be written in the form

$$f_0(\rho) = a_p[e^{i\rho} - \xi_0][e^{i\rho} - \eta_0]$$

for $\rho \in \mathbb{C}$, and we see immediately that the zeros of f_0 are given by

$$\mu'_k = (2\pi k + \operatorname{Arg} \xi_0) - i \ln |\xi_0|, \qquad k = 0, \pm 1, \pm 2, \ldots,$$
$$\mu''_k = (2\pi k + \operatorname{Arg} \eta_0) - i \ln |\eta_0|, \qquad k = 0, \pm 1, \pm 2, \ldots,$$

with $|\eta_0| = 1/|\xi_0|$ and $\arg \eta_0 = -\arg \xi_0 - 2\pi p/n + \pi$. From the above estimates for f_0 and f_1, the zeros μ'_k, μ''_k must all lie in the interior of the horizontal strip $|\operatorname{Im} \rho| \leq d$, i.e., $-d < \ln |\xi_0| < d$. In case $\xi_0 \neq \eta_0$, we have $\mu'_k \neq \mu''_l$ for all k, l, with each μ'_k and each μ''_k a zero of order 1 of f_0. In the special case $\xi_0 = \eta_0$, where Q has a double root, we have $b_p^2 = -4a_p^2/\omega_p$,

$$\xi_0 = \eta_0 = -\frac{b_p}{2a_p} = \pm \frac{i}{\sqrt{\omega_p}} \quad \text{and} \quad |\xi_0| = |\eta_0| = 1,$$

and

$$\mu'_k = \mu''_k = 2\pi k + \operatorname{Arg} \xi_0 := \mu_k \quad \text{for } k = 0, \pm 1, \pm 2, \ldots;$$

in this case each μ_k is a real zero of order 2 of f_0, and the relation $b_p^2 = -4a_p^2/\omega_p$ implies that $b_p \neq 0$. In both cases we will show that the zeros of Δ_0 and $f_0 + g_0$ in the strip Γ appear as perturbations of the μ'_k, μ''_k.

Since $-\pi < \operatorname{Arg} \xi_0 \leq \pi$ and $-\pi < \operatorname{Arg} \eta_0 \leq \pi$, we can select a constant $\omega \geq \pi$ such that $\omega - 2\pi < \operatorname{Arg} \xi_0 < \omega$ and $\omega - 2\pi < \operatorname{Arg} \eta_0 < \omega$. Then for $k = 1, 2, \ldots$ we introduce the rectangles

$$R'_k := \{\rho \in \mathbb{C} \mid \omega - 2\pi \leq \operatorname{Re} \rho \leq \omega + 2\pi(k-1) \text{ and } |\operatorname{Im} \rho| \leq d\}.$$

Clearly these rectangles lie in the horizontal strip Γ, and hence, they lie in the interior of the sector T_0, and the zeros μ'_0, μ''_0 lie in the interior of the rectangle R'_1. Choose a constant δ with $0 < \delta \leq \pi/4$ such that the two closed disks $|\rho - \mu'_0| \leq \delta$ and $|\rho - \mu''_0| \leq \delta$ both lie in the interior of R'_1 and such that these two disks are disjoint in the case $\xi_0 \neq \eta_0$. For $k = 0, \pm 1, \pm 2, \ldots$ form the circles

$$\Gamma'_k := \{\rho \in \mathbb{C} \mid |\rho - \mu'_k| = \delta\}, \qquad \Gamma''_k := \{\rho \in \mathbb{C} \mid |\rho - \mu''_k| = \delta\},$$

where in the case $\xi_0 = \eta_0$ we set

$$\Gamma'_k = \Gamma''_k := \Gamma_k \quad \text{for } k = 0, \pm 1, \pm 2, \ldots .$$

The following properties are obvious from these definitions: (i) the circles Γ'_k, Γ''_k, $k \geq 0$, lie in the interior of the horizontal strip Γ; (ii) the Γ'_k, Γ''_l and the points inside them do not overlap each other in the case $\xi_0 \neq \eta_0$; (iii) the Γ_k and the points inside them do not overlap each other in the case $\xi_0 = \eta_0$; and (iv) for each positive integer k_0 the circles Γ'_k, Γ''_k, $0 \leq k < k_0$, lie in the interior of the rectangle R'_{k_0}, the circles Γ'_k, Γ''_k, $k \geq k_0$, lie in the exterior of R'_{k_0} and to the right of R'_{k_0}, and the circles Γ'_k, Γ''_k, $k < 0$, lie in the exterior of R'_{k_0} and to the left of R'_{k_0}.

To complete the geometry, let Γ_* be the subset of T_0 defined by

$$\Gamma_* := \{\rho = a + ib \in \Gamma \mid \rho \text{ is not inside any of the circles } \Gamma'_k, \Gamma''_k\}.$$

In the sequel we refer to Γ_* as a *punctured horizontal strip*.

It is clear that $f_0(\rho) \neq 0$ for all $\rho \in R'_1$ not in the circles Γ'_0, Γ''_0. Set

$$m_* := \min\{\, |f_0(\rho)| \mid \rho \in R'_1 \text{ with } \rho \text{ not in } \Gamma'_0, \Gamma''_0 \,\} > 0.$$

Since $f_0(\rho + 2\pi) = f_0(\rho)$ for all $\rho \in \mathbb{C}$, it follows that $|f_0(\rho)| \geq m_*$ for all points $\rho = a + ib \in \mathbb{C}$ with $|b| \leq d$ and with ρ not in any of the circles Γ'_k, Γ''_k. If we set $m_0 := \min\{3|a_p|/4,\, m_*\} > 0$, then our estimates for f_0 combine to yield the result

$$|f_0(\rho)| \geq m_0 > 0 \tag{7.11}$$

for all $\rho \in \mathbb{C}$ with ρ not in any of the circles Γ'_k, Γ''_k.

Select a positive integer k_0 such that the constant $y_0 := \omega + 2\pi(k_0 - 1)$ has the following properties: $y_0 \geq r_1$, $y_0 \geq \omega$, and

$$\frac{\gamma_0}{|\rho|} \leq \frac{m_0}{2} \quad \text{for all } \rho \in \mathbb{C} \text{ with } |\rho| \geq y_0,$$

where γ_0 is the constant introduced above in the estimates for g_0 and g_1. Clearly $y_0 > 1$. Then for any point $\rho = a + ib \in \Gamma$ with $|\rho| \geq y_0$, we have $|\rho| \geq r_1 > R_0$, so $\rho \in G_0$, and by our previous estimate for g_0,

$$|g_0(\rho)| \leq \frac{\gamma_0}{|\rho|} \leq \frac{m_0}{2}. \tag{7.12}$$

Combining (7.11) and (7.12), we conclude that

$$|g_0(\rho)| \leq \frac{m_0}{2} < m_0 \leq |f_0(\rho)| \tag{7.13}$$

and

$$|f_0(\rho) + g_0(\rho)| \geq \frac{m_0}{2} \tag{7.14}$$

for all $\rho = a + ib \in \Gamma_*$ with $|\rho| \geq y_0$, and hence, by (7.6)

$$|\Delta_0(\rho)| \geq \frac{m_0}{2}\, |\rho|^p > 0 \tag{7.15}$$

for all $\rho = a + ib \in \Gamma_*$ with $|\rho| \geq y_0$.

The estimate (7.15) is our principal result for the growth rate of the characteristic determinant Δ_0 on the punctured horizontal strip Γ_* for the case $p = q$. At this point we divide the discussion into the two cases where $\xi_0 \neq \eta_0$ and $\xi_0 = \eta_0$.

7.1. Case 1. $p = q$, $\xi_0 \neq \eta_0$

Assume that $p = q$ and $\xi_0 \neq \eta_0$. Suppose $\rho = a + ib$ is a zero of Δ_0 in G_0 with $a \geq -\pi$, $b \geq -d$, and $|\rho| \geq y_0$. What can we say about the location of ρ? By Theorem 7.1 and the inequality (7.15), ρ must lie in one of the circles Γ'_k, Γ''_k for some integer k.

Let us consider the circles Γ'_k, Γ''_k for $k \geq k_0$, which lie in the interior of T_0 and in the interior of the horizontal strip Γ and to the right of the rectangle $R'_{k_0} = [\omega - 2\pi, y_0] \times [-d, d]$. From (7.13) we have $|g_0(\rho)| < |f_0(\rho)|$ for all points ρ on Γ'_k, Γ''_k for $k \geq k_0$, and hence, by Rouché's Theorem Δ_0 and $f_0 + g_0$ have precisely the same number of zeros as f_0 inside Γ'_k and Γ''_k for all $k \geq k_0$. But f_0 has only the single zero μ'_k of order 1 inside Γ'_k and only the single zero μ''_k of order 1 inside Γ''_k, implying that Δ_0 has exactly one zero ρ'_k inside Γ'_k with ρ'_k having order 1 for $k \geq k_0$, and Δ_0 has exactly one zero ρ''_k inside Γ''_k with ρ''_k having order 1 for $k \geq k_0$.

Setting

$$\lambda'_k := (\rho'_k)^n, \qquad k = k_0, k_0 + 1, \ldots,$$
$$\lambda''_k := (\rho''_k)^n, \qquad k = k_0, k_0 + 1, \ldots,$$

it follows that the λ'_k, λ''_k, $k = k_0, k_0 + 1, \ldots$, are eigenvalues of L. Since the points ρ'_k, ρ''_k, $k = k_0, k_0 + 1, \ldots$, are zeros of order 1 of Δ_0, applying our earlier work the corresponding algebraic multiplicities and ascents are

$$\begin{aligned} \nu(\lambda'_k) &= m(\lambda'_k) = 1, \qquad k = k_0, k_0 + 1, \ldots, \\ \nu(\lambda''_k) &= m(\lambda''_k) = 1, \qquad k = k_0, k_0 + 1, \ldots. \end{aligned} \tag{7.16}$$

See Theorem 2.1 of Chapter 4 in [**34**]; the proof given for the principal part T also works for the differential operator L.

Now suppose that λ_0 is any eigenvalue of L distinct from the λ'_k, λ''_k, $k = k_0, k_0+1, \ldots$. Choose $\rho_0 \in S_0 \cup S_1$ such that $\lambda_0 = (\rho_0)^n$. Clearly $\rho_0 \in \operatorname{Int} T_0 \cup \operatorname{Int} T_1$. There are two possible locations for the point ρ_0: either ρ_0 lies in the disk $|\rho| < y_0$, or $|\rho_0| \geq y_0$. In the former case only a finite number of such ρ_0 are possible because the spectrum $\sigma(L)$ is a countable set having no limit points in $\mathbb{C}$. Assume that ρ_0 belongs to the latter case, so $|\rho_0| \geq y_0$. Let $\rho_0 = a_0 + ib_0$. First, if $b_0 \geq d$, then we would have $\rho_0 \in S_0$ with $|\rho_0| \geq r_1$, and hence, ρ_0 would belong to the open set G_0 with $|\rho_0| \geq r_1$; but Theorem 7.1 would then place λ_0 in $\rho(L)$ — a contradiction. Thus, we must have $b_0 < d$. Second, if $b_0 \leq -d$, then we would have $\rho_0 \in S_1$ with $|\rho_0| \geq r_1$, so ρ_0 would belong to the open set G_1 with $|\rho_0| \geq r_1$, and Theorem 7.1 would again place λ_0 in $\rho(L)$ — a contradiction. Therefore, for the imaginary part of ρ_0 we must have $-d < b_0 < d$.

Now for the real part, clearly $a_0 \geq 0$ from the simple geometry of the sector $S_0 \cup S_1$, and hence, ρ_0 lies in the interior of the horizontal strip Γ and ρ_0 lies in $\operatorname{Int} T_0$. Suppose $a_0 > y_0$. We know that ρ_0 does not lie in any of the circles Γ'_k, Γ''_k for $k \geq k_0$ because λ_0 is distinct from the λ'_k, λ''_k. The circles Γ'_k, Γ''_k, $-\infty < k < k_0$, either lie in the interior of the rectangle R'_{k_0} or lie to the left of R'_{k_0}, and hence, ρ_0 can not be in any of these circles. This implies that ρ_0 belongs to the punctured horizontal strip Γ_* with $|\rho_0| \geq a_0 > y_0$, and by inequality (7.15) we have $\Delta_0(\rho_0) \neq 0$ — again putting λ_0 in $\rho(L)$. This contradiction shows that we must have $a_0 \leq y_0$, and hence, ρ_0 must lie in the rectangle $[0, y_0] \times [-d, d]$. Only a finite number of such ρ_0 are possible. Thus, the λ'_k, λ''_k, $k = k_0, k_0 + 1, \ldots$, account for all but a finite number of the eigenvalues of L.

To complete this case, we derive asymptotic formulas for the zeros ρ_k', ρ_k'' of Δ_0. Indeed, let G be the entire function defined by $G(\rho) := a_p[e^{i\rho} - \eta_0]$, and set

$$M_0 := \min\{\, |G(\rho)| \mid \rho \in \mathbb{C} \text{ with } |\rho - \mu_0'| \leq \delta\} > 0.$$

Because G has period 2π, it follows that $|G(\rho_k')| \geq M_0$ for $k = k_0, k_0 + 1, \ldots$. If we set $\zeta_k' := -g_0(\rho_k')/\xi_0 G(\rho_k')$, $k = k_0, k_0 + 1, \ldots$, then we can write the equation $f_0(\rho_k') + g_0(\rho_k') = 0$ as $e^{i\rho_k'} = \xi_0 + \xi_0\zeta_k'$, and upon dividing by $e^{i\mu_k'} = \xi_0$, it becomes

$$e^{i(\rho_k' - \mu_k')} = 1 + \zeta_k'.$$

But $|\operatorname{Re}(\rho_k' - \mu_k')| \leq |\rho_k' - \mu_k'| < \delta \leq \pi/4$, so

$$\rho_k' - \mu_k' = -i \operatorname{Log}[1 + \zeta_k'], \qquad k = k_0, k_0 + 1, \ldots. \tag{7.17}$$

For each integer $k \geq k_0$ we have

$$\begin{aligned} |\rho_k'| &\geq |\mu_k'| - |\rho_k' - \mu_k'| \geq 2\pi k + \operatorname{Arg} \xi_0 - \delta \\ &\geq 2\pi k - \pi - \frac{\pi}{4} \geq 6k - 5 \geq k \end{aligned}$$

and

$$|\zeta_k'| = \frac{|g_0(\rho_k')|}{|\xi_0||G(\rho_k')|} \leq \frac{\gamma_0}{|\xi_0| M_0 |\rho_k'|} \leq \frac{\gamma}{k}. \tag{7.18}$$

Since

$$-i \operatorname{Log}[1 + z] = zH(z) \quad \text{for } |z| < 1,$$

with H analytic on the disk $|z| < 1$, from (7.17) and (7.18) we obtain the estimate

$$|\rho_k' - \mu_k'| \leq \frac{\gamma}{k}, \qquad k = k_0, k_0 + 1, \ldots, \tag{7.19}$$

for an appropriate constant $\gamma > 0$. A similar argument shows that

$$|\rho_k'' - \mu_k''| \leq \frac{\gamma}{k}, \qquad k = k_0, k_0 + 1, \ldots. \tag{7.20}$$

The estimates (7.19) and (7.20) are the desired asymptotic formulas for Case 1.

We summarize these results for the eigenvalues as a theorem. This theorem was stated previously in Chapter 1 as Theorem 1.3.

THEOREM 7.2. *Let the differential operator L belong to Case 1, where the integers p and q satisfy the conditions $-\infty < p = q \leq p_0$ and where ξ_0 and η_0 are the roots of the quadratic polynomial $Q(z) = a_p z^2 + b_p z + c_p$ with $\xi_0 \neq \eta_0$ (so $|\eta_0| = 1/|\xi_0|$ and $\arg \eta_0 = -\arg \xi_0 - 2\pi p/n + \pi$). Then the elements of the spectrum $\sigma(L)$ can be listed as two distinct sequences*

$$\lambda_k' = (\rho_k')^n, \quad k = k_0, k_0 + 1, \ldots, \qquad \lambda_k'' = (\rho_k'')^n, \quad k = k_0, k_0 + 1, \ldots,$$

plus a finite number of additional points, where k_0 is a positive integer and

$$\begin{aligned} \rho_k' &= (2\pi k + \operatorname{Arg} \xi_0) - i \ln |\xi_0| + \epsilon_k', \quad k = k_0, k_0 + 1, \ldots, \\ \rho_k'' &= (2\pi k + \operatorname{Arg} \eta_0) + i \ln |\xi_0| + \epsilon_k'', \quad k = k_0, k_0 + 1, \ldots, \end{aligned}$$

with $|\epsilon_k'| \leq \gamma/k$ and $|\epsilon_k''| \leq \gamma/k$ for $k = k_0, k_0 + 1, \ldots$. Moreover, the corresponding algebraic multiplicities and ascents are

$$\begin{aligned} \nu(\lambda_k') = m(\lambda_k') = 1, \quad k = k_0, k_0 + 1, \ldots, \\ \nu(\lambda_k'') = m(\lambda_k'') = 1, \quad k = k_0, k_0 + 1, \ldots. \end{aligned}$$

7.2. Case 2. $p = q$, $\xi_0 = \eta_0$

Assume that $p = q$ and $\xi_0 = \eta_0$. Then the polynomial $Q(z) = a_p z^2 + b_p z + c_p$ has the double root $\xi_0 = \eta_0$, and multiple eigenvalues are possible. If $\rho = a + ib \in G_0$ is a zero of Δ_0 with $a \geq -\pi$, $b \geq -d$, and $|\rho| \geq y_0$, then by Theorem 7.1 and (7.15) ρ must lie in one of the circles $\Gamma_k = \Gamma_k' = \Gamma_k''$ for some integer k. Let us consider the circles Γ_k, $k \geq k_0$, which lie in the interior of T_0 and in the interior of the horizontal strip Γ and to the right of the rectangle $R_{k_0}' = [\omega - 2\pi, y_0] \times [-d, d]$. From (7.13) we have $|g_0(\rho)| < |f_0(\rho)|$ for all points ρ on Γ_k for $k \geq k_0$, and hence, by Rouché's Theorem Δ_0 and $f_0 + g_0$ have precisely the same number of zeros as f_0 inside Γ_k for all $k \geq k_0$. But f_0 has only the single zero μ_k of order 2 inside Γ_k, implying that Δ_0 has two zeros ρ_k' and ρ_k'' inside Γ_k for $k \geq k_0$, where either $\rho_k' \neq \rho_k''$ with ρ_k' and ρ_k'' both being zeros of order 1 or $\rho_k' = \rho_k''$ with ρ_k' being a zero of order 2.

Setting

$$\lambda_k' := (\rho_k')^n, \qquad k = k_0, k_0 + 1, \ldots,$$
$$\lambda_k'' := (\rho_k'')^n, \qquad k = k_0, k_0 + 1, \ldots,$$

it follows that the λ_k', λ_k'', $k = k_0, k_0 + 1, \ldots$, are eigenvalues of L. In addition, if $\rho_k' \neq \rho_k''$, then $\lambda_k' \neq \lambda_k''$ and by our earlier work the algebraic multiplicities and ascents are

$$\nu(\lambda_k') = m(\lambda_k') = 1, \qquad \nu(\lambda_k'') = m(\lambda_k'') = 1; \tag{7.21}$$

if $\rho_k' = \rho_k''$, then $\lambda_k' = \lambda_k''$, and again by our previous work the algebraic multiplicities and ascents are

$$\nu(\lambda_k') = 2, \qquad m(\lambda_k') = 1 \quad \text{or} \quad m(\lambda_k') = 2. \tag{7.22}$$

Let λ_0 be any eigenvalue of L distinct from the λ_k', λ_k'', $k = k_0, k_0 + 1, \ldots$. Choose $\rho_0 \in S_0 \cup S_1$ such that $\lambda_0 = (\rho_0)^n$. Then either ρ_0 lies in the disk $|\rho| < y_0$, or the modulus satisfies the condition $|\rho_0| \geq y_0$ with $|\operatorname{Im} \rho_0| < d$ by Theorem 7.1. In the latter case (7.15) implies that ρ_0 belongs to the rectangle $[0, y_0] \times [-d, d]$. In either case only a finite number of such ρ_0 are possible because the spectrum $\sigma(L)$ is a countable set having no limit points in $\mathbb{C}$. Thus, we conclude that the λ_k', λ_k'', $k = k_0, k_0 + 1, \ldots$, account for all but a finite number of the eigenvalues of L.

Next, let us derive asymptotic formulas for the zeros ρ_k', ρ_k'' of Δ_0. Fix any index $k \geq k_0$. We know that $f_0(\rho_k') + g_0(\rho_k') = 0$, so $\left[e^{i\rho_k'} - \xi_0\right]^2 = -(1/a_p) g_0(\rho_k')$. By constructing an appropriate analytic branch of the square root (depending on k), the last equation can be rewritten as

$$e^{i\rho_k'} = \xi_0 + \sqrt{-\frac{1}{a_p} g_0(\rho_k')},$$

and upon dividing by $e^{i\mu_k} = \xi_0$, it becomes

$$e^{i(\rho_k' - \mu_k)} = 1 + \underbrace{\frac{1}{\xi_0}\sqrt{-\frac{1}{a_p} g_0(\rho_k')}}_{\zeta_k'}.$$

Since $|\operatorname{Re}(\rho_k' - \mu_k)| \leq |\rho_k' - \mu_k| < \delta \leq \pi/4$, we get

$$\rho_k' - \mu_k = -i \operatorname{Log}[1 + \zeta_k'].$$

Now $|\rho_k'| \geq |\mu_k| - |\rho_k' - \mu_k| \geq 2\pi k + \operatorname{Arg}\xi_0 - \frac{\pi}{4} \geq 6k - 5 \geq k$ and

$$|\zeta_k'| = \frac{1}{|\xi_0|}\sqrt{\frac{1}{|a_p|}|g_0(\rho_k')|} \leq \frac{1}{|\xi_0|}\sqrt{\frac{\gamma_0}{|a_p||\rho_k'|}} \leq \frac{\gamma}{\sqrt{k}}\,.$$

As in Case 1 this leads to the estimate

$$|\rho_k' - \mu_k| \leq \frac{\gamma}{\sqrt{k}}\,, \qquad k = k_0, k_0+1, \ldots, \tag{7.23}$$

for an appropriate constant $\gamma > 0$. The same argument shows that

$$|\rho_k'' - \mu_k| \leq \frac{\gamma}{\sqrt{k}}\,, \qquad k = k_0, k_0+1, \ldots. \tag{7.24}$$

These last two results are our asymptotic formulas for the zeros in Case 2.

We now summarize the above results as a theorem. Previously this theorem was stated in Chapter 1 as Theorem 1.4.

THEOREM 7.3. *Let the differential operator L belong to Case 2, where the integers p and q satisfy $-\infty < p = q \leq p_0$ and where $\xi_0 = \eta_0$ is the double root of the quadratic polynomial $Q(z) = a_p z^2 + b_p z + c_p$ (so $\xi_0 = \eta_0 = \pm i/\sqrt{\omega_p}$). Then the elements of the spectrum $\sigma(L)$ can be listed as two sequences*

$$\lambda_k' = (\rho_k')^n, \quad k = k_0, k_0+1, \ldots, \qquad \lambda_k'' = (\rho_k'')^n, \quad k = k_0, k_0+1, \ldots,$$

plus a finite number of additional points, where k_0 is a positive integer and

$$\rho_k' = 2\pi k + \operatorname{Arg}\xi_0 + \epsilon_k', \quad k = k_0, k_0+1, \ldots,$$
$$\rho_k'' = 2\pi k + \operatorname{Arg}\xi_0 + \epsilon_k'', \quad k = k_0, k_0+1, \ldots,$$

with $|\epsilon_k'| \leq \gamma/\sqrt{k}$ and $|\epsilon_k''| \leq \gamma/\sqrt{k}$ for $k = k_0, k_0+1, \ldots$. For each $k \geq k_0$ if $\rho_k' \neq \rho_k''$, then $\lambda_k' \neq \lambda_k''$ and the algebraic multiplicities and ascents are

$$\nu(\lambda_k') = m(\lambda_k') = 1 \quad \text{and} \quad \nu(\lambda_k'') = m(\lambda_k'') = 1,$$

while if $\rho_k' = \rho_k''$, then $\lambda_k' = \lambda_k''$ and the algebraic multiplicities and ascents are

$$\nu(\lambda_k') = 2 \quad \text{and} \quad m(\lambda_k') = 1 \ \text{or}\ m(\lambda_k') = 2.$$

7.3. Case 3. $p < q$

Assume that $p < q$. This case becomes Case 3, the so-called *logarithmic case*. For this case the differential operator L is always simply irregular. For the functions that appear in equation (7.1), we know that the coefficients satisfy $a_p \neq 0$, $b_q \neq 0$, and $c_p \neq 0$. Set $n_0 := q - p > 0$. By (7.1) and (7.2) we can write the characteristic determinant Δ_0 in the form

$$\Delta_0(\rho) = \rho^p\big\{a_p e^{2i\rho}[1 + \phi_{02}(\rho)] + b_q \rho^{n_0} e^{i\rho}[1 + \phi_{01}(\rho)] + c_p[1 + \phi_{00}(\rho)]\big\} \tag{7.25}$$

for $\rho \in G_0$, where

$$\phi_{02}(\rho) := \sum_{\kappa=-(m-p_0-1)}^{p-1} \frac{a_\kappa}{a_p \rho^{p-\kappa}} + \frac{1}{a_p \rho^p} \Phi_{02}(\rho),$$

$$\phi_{01}(\rho) := \sum_{\kappa=-(m-p_0-1)}^{q-1} \frac{b_\kappa}{b_q \rho^{q-\kappa}} + \frac{1}{b_q \rho^q} \Phi_{01}(\rho),$$

$$\phi_{00}(\rho) := \sum_{\kappa=-(m-p_0-1)}^{p-1} \frac{c_\kappa}{c_p \rho^{p-\kappa}} + \frac{1}{c_p \rho^p} \Phi_{00}(\rho)$$

for $\rho \in G_0$. The functions ϕ_{0i}, $i = 0, 1, 2$, are analytic on the open set G_0, and recalling that $m - p_0 + p \geq 1$ and $m - p_0 + q \geq 1$, by (7.3) we obtain the growth rates

$$|\phi_{02}(\rho)| \leq \frac{\gamma_0}{|\rho|}, \qquad |\phi_{01}(\rho)| \leq \frac{\gamma_0}{|\rho|}, \qquad |\phi_{00}(\rho)| \leq \frac{\gamma_0}{|\rho|} \tag{7.26}$$

for $\rho \in G_0$. Choose a constant $r_1 > R_0$ such that

$$1/2 \leq |1 + \phi_{02}(\rho)| \leq 2, \quad 1/2 \leq |1 + \phi_{01}(\rho)| \leq 2, \quad 1/2 \leq |1 + \phi_{00}(\rho)| \leq 2$$

for $\rho \in G_0$ with $|\rho| \geq r_1$.

Similarly, by (7.1) and (7.4) the characteristic determinant Δ_1 can be expressed as

$$\Delta_1(\rho) = \rho^p \{ a_p[1 + \phi_{12}(\rho)] + b_q \rho^{n_0} e^{-i\rho}[1 + \phi_{11}(\rho)] + c_p e^{-2i\rho}[1 + \phi_{10}(\rho)] \} \tag{7.27}$$

for $\rho \in G_1$, where

$$\phi_{12}(\rho) := \sum_{\kappa=-(m-p_0-1)}^{p-1} \frac{a_\kappa}{a_p \rho^{p-\kappa}} + \frac{1}{a_p \rho^p} \Phi_{12}(\rho),$$

$$\phi_{11}(\rho) := \sum_{\kappa=-(m-p_0-1)}^{q-1} \frac{b_\kappa}{b_q \rho^{q-\kappa}} + \frac{1}{b_q \rho^q} \Phi_{11}(\rho),$$

$$\phi_{10}(\rho) := \sum_{\kappa=-(m-p_0-1)}^{p-1} \frac{c_\kappa}{c_p \rho^{p-\kappa}} + \frac{1}{c_p \rho^p} \Phi_{10}(\rho)$$

for $\rho \in G_1$. The functions ϕ_{1i}, $i = 0, 1, 2$, are analytic on the open set G_1 with

$$|\phi_{12}(\rho)| \leq \frac{\gamma_0}{|\rho|}, \qquad |\phi_{11}(\rho)| \leq \frac{\gamma_0}{|\rho|}, \qquad |\phi_{10}(\rho)| \leq \frac{\gamma_0}{|\rho|} \tag{7.28}$$

for $\rho \in G_1$. Without loss of generality we can assume that the constant r_1 chosen above also produces the inequalities

$$1/2 \leq |1 + \phi_{12}(\rho)| \leq 2, \quad 1/2 \leq |1 + \phi_{11}(\rho)| \leq 2, \quad 1/2 \leq |1 + \phi_{10}(\rho)| \leq 2$$

for $\rho \in G_1$ with $|\rho| \geq r_1$.

Set $\mu_0 := -b_q/c_p$, a nonzero complex constant, and let ω be the positive real number defined by the equation $1/\omega := 1/|\mu_0|^{1/n_0} + n_0$. Choose real numbers α and β with $0 < \alpha < [1/(2|\mu_0|)]^{1/n_0}$, $\beta > [2/|\mu_0|]^{1/n_0}$, and

$$2|\mu_0|(2\alpha)^{n_0} \leq \frac{1}{8}, \qquad \frac{4}{|\mu_0|\beta^{n_0}} \leq \frac{1}{4}.$$

Clearly $1/\beta < |\mu_0|^{1/n_0} < 1/\alpha$. We will first study the characteristic determinant Δ_0 on the sector S_0, which lies in Quadrant I. Note that if ρ is any point in S_0 with $|\rho| > R_0$, then ρ belongs to the open set G_0, and hence, we will be working in a region where Δ_0 is analytic. In terms of the constants α and β, form the region

$$\Omega_0 := \{\rho = a + ib \in S_0 \mid \alpha e^{b/n_0} \le a \le \beta e^{b/n_0}\}$$

and the two complementary regions

$$\Omega_{0\infty} := \{\rho = a + ib \in S_0 \mid a \le \alpha e^{b/n_0}\},$$

$$\Omega_{0\square} := \{\rho = a + ib \in S_0 \mid \beta e^{b/n_0} \le a\}.$$

Clearly these three regions lie in Quadrant I. An equivalent definition of Ω_0 is

$$\Omega_0 = \{\rho = a + ib \in S_0 \mid a \ge \alpha \text{ and } n_0 \ln[a/\beta] \le b \le n_0 \ln[a/\alpha]\},$$

so in the sequel Ω_0 is often referred to as a *logarithmic strip*. Also, observe that

$$n_0 \ln[a/\beta] < n_0 \ln[|\mu_0|^{1/n_0} a] < n_0 \ln[a/\alpha]$$

for $a \ge \beta$, and hence, the logarithmic curve $b = n_0 \ln[|\mu_0|^{1/n_0} a]$ runs down the 'middle' of Ω_0.

Let us begin by calculating the growth rate of Δ_0 on the region $\Omega_{0\infty}$. For any point $\rho = a + ib \in \Omega_{0\infty}$ we have

$$\begin{aligned} |\rho| &\le a + b \le \alpha e^{b/n_0} + n_0 e^{b/n_0} \\ &\le [1/|\mu_0|^{1/n_0} + n_0] e^{b/n_0} = [1/\omega] e^{b/n_0}, \end{aligned}$$

which shows that

$$n_0 \ln[\omega|\rho|] \le b \quad \text{for all } \rho = a + ib \in \Omega_{0\infty}. \tag{7.29}$$

Choose a constant $x_0 > 0$ such that $x \le \alpha e^{x/n_0}$ and $e^{-2x} \le 1/16$ for all $x \in \mathbb{R}$ with $x \ge x_0$, and then choose a second constant r_2 with $r_2 \ge r_1$ and $r_2 \ge \beta$ and with $x_0 \le n_0 \ln[\omega|\rho|]$ for all $\rho \in \mathbb{C}$ with $|\rho| \ge r_2$. If $\rho = a + ib \in \Omega_{0\infty}$ with $|\rho| \ge r_2$, then by (7.29) we have $b \ge x_0$, $b \le \alpha e^{b/n_0}$ and $e^{-2b} \le 1/16$, and

$$|\rho| \le a + b \le \alpha e^{b/n_0} + \alpha e^{b/n_0} = 2\alpha e^{b/n_0}. \tag{7.30}$$

Take any point $\rho = a + ib \in \Omega_{0\infty}$ with $|\rho| \ge r_2$. Clearly $b \ge x_0 > 0$. Combining (7.25) with (7.30), we have

$$\Delta_0(\rho) = c_p \rho^p \Big\{ [1 + \phi_{00}(\rho)] - \mu_0 \rho^{n_0} [1 + \phi_{01}(\rho)] e^{i\rho} + \frac{a_p}{c_p} [1 + \phi_{02}(\rho)] e^{2i\rho} \Big\}$$

and (recall $|a_p| = |c_p|$ by (3.30))

$$\begin{aligned} |\Delta_0(\rho)| &\ge |c_p||\rho|^p \left\{ \frac{1}{2} - 2|\mu_0||\rho|^{n_0} e^{-b} - 2e^{-2b} \right\} \\ &\ge |c_p||\rho|^p \left\{ \frac{1}{2} - 2|\mu_0| \cdot (2\alpha)^{n_0} e^b \cdot e^{-b} - 2 \cdot \frac{1}{16} \right\} \ge \frac{1}{4} |c_p||\rho|^p, \end{aligned}$$

or

$$|\Delta_0(\rho)| \ge \frac{1}{4} |a_p||\rho|^p > 0 \tag{7.31}$$

for all $\rho = a + ib \in \Omega_{0\infty}$ with $|\rho| \ge r_2$.

To determine the growth rate of Δ_0 on the region $\Omega_{0\square}$, we first use (7.25) to express Δ_0 in the form

$$\Delta_0(\rho) = b_q\rho^q e^{i\rho}\Big\{[1+\phi_{01}(\rho)] + \frac{a_p}{b_q\rho^{n_0}}[1+\phi_{02}(\rho)]e^{i\rho} + \frac{c_p}{b_q\rho^{n_0}}[1+\phi_{00}(\rho)]e^{-i\rho}\Big\}$$

for $\rho \in G_0$. Then for any point $\rho = a+ib \in \Omega_{0\square}$ with $|\rho| \geq r_2$ we have

$$|\rho| \geq a \geq \beta e^{b/n_0}, \tag{7.32}$$

and hence,

$$\begin{aligned}|\Delta_0(\rho)| &\geq |b_q||\rho|^q e^{-b}\Big\{\frac{1}{2} - \frac{2}{|\mu_0||\rho|^{n_0}}\left[e^{-b}+e^{b}\right]\Big\}\\ &\geq |b_q||\rho|^q e^{-b}\Big\{\frac{1}{2} - \frac{4e^b}{|\mu_0|\cdot\beta^{n_0}e^b}\Big\} \geq \frac{1}{4}|b_q||\rho|^q e^{-b}\\ &= \frac{1}{4}|b_q||\rho|^p\cdot|\rho|^{n_0}e^{-b} \geq \frac{\beta^{n_0}}{4}|b_q||\rho|^p \geq \frac{4}{|\mu_0|}|b_q||\rho|^p > 0,\end{aligned}$$

or

$$|\Delta_0(\rho)| \geq \frac{1}{4}|b_q||\rho|^q e^{-b} \geq \frac{4}{|\mu_0|}|b_q||\rho|^p > 0 \tag{7.33}$$

for all $\rho = a+ib \in \Omega_{0\square}$ with $|\rho| \geq r_2$.

Next, we rework the above material for the characteristic determinant Δ_1 on the sector S_1. For $\rho \in S_1$ with $|\rho| > R_0$, we have $\rho \in G_1$, and we are working in a region of analyticity for Δ_1. Now let us form the region

$$\Omega_1 := \big\{\rho = a+ib \in S_1 \mid \alpha e^{-b/n_0} \leq a \leq \beta e^{-b/n_0}\big\}$$

and the two complementary regions

$$\Omega_{1\infty} := \big\{\rho = a+ib \in S_1 \mid a \leq \alpha e^{-b/n_0}\big\},$$

$$\Omega_{1\square} := \big\{\rho = a+ib \in S_1 \mid \beta e^{-b/n_0} \leq a\big\}.$$

These three regions lie in Quadrant IV, and Ω_1 can be expressed as

$$\Omega_1 = \big\{\rho = a+ib \in S_1 \mid a \geq \alpha \text{ and } -n_0\ln[a/\alpha] \leq b \leq -n_0\ln[a/\beta]\big\}.$$

For the region $\Omega_{1\infty}$ we obtain the estimates

$$n_0\ln[\omega|\rho|] \leq |b| \quad \text{for } \rho = a+ib \in \Omega_{1\infty} \tag{7.34}$$

and

$$|\rho| \leq 2\alpha e^{|b|/n_0} \quad \text{for } \rho = a+ib \in \Omega_{1\infty} \text{ with } |\rho| \geq r_2. \tag{7.35}$$

Rewriting (7.27) in the form

$$\Delta_1(\rho) = a_p\rho^p\Big\{[1+\phi_{12}(\rho)] + \frac{b_q}{a_p}\rho^{n_0}[1+\phi_{11}(\rho)]e^{-i\rho} + \frac{c_p}{a_p}[1+\phi_{10}(\rho)]e^{-2i\rho}\Big\}$$

for $\rho \in G_1$, we obtain the growth rate

$$|\Delta_1(\rho)| \geq \frac{1}{4}|a_p||\rho|^p > 0 \tag{7.36}$$

for all $\rho = a + ib \in \Omega_{1\infty}$ with $|\rho| \geq r_2$. Compare this growth rate to (7.31). Also, for the region $\Omega_{1\square}$ we have the estimate

$$|\rho| \geq \beta e^{|b|/n_0} \quad \text{for } \rho = a + ib \in \Omega_{1\square} \text{ with } |\rho| \geq r_2, \tag{7.37}$$

and expressing (7.27) in the alternate form

$$\Delta_1(\rho) = b_q\rho^q e^{-i\rho}\Big\{[1+\phi_{11}(\rho)] + \frac{a_p}{b_q\rho^{n_0}}[1+\phi_{12}(\rho)]e^{i\rho} + \frac{c_p}{b_q\rho^{n_0}}[1+\phi_{10}(\rho)]e^{-i\rho}\Big\}$$

for $\rho \in G_1$, we obtain the growth rate

$$|\Delta_1(\rho)| \geq \frac{1}{4}\,|b_q||\rho|^q e^b \geq \frac{4}{|\mu_0|}\,|b_q||\rho|^p > 0 \tag{7.38}$$

for all $\rho = a + ib \in \Omega_{1\square}$ with $|\rho| \geq r_2$. Cf. (7.33).

As an application of these growth rates on the regions $\Omega_{0\infty}$, $\Omega_{0\square}$ and $\Omega_{1\infty}$, $\Omega_{1\square}$, we obtain the following *apriori estimates* for the eigenvalues of L.

THEOREM 7.4. *Assume $p < q$. For any point $\lambda = \rho^n \in \mathbb{C}$ with $\rho = a + ib \in S_0$ and $|\rho| \geq r_2$:*

(a) *If $\rho \in \Omega_{0\infty}$ or $\rho \in \Omega_{0\square}$, then $\Delta_0(\rho) \neq 0$ and $\lambda \in \rho(L)$.*

(b) *If λ is an eigenvalue of L, then $\Delta_0(\rho) = 0$ and ρ lies in the interior of the logarithmic strip Ω_0.*

Also, for any point $\lambda = \rho^n \in \mathbb{C}$ with $\rho = a + ib \in S_1$ and $|\rho| \geq r_2$:

(c) *If $\rho \in \Omega_{1\infty}$ or $\rho \in \Omega_{1\square}$, then $\Delta_1(\rho) \neq 0$ and $\lambda \in \rho(L)$.*

(d) *If λ is an eigenvalue of L, then $\Delta_1(\rho) = 0$ and ρ lies in the interior of the logarithmic strip Ω_1.*

The theorem implies that the resolvent set $\rho(L)$ is nonempty, and hence, by our earlier remarks the spectrum $\sigma(L)$ is a countable set having no limit points in $\mathbb{C}$.

Next, we calculate the eigenvalues of L that correspond to the zeros of Δ_0 in the logarithmic strip Ω_0. Following this we compute the eigenvalues corresponding to the zeros of Δ_1 in the logarithmic strip Ω_1. Set $\xi := 1 + n_0/\alpha$ and $\eta := \beta + n_0$. Then for $\rho = a + ib \in \Omega_0$ we have

$$|\rho| \leq a + b \leq a + n_0 \ln[a/\alpha] \leq a + (n_0/\alpha)a = \xi a \tag{7.39}$$

and

$$|\rho| \leq a + b \leq \beta e^{b/n_0} + b \leq \beta e^{b/n_0} + n_0 e^{b/n_0} = \eta e^{b/n_0}. \tag{7.40}$$

In studying the characteristic determinant on the logarithmic strip Ω_0, we use (7.25) to write Δ_0 in the form

$$\Delta_0(\rho) = c_p\rho^p e^{i\rho}\{e^{-i\rho} - \mu_0\rho^{n_0}[1 + h_0(\rho)]\} \tag{7.41}$$

for $\rho \in G_0$, where h_0 is the function given by

$$h_0(\rho) := -\frac{e^{-i\rho}}{\mu_0\rho^{n_0}}\,\phi_{00}(\rho) + \phi_{01}(\rho) + \frac{a_p}{b_q\rho^{n_0}}\,[1 + \phi_{02}(\rho)]e^{i\rho}$$

for $\rho \in G_0$. The function h_0 is analytic on the open set G_0. Observe that for any point $\rho = a + ib \in \Omega_0$ with $|\rho| \geq r_2$, the inequalities defining Ω_0 yield

$$\left|\frac{e^{-i\rho}}{\mu_0\rho^{n_0}}\right| = \frac{e^b}{|\mu_0||\rho|^{n_0}} \leq \frac{1}{|\mu_0||\rho|^{n_0}}\cdot\frac{a^{n_0}}{\alpha^{n_0}} \leq \frac{1}{|\mu_0|\alpha^{n_0}},$$

and clearly $|1+\phi_{02}(\rho)||e^{i\rho}| = |1+\phi_{02}(\rho)|e^{-b} \le 2$ because $b \ge 0$ for $\rho \in S_0$, and hence, by the inequalities in (7.26)

$$|h_0(\rho)| \le \frac{\gamma_1}{|\rho|} \quad \text{for } \rho \in \Omega_0 \text{ with } |\rho| \ge r_2. \tag{7.42}$$

Set

$$\Omega' := \{\rho = a+ib \in S_0 \mid a \ge r_2,\ n_0 \ln[a/\beta] \le b \le n_0 \ln[a/\alpha]\},$$

which is a subset of both G_0 and Ω_0 (recall that $r_2 \ge \beta > \alpha$). Fix a real number δ with $0 < \delta \le \pi/4$ and $0 < \delta < (\ln 2)/(1+n_0)$, and then for $k = 1, 2, \ldots$ define

$$\begin{cases} \alpha'_k := 2\pi k - \operatorname{Arg}\mu_0, & \beta'_k := n_0 \ln[|\mu_0|^{1/n_0}\alpha'_k], \\ \mu'_k := \alpha'_k + i\beta'_k, \end{cases}$$

and introduce the circles

$$\Gamma'_k := \{\rho \in \mathbb{C} \mid |\rho - \mu'_k| = \delta\}.$$

Choose an integer $k_1 \ge 2$ such that the real number $y'_1 := \alpha'_{k_1} - \pi$ satisfies the inequality $y'_1 \ge r_2$. Note that $\alpha'_k - \pi \ge r_2 \ge \beta$ and $\alpha'_k \ge 3\pi$ for $k = k_1, k_1+1, \ldots$. Also, introduce the *logarithmic rectangles*

$$R'_k := \{\rho = a+ib \in \mathbb{C} \mid \alpha'_k - \pi \le a \le \alpha'_k + \pi,\ n_0 \ln[a/\beta] \le b \le n_0 \ln[a/\alpha]\}$$

for $k = k_1, k_1+1, \ldots$. Without loss of generality we can assume that k_1 is sufficiently large to guarantee that each R'_k is contained in S_0, and hence, for $k = k_1, k_1+1, \ldots$ the point μ'_k lies in the interior of R'_k with R'_k a subset of Ω'.

Fix any index $k \ge k_1$, and take any point $\rho = a+ib \in \mathbb{C}$ with $|\rho - \mu'_k| \le \delta$. We assert that ρ lies in the interior of R'_k. Indeed, we clearly have $|a - \alpha'_k| \le \delta < \pi$ and $|b - n_0 \ln[|\mu_0|^{1/n_0}\alpha'_k]| \le \delta$, so

$$\begin{aligned} |b - n_0 \ln[|\mu_0|^{1/n_0} a]| &\le |b - n_0 \ln[|\mu_0|^{1/n_0}\alpha'_k]| \\ &\quad + |n_0 \ln[|\mu_0|^{1/n_0} a] - n_0 \ln[|\mu_0|^{1/n_0}\alpha'_k]| \\ &\le \delta + n_0|a - \alpha'_k| \le \delta(1+n_0) < \ln 2. \end{aligned}$$

It follows that $n_0 \ln[(|\mu_0|/2)^{1/n_0} a] < b < n_0 \ln[(2|\mu_0|)^{1/n_0} a]$ and

$$n_0 \ln[a/\beta] < b < n_0 \ln[a/\alpha].$$

This establishes the assertion, and it is immediate that the circle Γ'_k lies in the interior of the logarithmic rectangle R'_k for $k = k_1, k_1+1, \ldots$. To complete the setup of the geometry, for $k = k_1, k_1+1, \ldots$ let Ω'_k be the *punctured logarithmic rectangle* formed from R'_k by removing all the points inside Γ'_k.

The next step is to establish the growth rate of Δ_0 on each of the regions Ω'_k. Note that

$$e^{-i\mu'_k} = \mu_0(\alpha'_k)^{n_0}$$

for $k = k_1, k_1+1, \ldots$. Let f_k, $k = k_1, k_1+1, \ldots$, and g_k, $k = k_1, k_1+1, \ldots$, be the sequences of functions defined by

$$f_k(\rho) := e^{-i(\rho - \mu'_k)} - 1 \quad \text{for } \rho \in \mathbb{C},$$

$$g_k(\rho) := -h_0(\rho) - \sum_{j=1}^{n_0} \binom{n_0}{j} \left[\frac{1}{\alpha'_k}(\rho - \mu'_k) + \frac{i\beta'_k}{\alpha'_k}\right]^j [1 + h_0(\rho)] \quad \text{for } \rho \in G_0.$$

The functions g_k are analytic on the open set G_0. We can use (7.41) to write Δ_0 in its final form

$$\Delta_0(\rho) = c_p\mu_0(\alpha'_k)^{n_0}\rho^p e^{i\rho}\Big\{e^{-i\rho}\cdot e^{i\mu'_k} - \Big[\frac{\rho}{\alpha'_k}\Big]^{n_0}[1+h_0(\rho)]\Big\}$$

$$= c_p\mu_0(\alpha'_k)^{n_0}\rho^p e^{i\rho}\Big\{e^{-i(\rho-\mu'_k)} - \Big[\frac{1}{\alpha'_k}(\rho-\mu'_k) + \frac{i\beta'_k}{\alpha'_k} + 1\Big]^{n_0}[1+h_0(\rho)]\Big\},$$

or

$$\Delta_0(\rho) = c_p\mu_0(\alpha'_k)^{n_0}\rho^p e^{i\rho}[f_k(\rho) + g_k(\rho)] \tag{7.43}$$

for $\rho \in G_0$ and for $k = k_1, k_1+1, \ldots$. Here we have a family of representations for Δ_0 depending on the integer k. We will use the kth representation to determine the growth rate of Δ_0 on the kth region Ω'_k.

In terms of the constants α, β, δ, choose $d_0 > 0$ such that

$$n_0\ln[2/(|\mu_0|^{1/n_0}\alpha)] \le d_0, \qquad n_0\ln[2|\mu_0|^{1/n_0}\beta] \le d_0,$$

and $\delta < d_0$, and then form the punctured rectangle

$$R_* := \{\rho = a + ib \in \mathbb{C} \mid -\pi \le a \le \pi,\ -d_0 \le b \le d_0, \text{ and } |\rho| \ge \delta\}.$$

Set $m_0 := \min\{\,|e^{-i\rho} - 1| \mid \rho \in R_*\} = \min\{\,|e^{i\rho} - 1| \mid \rho \in R_*\} > 0$. Fix any index $k \ge k_1$ and any point $\rho = a + ib \in \Omega'_k$. Then it follows that $-\pi \le a - \alpha'_k \le \pi$, $\pi/\alpha'_k \le 1/2$ because $\alpha'_k \ge 3\pi$,

$$\frac{a}{\alpha'_k} \le \frac{\alpha'_k + \pi}{\alpha'_k} \le 2, \qquad \frac{a}{\alpha'_k} \ge \frac{\alpha'_k - \pi}{\alpha'_k} \ge \frac{1}{2},$$

and

$$b - \beta'_k \le n_0\ln[a/\alpha] - n_0\ln[|\mu_0|^{1/n_0}\alpha'_k]$$
$$= n_0\ln[a/(|\mu_0|^{1/n_0}\alpha\alpha'_k)] \le n_0\ln[2/(|\mu_0|^{1/n_0}\alpha)] \le d_0,$$

$$b - \beta'_k \ge n_0\ln[a/\beta] - n_0\ln[|\mu_0|^{1/n_0}\alpha'_k]$$
$$= n_0\ln[a/(|\mu_0|^{1/n_0}\beta\alpha'_k)] \ge n_0\ln[1/(2|\mu_0|^{1/n_0}\beta)] \ge -d_0.$$

Thus, the point $\rho - \mu'_k$ belongs to the punctured rectangle R_* with $|\rho - \mu'_k| \le \pi + d_0$. We conclude that

$$|f_k(\rho)| \ge m_0 \tag{7.44}$$

for $k \ge k_1$ and for $\rho \in \Omega'_k$. Note that the constant m_0 is independent of the index k.

Clearly $\lim_{k\to\infty} 1/\alpha'_k = \lim_{k\to\infty}\beta'_k/\alpha'_k = 0$. In terms of the constant γ_1 that appears in the inequality (7.42), select an integer $k_0 \ge k_1$ such that the real number $y'_0 := \alpha'_{k_0} - \pi$ satisfies the condition $y'_0 \ge r_2$, such that

$$\frac{\gamma_1}{|\rho|} \le \frac{m_0}{4} \quad \text{for all } \rho \in \mathbb{C} \text{ with } |\rho| \ge y'_0,$$

and such that

$$\sum_{j=1}^{n_0}\binom{n_0}{j}\Big[\frac{\pi + d_0}{\alpha'_k} + \frac{\beta'_k}{\alpha'_k}\Big]^j\Big[1 + \frac{\gamma_1}{r_2}\Big] \le \frac{m_0}{4} \quad \text{for all } k \ge k_0.$$

Then for $k \ge k_0$ and for $\rho = a + ib \in \Omega'_k$, we have $|\rho| \ge a \ge \alpha'_k - \pi \ge y'_0 \ge r_2$, and hence, by (7.42) and the definition of the integer k_0:

$$|g_k(\rho)| \le \frac{m_0}{2} < m_0 \le |f_k(\rho)|. \tag{7.45}$$

Also, since $\alpha_k' \geq 3\pi$, we have $a \geq \alpha_k' - \pi \geq 2\pi$ or $a/2 \geq \pi$, and $\alpha_k' \geq a - \pi \geq a/2 \geq |\rho|/(2\xi)$ by (7.39). Therefore, from (7.43) and (7.45) we conclude that

$$|\Delta_0(\rho)| \geq |c_p||\mu_0|(\alpha_k')^{n_0}|\rho|^p e^{-b} \cdot \frac{m_0}{2} \geq \frac{m_0|a_p||\mu_0|}{2(2\xi)^{n_0}} |\rho|^q e^{-b},$$

or

$$|\Delta_0(\rho)| \geq \frac{m_0|a_p||\mu_0|}{2(2\xi)^{n_0}} |\rho|^q e^{-b} > 0 \tag{7.46}$$

for $k \geq k_0$ and for $\rho = a + ib \in \Omega_k'$.

The estimate (7.45) is local in character in that it depends on k: it is valid only on the region Ω_k'. In contrast, the estimate (7.46) is global because the constant on the right is independent of k. If we introduce the *punctured logarithmic strip*

$$\Omega_*' := \bigcup_{k=k_0}^{\infty} \Omega_k',$$

then we see that Ω_*' consists of all points $\rho = a + ib \in \Omega_0$ which satisfy $a \geq y_0'$ and which do not lie inside any of the circles Γ_k' for $k \geq k_0$, and from (7.46) and the definition of Ω_0:

$$|\Delta_0(\rho)| \geq \frac{m_0|a_p||\mu_0|}{2(2\xi)^{n_0}} |\rho|^q e^{-b} \geq \frac{m_0|a_p||\mu_0|\alpha^{n_0}}{2(2\xi)^{n_0}} |\rho|^p > 0 \tag{7.47}$$

for all $\rho = a + ib \in \Omega_*'$.

With the basic estimates (7.45) and (7.47) in place, consider one of the circles Γ_k' for $k \geq k_0$. Since (7.45) is valid for each point ρ on Γ_k', it follows by Rouché's Theorem that Δ_0 and $f_k + g_k$ have the same number of zeros as f_k inside Γ_k'. But μ_k' is the only zero of f_k inside Γ_k', μ_k' being a zero of order 1. Consequently, Δ_0 has a unique zero ρ_k' inside Γ_k' with ρ_k' having order 1 for $k \geq k_0$. Setting

$$\lambda_k' := (\rho_k')^n, \qquad k = k_0, k_0 + 1, \ldots,$$

the complex numbers λ_k' are eigenvalues of L, and by our earlier work the corresponding algebraic multiplicities and ascents are given by

$$\nu(\lambda_k') = m(\lambda_k') = 1, \qquad k = k_0, k_0 + 1, \ldots. \tag{7.48}$$

It is also easy to derive asymptotic formulas for the zeros ρ_k' of Δ_0. Indeed, set $\zeta_k' := -g_k(\rho_k')$ for $k = k_0, k_0 + 1, \ldots$. Then $e^{-i(\rho_k' - \mu_k')} = 1 + \zeta_k'$ and

$$\rho_k' - \mu_k' = i \operatorname{Log}[1 + \zeta_k'] \tag{7.49}$$

for $k = k_0, k_0 + 1, \ldots$, where

$$|\zeta_k'| \leq \frac{\gamma_1}{|\rho_k'|} + \sum_{j=1}^{n_0} \binom{n_0}{j} \left[\frac{\delta}{\alpha_k'} + \frac{\beta_k'}{\alpha_k'}\right]^j \left[1 + \frac{\gamma_1}{r_2}\right].$$

Now for each $k \geq k_0$, $\alpha_k' \geq 2\pi k - \pi \geq k$, $\beta_k' \leq n_0 \ln[|\mu_0|^{1/n_0}(2\pi k + \pi)] \leq \gamma_2 \ln k$, and $|\rho_k'| \geq |\mu_k'| - |\rho_k' - \mu_k'| \geq \alpha_k' - \delta \geq 6k - 5 \geq k$, which yields $|\zeta_k'| \leq \gamma_3 \ln k/k$ and

$$|\rho_k' - \mu_k'| \leq \frac{\gamma \ln k}{k}, \qquad k = k_0, k_0 + 1, \ldots. \tag{7.50}$$

To compute the zeros of Δ_1 in the logarithmic strip Ω_1, we begin by using (7.27) to express Δ_1 in the form

$$\Delta_1(\rho) = a_p\rho^p e^{-i\rho}\{e^{i\rho} - \mu_1\rho^{n_0}[1 + h_1(\rho)]\} \tag{7.51}$$

for $\rho \in G_1$, where $\mu_1 := -b_q/a_p = -\mu_0/\omega_p$, $|\mu_1| = |\mu_0|$, and h_1 is the analytic function given by

$$h_1(\rho) := -\frac{e^{i\rho}}{\mu_1 \rho^{n_0}} \phi_{12}(\rho) + \phi_{11}(\rho) + \frac{c_p}{b_q \rho^{n_0}} [1 + \phi_{10}(\rho)] e^{-i\rho}$$

for $\rho \in G_1$. The function h_1 satisfies the growth rate

$$|h_1(\rho)| \leq \frac{\gamma_1}{|\rho|} \quad \text{for } \rho \in \Omega_1 \text{ with } |\rho| \geq r_2. \tag{7.52}$$

We concentrate the search for zeros in the region

$$\Omega'' := \{\rho = a + ib \in S_1 \mid a \geq r_2, \; -n_0 \ln[a/\alpha] \leq b \leq -n_0 \ln[a/\beta]\},$$

which is a subset of both G_1 and Ω_1. Using the real number δ defined above, for $k = 1, 2, \ldots$ define

$$\begin{cases} \alpha_k'' := 2\pi k + \operatorname{Arg} \mu_1, & \beta_k'' := -n_0 \ln[|\mu_1|^{1/n_0} \alpha_k''], \\ \mu_k'' := \alpha_k'' + i\beta_k'', \end{cases}$$

and introduce the circles

$$\Gamma_k'' := \{\rho \in \mathbb{C} \mid |\rho - \mu_k''| = \delta\}.$$

Assume the positive integer k_1 also satisfies the condition $y_1'' := \alpha_{k_1}'' - \pi \geq r_2$. Let us introduce the *logarithmic rectangles*

$$R_k'' := \{\rho = a + ib \in \mathbb{C} \mid \alpha_k'' - \pi \leq a \leq \alpha_k'' + \pi, \; -n_0 \ln[a/\alpha] \leq b \leq -n_0 \ln[a/\beta]\}$$

for $k = k_1, k_1+1, \ldots$. Without loss of generality we can assume that k_1 is sufficiently large to guarantee that each R_k'' is contained in S_1, and hence, for $k = k_1, k_1+1, \ldots$ the point μ_k'' lies in the interior of R_k'' with R_k'' a subset of Ω''. As above the circle Γ_k'' lies in the interior of the logarithmic rectangle R_k'' for $k = k_1, k_1+1, \ldots$. Finally, for $k = k_1, k_1+1, \ldots$ let Ω_k'' be the *punctured logarithmic rectangle* formed from R_k'' by removing all the points inside Γ_k''.

We next determine the growth rate of Δ_1 on each of the regions Ω_k''. Observe that the points μ_k'' satisfy the equation

$$e^{i\mu_k''} = \mu_1 (\alpha_k'')^{n_0}$$

for $k = k_1, k_1+1, \ldots$. If F_k, $k = k_1, k_1+1, \ldots$, and G_k, $k = k_1, k_1+1, \ldots$, are the sequences of analytic functions defined by

$$F_k(\rho) = e^{i(\rho - \mu_k'')} - 1 \quad \text{for } \rho \in \mathbb{C},$$

$$G_k(\rho) = -h_1(\rho) - \sum_{j=1}^{n_0} \binom{n_0}{j} \Big[\frac{1}{\alpha_k''} (\rho - \mu_k'') + \frac{i\beta_k''}{\alpha_k''} \Big]^j [1 + h_1(\rho)] \quad \text{for } \rho \in G_1,$$

then we can rewrite Δ_1 in the form

$$\Delta_1(\rho) = a_p \mu_1 (\alpha_k'')^{n_0} \rho^p e^{-i\rho} [F_k(\rho) + G_k(\rho)] \tag{7.53}$$

for $\rho \in G_1$ and for $k = k_1, k_1+1, \ldots$. In the last equation we actually have a family of representations for Δ_1 depending on the integer k.

Using the constant d_0, the punctured rectangle R_*, and the constant m_0 defined above, it follows that

$$|F_k(\rho)| \geq m_0 \tag{7.54}$$

for $k \ge k_1$ and for $\rho \in \Omega_k''$. Since $\lim_{k\to\infty} 1/\alpha_k'' = \lim_{k\to\infty} \beta_k''/\alpha_k'' = 0$, in terms of (7.52) it can be assumed that the previous integer k_0 also guarantees that the real number $y_0'' := \alpha_{k_0}'' - \pi$ satisfies the inequality $y_0'' \ge r_2$, that

$$\frac{\gamma_1}{|\rho|} \le \frac{m_0}{4} \quad \text{for all } \rho \in \mathbb{C} \text{ with } |\rho| \ge y_0'',$$

and that

$$\sum_{j=1}^{n_0} \binom{n_0}{j} \left[\frac{\pi + d_0}{\alpha_k''} + \frac{|\beta_k''|}{\alpha_k''}\right]^j \left[1 + \frac{\gamma_1}{r_2}\right] \le \frac{m_0}{4} \quad \text{for all } k \ge k_0.$$

Then for each $k \ge k_0$ and each $\rho = a + ib \in \Omega_k''$, we have $|\rho| \ge a \ge \alpha_k'' - \pi \ge y_0'' \ge r_2$, and by (7.52) and the defining properties of k_0:

$$|G_k(\rho)| \le \frac{m_0}{2} < m_0 \le |F_k(\rho)|. \tag{7.55}$$

Also, $\alpha_k'' \ge a - \pi \ge a/2 \ge |\rho|/(2\xi)$, and from (7.53) and (7.55)

$$|\Delta_1(\rho)| \ge |a_p||\mu_1|(\alpha_k'')^{n_0}|\rho|^p e^b \cdot \frac{m_0}{2} \ge \frac{m_0|a_p||\mu_0|}{2(2\xi)^{n_0}} |\rho|^q e^b > 0 \tag{7.56}$$

for $k \ge k_0$ and for $\rho = a + ib \in \Omega_k''$. Cf. the estimate (7.46).

The estimate (7.55) is local as it depends on k, while the estimate (7.56) is global being independent of k. Introducing the *punctured logarithmic strip*

$$\Omega_*'' := \bigcup_{k=k_0}^{\infty} \Omega_k'',$$

we see that Ω_*'' consists of all points $\rho = a + ib \in \Omega_1$ which satisfy $a \ge y_0''$ and which do not lie inside any of the circles Γ_k'' for $k \ge k_0$, and from (7.56) and the definition of Ω_1:

$$|\Delta_1(\rho)| \ge \frac{m_0|a_p||\mu_0|}{2(2\xi)^{n_0}} |\rho|^q e^b \ge \frac{m_0|a_p||\mu_0|\alpha^{n_0}}{2(2\xi)^{n_0}} |\rho|^p > 0 \tag{7.57}$$

for all $\rho = a + ib \in \Omega_*''$. Cf. (7.47).

Consider one of the circles Γ_k'' for $k \ge k_0$. Inequality (7.55) is valid for each point ρ on the circle Γ_k'', so by Rouché's Theorem Δ_1 and $F_k + G_k$ have the same number of zeros as F_k inside Γ_k''. The point μ_k'' is the only zero of F_k inside Γ_k'' with μ_k'' being a zero of order 1. We conclude that Δ_1 has a unique zero ρ_k'' inside Γ_k'' of order 1 for $k \ge k_0$. Setting

$$\lambda_k'' := (\rho_k'')^n, \qquad k = k_0, k_0 + 1, \ldots,$$

the λ_k'' are all eigenvalues of L with algebraic multiplicities and ascents

$$\nu(\lambda_k'') = m(\lambda_k'') = 1, \qquad k = k_0, k_0 + 1, \ldots. \tag{7.58}$$

To derive asymptotic formulas, set $\zeta_k'' := -G_k(\rho_k'')$ for $k = k_0, k_0 + 1, \ldots$. It is immediate that $e^{i(\rho_k'' - \mu_k'')} = 1 + \zeta_k''$ and

$$\rho_k'' - \mu_k'' = -i \operatorname{Log}[1 + \zeta_k''] \tag{7.59}$$

for $k = k_0, k_0 + 1, \ldots$, where

$$|\zeta_k''| \le \frac{\gamma_1}{|\rho_k''|} + \sum_{j=1}^{n_0} \binom{n_0}{j} \left[\frac{\delta}{\alpha_k''} + \frac{|\beta_k''|}{\alpha_k''}\right]^j \left[1 + \frac{\gamma_1}{r_2}\right].$$

For each $k \geq k_0$ we have $\alpha_k'' \geq k$, $|\beta_k''| \leq \gamma_2 \ln k$, and $|\rho_k''| \geq k$, which yields the estimates $|\zeta_k''| \leq \gamma_3 \ln k/k$ and

$$|\rho_k'' - \mu_k''| \leq \frac{\gamma \ln k}{k}, \qquad k = k_0, k_0+1, \ldots. \tag{7.60}$$

Finally, we assert that the λ_k', λ_k'', $k = k_0, k_0+1, \ldots$, account for all but a finite number of the eigenvalues of L. Indeed, suppose $\lambda_0 = (\rho_0)^n$ is any eigenvalue of L distinct from the λ_k', λ_k'', with $\rho_0 = a_0 + ib_0$ belonging to the sector $S_0 \cup S_1$. Only a finite number of such ρ_0 are possible in the disk $|\rho| < r_2$. Assume that $|\rho_0| \geq r_2$.

First, consider the case where $\rho_0 \in S_0$. By Theorem 7.4 we know that $\Delta_0(\rho_0) = 0$ and ρ_0 lies in the interior of the logarithmic strip Ω_0. Now ρ_0 does not lie in any of the circles Γ_k' for $k \geq k_0$ since λ_0 is distinct from the λ_k', and ρ_0 does not lie in the punctured logarithmic strip Ω_*' by virtue of (7.47). Thus, we must have $\rho_0 \in \Omega_0$ with $\alpha \leq a_0 < y_0'$. But this implies that $0 \leq b_0 \leq n_0 \ln[a_0/\alpha] < n_0 \ln[y_0'/\alpha]$, so these ρ_0 lie in a bounded region of the ρ plane, and again only a finite number of such ρ_0 are possible.

Second, consider the other possible case where $\rho_0 \in S_1$. By Theorem 7.4 we have $\Delta_1(\rho_0) = 0$ with ρ_0 lying in the interior of the logarithmic strip Ω_1. Following the same argument as in the previous case, only a finite number of such ρ_0 are possible. This establishes the assertion.

We summarize the above results for this logarithmic case in the following theorem. This theorem was stated previously in Chapter 1 as Theorem 1.5.

THEOREM 7.5. *Let the differential operator L belong to Case 3, a logarithmic case, where the integers p and q satisfy the conditions $-\infty < p < q \leq p_0$, and let $n_0 = q - p$, $\mu_0 = -b_q/c_p \neq 0$, and $\mu_1 = -b_q/a_p \neq 0$ (so $|\mu_1| = |\mu_0|$ and $\arg \mu_1 = \arg \mu_0 - 2\pi p/n + \pi$). Then the elements of the spectrum $\sigma(L)$ can be listed as two distinct sequences*

$$\lambda_k' = (\rho_k')^n, \quad k = k_0, k_0+1, \ldots, \qquad \lambda_k'' = (\rho_k'')^n, \quad k = k_0, k_0+1, \ldots,$$

plus a finite number of additional points, where k_0 is a positive integer and

$$\rho_k' = (2\pi k - \operatorname{Arg} \mu_0) + in_0 \ln[|\mu_0|^{1/n_0}(2\pi k - \operatorname{Arg} \mu_0)] + \epsilon_k', \qquad k = k_0, k_0+1, \ldots,$$

$$\rho_k'' = (2\pi k + \operatorname{Arg} \mu_1) - in_0 \ln[|\mu_0|^{1/n_0}(2\pi k + \operatorname{Arg} \mu_1)] + \epsilon_k'', \qquad k = k_0, k_0+1, \ldots,$$

with $|\epsilon_k'| \leq \gamma \ln k/k$ and $|\epsilon_k''| \leq \gamma \ln k/k$ for $k = k_0, k_0+1, \ldots$. Moreover, the corresponding algebraic multiplicities and ascents are

$$\nu(\lambda_k') = m(\lambda_k') = 1, \quad k = k_0, k_0+1, \ldots,$$
$$\nu(\lambda_k'') = m(\lambda_k'') = 1, \quad k = k_0, k_0+1, \ldots.$$

CHAPTER 8

The Eigenvalues for n Odd

In this chapter we compute the eigenvalues of the differential operator L for the case n odd. Throughout the hypotheses of Chapters 3–5 are assumed: (i) $n = 2\nu - 1 \geq 3$; (ii) the differential operator L is either regular or simply irregular; (iii) the integers p and q have been determined with $-\infty < p, q \leq p_0$ and with $a_p \neq 0$, $b_q \neq 0$, and $a_\kappa = 0$ for $\kappa = p+1, \ldots, p_0$ and $b_\kappa = 0$ for $\kappa = q+1, \ldots, p_0$; (iv) the translated sectors T_0 and T_1 have been selected; (v) the integer m has been fixed with $m > n$, $m > p_0$, and $-(m - p_0 - 1) \leq p, q \leq p_0$; and (vi) the functions π_i, $i = 0, 1$, and the functions π_i', $i = 0, 1$, have been determined as per Chapter 3 or equations (5.47) and (5.54):

$$\pi_1(\rho) = \sum_{\kappa=-(m-p_0-1)}^{p} a_\kappa \rho^\kappa, \qquad \pi_0(\rho) = \sum_{\kappa=-(m-p_0-1)}^{q} b_\kappa \rho^\kappa, \tag{8.1}$$

$$\pi_1'(\rho) = \sum_{\kappa=-(m-p_0-1)}^{q} a_\kappa' \rho^\kappa, \qquad \pi_0'(\rho) = \sum_{\kappa=-(m-p_0-1)}^{p} b_\kappa' \rho^\kappa \tag{8.2}$$

for $\rho \neq 0$ in $\mathbb{C}$. The leading coefficients in these representations are related by the equations $|a_q'| = |b_q|$ and $|b_p'| = |a_p|$.

To determine the eigenvalues of L, we calculate the zeros of the characteristic determinant Δ_0 in the open set G_0 and the zeros of the characteristic determinant Δ_1 in the open set G_1. The basic properties of Δ_0 and Δ_1 have been developed previously in Theorem 5.3 and Theorem 5.4. Our analysis divides naturally into the three cases where $p = q$, $p < q$, and $p > q$. The latter two cases are logarithmic cases. The development of the mathematics closely parallels that for the case n even. Consequently, we will only indicate the main features of the theory.

8.1. Case 1. $p = q$

Assume that $p = q$. This first case is a case with simple eigenvalues; it includes all of the regular differential operators and many of the simply irregular ones. We begin by working on the sector T_0 and the open set $G_0 = \{\rho \in \operatorname{Int} T_0 \mid |\rho| > R_0\}$. Let

$$f_0(\rho) := a_p e^{i\rho} + b_p = a_p\big[e^{i\rho} - \xi_0\big]$$

for $\rho \in \mathbb{C}$, where $\xi_0 := -b_p/a_p \neq 0$, and let

$$g_0(\rho) := \sum_{\kappa=-(m-p_0-1)}^{p-1} \frac{a_\kappa}{\rho^{p-\kappa}} e^{i\rho} + \sum_{\kappa=-(m-p_0-1)}^{p-1} \frac{b_\kappa}{\rho^{p-\kappa}} + \frac{1}{\rho^p}\left[\Phi_{01}(\rho)e^{i\rho} + \Phi_{00}(\rho)\right]$$

for $\rho \in G_0$. The function f_0 is an entire function, while the function g_0 is analytic on the open set G_0. By equation (5.46) or Theorem 5.3, the characteristic determinant

Δ_0 can then be expressed as

$$\Delta_0(\rho) = \rho^p[f_0(\rho) + g_0(\rho)] \quad \text{for } \rho \in G_0. \tag{8.3}$$

For the case $p = q$ the constant $d > 0$ was chosen to satisfy condition (3.47):

$$|a_p|e^{-d} + |b_p|e^{-d} \leq \frac{1}{4}\min\{|a_p|, |b_p|\}, \tag{8.4}$$

which led to the companion result

$$|a_p'|e^{-d} + |b_p'|e^{-d} \leq \frac{1}{4}\min\{|a_p'|, |b_p'|\} = \frac{1}{4}\min\{|a_p|, |b_p|\}. \tag{8.5}$$

Using (8.4) and (5.48), we proceed to determine a constant $r_1 > R_0$ and to establish the growth rates

$$|\Delta_0(\rho)| \geq \frac{1}{2}|b_p||\rho|^p > 0 \tag{8.6}$$

for $\rho = a + ib \in G_0$ with $|\rho| \geq r_1$ and $b \geq d$, and

$$|\Delta_0(\rho)| \geq \frac{1}{2}e^{-b}|a_p||\rho|^p \geq \frac{1}{2}|a_p||\rho|^p > 0 \tag{8.7}$$

for $\rho = a + ib \in G_0$ with $|\rho| \geq r_1$ and $b \leq -d$.

Next, we establish similar growth rates for the characteristic determinant Δ_1 on the open set $G_1 = \{\rho \in \operatorname{Int} T_1 \mid |\rho| > R_0\}$. Let

$$f_1(\rho) := a_p'e^{-i\rho} + b_p' = a_p'\left[e^{-i\rho} - \eta_0\right]$$

for $\rho \in \mathbb{C}$, where by (5.56) $\eta_0 := -b_p'/a_p' = 1/(\omega_p\xi_0)$, and let

$$g_1(\rho) := \sum_{\kappa=-(m-p_0-1)}^{p-1} \frac{a_\kappa'}{\rho^{p-\kappa}}e^{-i\rho} + \sum_{\kappa=-(m-p_0-1)}^{p-1} \frac{b_\kappa'}{\rho^{p-\kappa}} + \frac{1}{\rho^p}\left[\Phi_{11}(\rho)e^{-i\rho} + \Phi_{10}(\rho)\right]$$

for $\rho \in G_1$. The function f_1 is an entire function, and the function g_1 is analytic on the open set G_1. From (5.53) or Theorem 5.4 we can express the characteristic determinant Δ_1 in the form

$$\Delta_1(\rho) = \rho^p[f_1(\rho) + g_1(\rho)] \quad \text{for } \rho \in G_1. \tag{8.8}$$

Using (8.5) and (5.55), we derive the growth rates

$$|\Delta_1(\rho)| \geq \frac{1}{2}|b_p'||\rho|^p > 0 \tag{8.9}$$

for $\rho = a + ib \in G_1$ with $|\rho| \geq r_1$ and $b \leq -d$, and

$$|\Delta_1(\rho)| \geq \frac{1}{2}e^{b}|a_p'||\rho|^p \geq \frac{1}{2}|a_p'||\rho|^p > 0 \tag{8.10}$$

for $\rho = a + ib \in G_1$ with $|\rho| \geq r_1$ and $b \geq d$.

As an immediate application of (8.6), (8.7) and (8.9), (8.10), we obtain the following *apriori estimates* for the eigenvalues of L.

THEOREM 8.1. *Assume that $p = q$. Let $\lambda = \rho^n \in \mathbb{C}$ with $\rho = a + ib \in G_0$ and with $|\rho| \geq r_1$.*

(a) *If $|b| \geq d$, then $\Delta_0(\rho) \neq 0$ and $\lambda \in \rho(L)$.*

(b) *If λ is an eigenvalue of L, then $\Delta_0(\rho) = 0$ and $|b| < d$.*

In addition, let $\lambda = \rho^n \in \mathbb{C}$ with $\rho = a + ib \in G_1$ and with $|\rho| \geq r_1$.

(c) *If $|b| \geq d$, then $\Delta_1(\rho) \neq 0$ and $\lambda \in \rho(L)$.*

(d) *If λ is an eigenvalue of L, then $\Delta_1(\rho) = 0$ and $|b| < d$.*

By the last theorem the resolvent set $\rho(L)$ is nonempty. Hence, the differential operator L is a Fredholm operator with Fredholm set $\Phi(L) = \mathbb{C}$ and with resolvent set $\rho(L) \neq \emptyset$; this implies that the spectrum $\sigma(L)$ is a countable set having no limit points in $\mathbb{C}$. See [**34**, p. 58 or p. 60].

We next focus our search for the zeros of Δ_0 on the horizontal strip

$$\Gamma_0 = \{\rho = a + ib \in \mathbb{C} \mid a \geq -\pi \text{ and } |b| \leq d\}.$$

Recall that the sector T_0 was selected in Chapter 3 so that the horizontal strip Γ_0 lies in the interior of T_0. Clearly the zeros of f_0 are given by the sequence

$$\mu_k' = (2\pi k + \operatorname{Arg} \xi_0) - i \ln |\xi_0|, \qquad k = 0, \pm 1, \pm 2, \ldots,$$

where each μ_k' is a zero of order 1 of f_0. The μ_k' all lie in the interior of the horizontal strip $|\operatorname{Im} \rho| \leq d$, i.e., $-d < \ln |\xi_0| < d$. Choose a real number $\omega \geq \pi$ such that $\omega - 2\pi < \operatorname{Arg} \xi_0 < \omega$, and then for $k = 1, 2, \ldots$ introduce the rectangles

$$R_k' := \{\rho \in \mathbb{C} \mid \omega - 2\pi \leq \operatorname{Re} \rho \leq \omega + 2\pi(k-1) \text{ and } |\operatorname{Im} \rho| \leq d\}.$$

These rectangles lie in the horizontal strip Γ_0, and hence, they lie in the interior of the sector T_0, and the zero μ_0' lies in the interior of the rectangle R_1'. Choose a real number δ with $0 < \delta \leq \pi/4$ such that the disk $|\rho - \mu_0'| \leq \delta$ lies in the interior of R_1'. For $k = 0, \pm 1, \pm 2, \ldots$ form the circles

$$\Gamma_k' := \{\rho \in \mathbb{C} \mid |\rho - \mu_k'| = \delta\}.$$

To complete the geometry, let Ω_0 be the subset of the sector T_0 defined by

$$\Omega_0 := \{\rho = a + ib \in \Gamma_0 \mid \rho \text{ is not inside any of the circles } \Gamma_k'\},$$

a *punctured horizontal strip.*

Let $m_* := \min\{|f_0(\rho)| \mid \rho \in R_1' \text{ with } \rho \text{ not in } \Gamma_0'\} > 0$, and set

$$m_0 := \min\left\{\frac{3}{4}|a_p|, \frac{3}{4}|b_p|, m_*\right\} > 0.$$

We then choose a positive integer k_0 and a real number $y_0 := \omega + 2\pi(k_0 - 1)$ sufficiently large to produce the inequalities

$$|g_0(\rho)| \leq \frac{m_0}{2} < m_0 \leq |f_0(\rho)| \tag{8.11}$$

and

$$|\Delta_0(\rho)| \geq \frac{m_0}{2} |\rho|^p > 0 \tag{8.12}$$

for all $\rho = a + ib \in \Omega_0$ with $|\rho| \geq y_0$. From (8.12) it is immediate that Δ_0 has no zeros in Ω_0 when $|\rho| \geq y_0$.

Now let us consider the circles Γ_k' for $k \geq k_0$, which lie in the interior of T_0 and in the interior of the horizontal strip Γ_0 and to the right of the rectangle $R_{k_0}' = [\omega - 2\pi, y_0] \times [-d, d]$. From (8.11) we have $|g_0(\rho)| < |f_0(\rho)|$ for all points ρ on Γ_k' for $k \geq k_0$, and hence, by Rouché's Theorem Δ_0 and $f_0 + g_0$ have precisely the same number of zeros as f_0 inside Γ_k' for all $k \geq k_0$. But f_0 has only the single zero μ_k' of order 1 inside Γ_k', implying that Δ_0 has exactly one zero ρ_k' inside Γ_k' with ρ_k' having order 1 for $k \geq k_0$. Setting

$$\lambda_k' := (\rho_k')^n, \qquad k = k_0, k_0 + 1, \ldots,$$

it follows that the λ'_k, $k = k_0, k_0+1, \ldots$, are eigenvalues of L, and by our earlier work the corresponding algebraic multiplicities and ascents are

$$\nu(\lambda'_k) = m(\lambda'_k) = 1, \qquad k = k_0, k_0+1, \ldots. \tag{8.13}$$

As in the case n even we obtain the asymptotic formulas

$$|\rho'_k - \mu'_k| \le \frac{\gamma}{k}, \qquad k = k_0, k_0+1, \ldots. \tag{8.14}$$

Next, we compute the zeros of Δ_1 in the horizontal strip

$$\Gamma_1 = \{\rho = a + ib \in \mathbb{C} \mid a \le \pi \text{ and } |b| \le d\}.$$

Recall that the sector T_1 was selected in Chapter 3 so that the horizontal strip Γ_1 lies in the interior of T_1. The zeros of the entire function f_1 are given by

$$\mu''_k = -(2\pi k + \operatorname{Arg}\eta_0) + i \ln|\eta_0|, \qquad k = 0, \pm 1, \pm 2, \ldots,$$

where $|\eta_0| = 1/|\xi_0|$ and $\arg\eta_0 = -\arg\xi_0 - 2\pi p/n$ and where each μ''_k is a zero of order 1 of f_1. The μ''_k all lie in the interior of the horizontal strip $|\operatorname{Im}\rho| \le d$, i.e., $-d < \ln|\eta_0| < d$. Choose a constant $\omega' \le \pi$ such that $\omega' - 2\pi < -\operatorname{Arg}\eta_0 < \omega'$, and then for $k = 1, 2, \ldots$ introduce the rectangles

$$R''_k := \{\rho \in \mathbb{C} \mid \omega' - 2\pi k \le \operatorname{Re}\rho \le \omega' \text{ and } |\operatorname{Im}\rho| \le d\}.$$

The zero μ''_0 lies in the interior of the rectangle R''_1, so we can choose a constant δ' with $0 < \delta' \le \pi/4$ such that the disk $|\rho - \mu''_0| \le \delta'$ lies in the interior of R''_1. Without loss of generality we can assume that $\delta = \delta'$. For $k = 0, \pm 1, \pm 2, \ldots$ form the circles

$$\Gamma''_k := \{\rho \in \mathbb{C} \mid |\rho - \mu''_k| = \delta\}.$$

Lastly, let Ω_1 be the *punctured horizontal strip* in the ρ plane defined by

$$\Omega_1 := \{\rho = a + ib \in \Gamma_1 \mid \rho \text{ is not inside any of the circles } \Gamma''_k\}.$$

Set $m^* := \min\{|f_1(\rho)| \mid \rho \in R''_1 \text{ with } \rho \text{ not in } \Gamma''_0\} > 0$, and then set

$$m_\diamond := \min\left\{\frac{3}{4}|a'_p|, \frac{3}{4}|b'_p|, m^*\right\} > 0.$$

Choosing a positive integer k_1 and a real number $x_0 := \omega' - 2\pi k_1$ with k_1 sufficiently large, we obtain the estimates

$$|g_1(\rho)| \le \frac{m_\diamond}{2} < m_\diamond \le |f_1(\rho)| \tag{8.15}$$

and

$$|\Delta_1(\rho)| \ge \frac{m_\diamond}{2}|\rho|^p > 0 \tag{8.16}$$

for $\rho = a + ib \in \Omega_1$ with $|\rho| \ge |x_0|$. It is immediate that Δ_1 has no zeros in the punctured horizontal strip Ω_1 when $|\rho| \ge |x_0|$.

Now let us consider the circles Γ''_k for $k \ge k_0$, which lie in Γ_1 and to the left of the rectangle $R''_{k_0} = [x_0, \omega'] \times [-d, d]$. From (8.15) we have $|g_1(\rho)| < |f_1(\rho)|$ for all points ρ on Γ''_k for $k \ge k_0$, and hence, by Rouché's Theorem Δ_1 and $f_1 + g_1$ have precisely the same number of zeros as f_1 inside Γ''_k for all $k \ge k_0$. But f_1 has only the single zero μ''_k of order 1 inside Γ''_k, implying that Δ_1 has exactly one zero ρ''_k inside Γ''_k with ρ''_k having order 1 for each $k \ge k_0$. Setting

$$\lambda''_k := (\rho''_k)^n, \qquad k = k_0, k_0+1, \ldots,$$

the λ_k'' are eigenvalues of L with algebraic multiplicities and ascents

$$\nu(\lambda_k'') = m(\lambda_k'') = 1, \qquad k = k_0, k_0+1, \ldots. \tag{8.17}$$

For the zeros ρ_k'', $k = k_0, k_0+1, \ldots$, we obtain the asymptotic formulas

$$|\rho_k'' - \mu_k''| \leq \frac{\gamma}{k}, \qquad k = k_0, k_0+1, \ldots. \tag{8.18}$$

As in the case n even, we can show that the λ_k', λ_k'', $k = k_0, k_0+1, \ldots$, account for all but a finite number of the eigenvalues of L.

The above results for the eigenvalues are summarized in the following theorem. This theorem was stated previously in Chapter 1 as Theorem 1.6.

THEOREM 8.2. *Let the differential operator L belong to Case 1, where the integers p and q satisfy the conditions $-\infty < p = q \leq p_0$ and where $\xi_0 = -b_p/a_p \neq 0$ and $\eta_0 = -b_p'/a_p' \neq 0$ (so $|\eta_0| = 1/|\xi_0|$, $\arg \eta_0 = -\arg \xi_0 - 2\pi p/n$). Then the elements of the spectrum $\sigma(L)$ can be listed as two sequences*

$$\lambda_k' = (\rho_k')^n, \quad k = k_0, k_0+1, \ldots, \qquad \lambda_k'' = (\rho_k'')^n, \quad k = k_0, k_0+1, \ldots,$$

plus a finite number of additional points, where k_0 is a positive integer and

$$\rho_k' = (2\pi k + \operatorname{Arg} \xi_0) - i \ln |\xi_0| + \epsilon_k', \qquad k = k_0, k_0+1, \ldots,$$
$$\rho_k'' = -(2\pi k + \operatorname{Arg} \eta_0) - i \ln |\xi_0| + \epsilon_k'', \quad k = k_0, k_0+1, \ldots,$$

with $|\epsilon_k'| \leq \gamma/k$ and $|\epsilon_k''| \leq \gamma/k$ for $k = k_0, k_0+1, \ldots$. Moreover, the corresponding algebraic multiplicities and ascents are

$$\nu(\lambda_k') = m(\lambda_k') = 1, \quad k = k_0, k_0+1, \ldots,$$
$$\nu(\lambda_k'') = m(\lambda_k'') = 1, \quad k = k_0, k_0+1, \ldots.$$

8.2. Case 2. $p < q$

Assume that $p < q$. This second case is a logarithmic case; all the differential operators belonging to this case are simply irregular. For the translated sectors T_0 and T_1 and the corresponding open sets

$$G_0 = \{\rho \in \operatorname{Int} T_0 \mid |\rho| > R_0\}, \qquad G_1 = \{\rho \in \operatorname{Int} T_1 \mid |\rho| > R_0\},$$

we can assume that the constant R_0 has been chosen sufficiently large to guarantee that $\operatorname{Re} \rho \geq 0$ for all $\rho \in G_0$ and $\operatorname{Re} \rho \leq 0$ for all $\rho \in G_1$.

Set $n_0 := q - p > 0$, and consider the sector T_0 and the open set G_0. By (5.46) or Theorem 5.3 and (8.1) we can write the characteristic determinant Δ_0 in the form

$$\Delta_0(\rho) = \rho^p \{a_p e^{i\rho}[1 + \phi_{01}(\rho)] + b_q \rho^{n_0}[1 + \phi_{00}(\rho)]\} \tag{8.19}$$

for $\rho \in G_0$, where

$$\phi_{01}(\rho) := \sum_{\kappa=-(m-p_0-1)}^{p-1} \frac{a_\kappa}{a_p \rho^{p-\kappa}} + \frac{1}{a_p \rho^p} \Phi_{01}(\rho),$$

$$\phi_{00}(\rho) := \sum_{\kappa=-(m-p_0-1)}^{q-1} \frac{b_\kappa}{b_q \rho^{q-\kappa}} + \frac{1}{b_q \rho^q} \Phi_{00}(\rho)$$

for $\rho \in G_0$. The functions ϕ_{01}, ϕ_{00} are analytic on the open set G_0.

Set $\mu_0 := -b_q/a_p \neq 0$. Choosing a constant $r_1 > R_0$ sufficiently large, we rewrite (8.19) in the form

$$\Delta_0(\rho) = b_q\rho^q\Big\{[1+\phi_{00}(\rho)] + \frac{a_p}{b_q\rho^{n_0}}[1+\phi_{01}(\rho)]e^{i\rho}\Big\}$$

for $\rho \in G_0$, and then establish the growth rate

$$|\Delta_0(\rho)| \geq \frac{1}{4}|b_q||\rho|^q > 0 \tag{8.20}$$

for $\rho = a + ib \in G_0$ with $|\rho| \geq r_1$ and $b \geq 0$. Consequently, as we search for the zeros of Δ_0 in the open set G_0, we will concentrate our search in Quadrant IV.

Let ω be the real number defined by the equation $1/\omega := 1/|\mu_0|^{1/n_0} + n_0$. Choose real numbers α and β with $0 < \alpha < [1/(2|\mu_0|)]^{1/n_0}$, $\beta > [2/|\mu_0|]^{1/n_0}$, and

$$2|\mu_0|(2\alpha)^{n_0} \leq \frac{1}{4}, \qquad \frac{2}{|\mu_0|\beta^{n_0}} \leq \frac{1}{4}.$$

Clearly $1/\beta < |\mu_0|^{1/n_0} < 1/\alpha$. In terms of these constants we form the *logarithmic strip*

$$\begin{aligned}\Omega_0 &:= \big\{\rho = a+ib \in S_0 \;\big|\; b \leq 0,\ \alpha e^{-b/n_0} \leq a \leq \beta e^{-b/n_0}\big\}\\ &= \big\{\rho = a+ib \in S_0 \;\big|\; a \geq \alpha,\ b \leq 0,\ -n_0 \ln[a/\alpha] \leq b \leq -n_0\ln[a/\beta]\big\}\end{aligned}$$

and the two complementary regions

$$\Omega_{0\infty} := \big\{\rho = a+ib \in S_0 \;\big|\; b \leq 0,\ a \leq \alpha e^{-b/n_0}\big\},$$

$$\Omega_{0\Box} := \big\{\rho = a+ib \in S_0 \;\big|\; b \leq 0,\ \beta e^{-b/n_0} \leq a\big\}.$$

These three regions are contained in Quadrant IV.

To calculate the growth rate of Δ_0 on the region $\Omega_{0\infty}$, we first choose a real number r_2 sufficiently large with $r_2 \geq r_1$ and $r_2 \geq \beta$ and such that

$$|\rho| \leq 2\alpha e^{|b|/n_0} \tag{8.21}$$

for $\rho = a + ib \in \Omega_{0\infty}$ with $|\rho| \geq r_2$. Then rewriting (8.19) in the form

$$\Delta_0(\rho) = a_p\rho^p e^{i\rho}\Big\{[1+\phi_{01}(\rho)] + \frac{b_q}{a_p}\rho^{n_0}[1+\phi_{00}(\rho)]e^{-i\rho}\Big\}$$

for $\rho \in G_0$, we proceed to establish the growth rate

$$|\Delta_0(\rho)| \geq \frac{1}{4}|a_p||\rho|^p e^{-b} \geq \frac{|a_p|}{4(2\alpha)^{n_0}}|\rho|^q > 0 \tag{8.22}$$

for all $\rho = a + ib \in \Omega_{0\infty}$ with $|\rho| \geq r_2$. Similarly, for the region $\Omega_{0\Box}$ we use (8.19) to express Δ_0 in the form

$$\Delta_0(\rho) = b_q\rho^q\Big\{[1+\phi_{00}(\rho)] + \frac{a_p}{b_q\rho^{n_0}}[1+\phi_{01}(\rho)]e^{i\rho}\Big\}$$

for $\rho \in G_0$, and then derive the growth rate

$$|\Delta_0(\rho)| \geq \frac{1}{4}|b_q||\rho|^q > 0 \tag{8.23}$$

for all $\rho = a + ib \in \Omega_{0\Box}$.

Relative to the sector T_1 and the open set G_1, we can use (5.53) or Theorem 5.4 and (8.2) to write the characteristic determinant Δ_1 in the form

$$\Delta_1(\rho) = \rho^q\Big\{a'_q e^{-i\rho}\big[1+\phi_{11}(\rho)\big] + \frac{b'_p}{\rho^{n_0}}\big[1+\phi_{10}(\rho)\big]\Big\} \tag{8.24}$$

for $\rho \in G_1$, where

$$\phi_{11}(\rho) := \sum_{\kappa=-(m-p_0-1)}^{q-1} \frac{a'_\kappa}{a'_q\rho^{q-\kappa}} + \frac{1}{a'_q\rho^q}\Phi_{11}(\rho),$$

$$\phi_{10}(\rho) := \sum_{\kappa=-(m-p_0-1)}^{p-1} \frac{b'_\kappa}{b'_p\rho^{p-\kappa}} + \frac{1}{b'_p\rho^p}\Phi_{10}(\rho)$$

for $\rho \in G_1$. The functions ϕ_{11}, ϕ_{10} are analytic on the open set G_1.

Set $\mu_1 := -a'_q/b'_p \neq 0$. From (5.56) we have $a'_q = \omega_\nu^q b_q$ and $b'_p = \omega_{\nu-1}^p a_p$, and hence,

$$\mu_1 = -\omega_\nu^q b_q/(\omega_{\nu-1}^p a_p) = \omega_{n_0\nu+p}\mu_0. \tag{8.25}$$

It follows that $|\mu_1| = |\mu_0|$ and $\arg\mu_1 = \arg\mu_0 + 2\pi(n_0\nu+p)/n$. Choosing a constant $r'_1 > R_0$ sufficiently large, we rewrite (8.24) in the form

$$\Delta_1(\rho) = a'_q\rho^q e^{-i\rho}\Big\{\big[1+\phi_{11}(\rho)\big] + \frac{b'_p}{a'_q\rho^{n_0}}\big[1+\phi_{10}(\rho)\big]e^{i\rho}\Big\}$$

for $\rho \in G_1$, and then derive the growth rate

$$|\Delta_1(\rho)| \geq \frac{1}{4}|a'_q||\rho|^q e^b > 0 \tag{8.26}$$

for $\rho = a+ib \in G_1$ with $|\rho| \geq r'_1$ and $b \geq 0$. In view of this result, our search for the zeros of the characteristic determinant Δ_1 in the open set G_1 will be concentrated in Quadrant III.

Set $\omega_1 := \omega$, $\alpha_1 := \alpha$, and $\beta_1 := \beta$. Then these constants satisfy the conditions $1/\omega_1 = 1/|\mu_1|^{1/n_0} + n_0$, $0 < \alpha_1 < [1/(2|\mu_1|)]^{1/n_0}$, $\beta_1 > [2/|\mu_1|]^{1/n_0}$, and

$$2|\mu_1|(2\alpha_1)^{n_0} \leq \frac{1}{4}, \qquad \frac{2}{|\mu_1|(\beta_1)^{n_0}} \leq \frac{1}{4}.$$

In terms of them we form the *logarithmic strip*

$$\begin{aligned}\Omega_1 :=& \big\{\rho = a+ib \in S_1 \;\big|\; b \leq 0,\ -\beta_1 e^{-b/n_0} \leq a \leq -\alpha_1 e^{-b/n_0}\big\}\\ =& \big\{\rho = a+ib \in S_1 \;\big|\; a \leq -\alpha_1,\ b \leq 0,\ -n_0\ln[-a/\alpha_1] \leq b \leq -n_0\ln[-a/\beta_1]\big\}\end{aligned}$$

and the two complementary regions

$$\Omega_{1\infty} := \big\{\rho = a+ib \in S_1 \;\big|\; b \leq 0,\ -\alpha_1 e^{-b/n_0} \leq a\big\},$$

$$\Omega_{1\square} := \big\{\rho = a+ib \in S_1 \;\big|\; b \leq 0,\ a \leq -\beta_1 e^{-b/n_0}\big\}.$$

These three regions are contained in Quadrant III.

To calculate the growth rate of Δ_1 on the region $\Omega_{1\infty}$, we choose a real number r'_2 sufficiently large with $r'_2 \geq r'_1$ and $r'_2 \geq \beta_1$ and such that

$$|\rho| \leq 2\alpha_1 e^{|b|/n_0} \tag{8.27}$$

for $\rho = a + ib \in \Omega_{1\infty}$ with $|\rho| \geq r_2'$. Without loss of generality we can assume that $r_2 = r_2'$. Rewriting (8.24) in the form

$$\Delta_1(\rho) = b_p'\rho^p\{[1 + \phi_{10}(\rho)] + (a_q'/b_p')\rho^{n_0}[1 + \phi_{11}(\rho)]e^{-i\rho}\}$$

for $\rho \in G_1$, we obtain the growth rate

$$|\Delta_1(\rho)| \geq \frac{1}{4}\,|b_p'||\rho|^p > 0 \tag{8.28}$$

for all $\rho = a + ib \in \Omega_{1\infty}$ with $|\rho| \geq r_2$. Similarly, for the region $\Omega_{1\square}$ we express the characteristic determinant Δ_1 in the form

$$\Delta_1(\rho) = a_q'\rho^q e^{-i\rho}\Big\{[1 + \phi_{11}(\rho)] + \frac{b_p'}{a_q'\rho^{n_0}}\,[1 + \phi_{10}(\rho)]e^{i\rho}\Big\}$$

for $\rho \in G_1$, and then establish the growth rate

$$|\Delta_1(\rho)| \geq \frac{1}{4}\,|a_q'||\rho|^q e^b \geq \frac{1}{4}\,|a_q'|(\beta_1)^{n_0}|\rho|^p > 0 \tag{8.29}$$

for all $\rho = a + ib \in \Omega_{1\square}$.

As an application of these growth rates on the regions $\Omega_{0\infty}$, $\Omega_{0\square}$ and $\Omega_{1\infty}$, $\Omega_{1\square}$, we obtain the following *apriori estimates* for the eigenvalues of L.

THEOREM 8.3. *Assume that $p < q$. Let $\lambda = \rho^n \in \mathbb{C}$ with $\rho = a + ib \in S_0$ and with $|\rho| \geq r_2$.*

(a) *If $\rho \in \Omega_{0\infty}$ or $\rho \in \Omega_{0\square}$, then $\Delta_0(\rho) \neq 0$ and $\lambda \in \rho(L)$.*

(b) *If λ is an eigenvalue of L, then $\Delta_0(\rho) = 0$ and ρ lies in the interior of the logarithmic strip Ω_0.*

In addition, let $\lambda = \rho^n \in \mathbb{C}$ with $\rho = a + ib \in S_1$ and with $|\rho| \geq r_2$.

(c) *If $\rho \in \Omega_{1\infty}$ or $\rho \in \Omega_{1\square}$, then $\Delta_1(\rho) \neq 0$ and $\lambda \in \rho(L)$.*

(d) *If λ is an eigenvalue of L, then $\Delta_1(\rho) = 0$ and ρ lies in the interior of the logarithmic strip Ω_1.*

By the theorem the resolvent set $\rho(L)$ is nonempty, and hence, by our earlier remarks the spectrum $\sigma(L)$ is a countable set having no limit points in $\mathbb{C}$.

Next, let us consider the behavior of Δ_0 on the logarithmic strip Ω_0. Setting $\xi := 1 + n_0/\alpha$, we see that

$$|\rho| \leq \xi|a| \quad \text{for all } \rho = a + ib \in \Omega_0. \tag{8.30}$$

Relative to the strip Ω_0 let Δ_0 be expressed in the form

$$\Delta_0(\rho) = a_p\rho^p\{e^{i\rho} - \mu_0\rho^{n_0}[1 + h_0(\rho)]\} \tag{8.31}$$

for $\rho \in G_0$, where h_0 is the analytic function given by

$$h_0(\rho) := -\frac{e^{i\rho}}{\mu_0\rho^{n_0}}\,\phi_{01}(\rho) + \phi_{00}(\rho)$$

for $\rho \in G_0$. The function h_0 satisfies the growth rate

$$|h_0(\rho)| \leq \frac{\gamma_1}{|\rho|} \tag{8.32}$$

for all $\rho = a + ib \in \Omega_0$ with $|\rho| \geq r_2$.

Fix a real number δ with $0 < \delta \le \pi/4$ and $0 < \delta < (\ln 2)/(1+n_0)$, and then for the integers $k = 1, 2, \ldots$ define

$$\begin{cases} \alpha'_k := 2\pi k + \operatorname{Arg}\mu_0, & \beta'_k := -n_0 \ln[|\mu_0|^{1/n_0}\alpha'_k], \\ \mu'_k := \alpha'_k + i\beta'_k, \end{cases}$$

and introduce the circles

$$\Gamma'_k := \{\rho \in \mathbb{C} \mid |\rho - \mu'_k| = \delta\}.$$

Choose an integer $k_1 \ge 2$ such that the real number $y'_1 := \alpha'_{k_1} - \pi$ satisfies the condition $y'_1 \ge r_2$. Also, introduce the *logarithmic rectangles*

$$R'_k := \{\rho = a + ib \in \mathbb{C} \mid \alpha'_k - \pi \le a \le \alpha'_k + \pi,\ -n_0 \ln[a/\alpha] \le b \le -n_0 \ln[a/\beta]\}$$

for $k = k_1, k_1+1, \ldots$. Without loss of generality we can assume that k_1 is sufficiently large to guarantee that each R'_k is contained in the sector S_0, and hence, for $k = k_1, k_1+1, \ldots$ the point μ'_k lies in the interior of R'_k with R'_k a subset of Ω_0. The circle Γ'_k also lies in the interior of the logarithmic rectangle R'_k for $k = k_1, k_1+1, \ldots$. To complete the setup of the geometry, for $k = k_1, k_1+1, \ldots$ let Ω'_k be the *punctured logarithmic rectangle* formed from R'_k by removing all the points inside Γ'_k.

The next step is to establish the growth rate of Δ_0 on each of the regions Ω'_k. Note that

$$e^{i\mu'_k} = \mu_0(\alpha'_k)^{n_0} \quad \text{for } k = k_1, k_1+1, \ldots.$$

Let f_k, $k = k_1, k_1+1, \ldots$, and g_k, $k = k_1, k_1+1, \ldots$, be the sequences of functions defined by

$$f_k(\rho) := e^{i(\rho - \mu'_k)} - 1 \quad \text{for } \rho \in \mathbb{C},$$

$$g_k(\rho) := -h_0(\rho) - \sum_{j=1}^{n_0} \binom{n_0}{j} \Big[\frac{1}{\alpha'_k}(\rho - \mu'_k) + \frac{i\beta'_k}{\alpha'_k}\Big]^j [1 + h_0(\rho)] \quad \text{for } \rho \in G_0.$$

Then we can use (8.31) to write Δ_0 in its final form:

$$\Delta_0(\rho) = a_p \mu_0 (\alpha'_k)^{n_0} \rho^p [f_k(\rho) + g_k(\rho)] \tag{8.33}$$

for $\rho \in G_0$ and for $k = k_1, k_1+1, \ldots$. Here we have a family of representations for Δ_0 depending on the integer k. We will use the kth representation to determine the growth rate of Δ_0 on the kth region Ω'_k.

In terms of the constants α, β, δ, choose $d_0 > 0$ such that

$$n_0 \ln[2/(|\mu_0|^{1/n_0}\alpha)] \le d_0, \qquad n_0 \ln[2|\mu_0|^{1/n_0}\beta] \le d_0,$$

and $\delta < d_0$, and then form the punctured rectangle

$$R_* := \{\rho = a + ib \in \mathbb{C} \mid -\pi \le a \le \pi,\ -d_0 \le b \le d_0, \text{ and } |\rho| \ge \delta\}.$$

Set $m_0 := \min\{|e^{i\rho} - 1| \mid \rho \in R_*\} > 0$. Then for any index $k \ge k_1$ and for any point $\rho = a + ib \in \Omega'_k$, the translate $\rho - \mu'_k$ belongs to R_* with $|\rho - \mu'_k| \le \pi + d_0$, and hence,

$$|f_k(\rho)| \ge m_0 \tag{8.34}$$

for $k \ge k_1$ and for $\rho \in \Omega'_k$. Note that the constant m_0 is independent of the index k.

Clearly $\lim_{k\to\infty} 1/\alpha'_k = \lim_{k\to\infty} \beta'_k/\alpha'_k = 0$. In terms of the constant γ_1 that appears in the inequality (8.32), select an integer $k_0 \ge k_1$ such that

$$\frac{\gamma_1}{|\rho|} \le \frac{m_0}{4} \quad \text{for all } \rho \in \mathbb{C} \text{ with } |\rho| \ge y'_0,$$

where $y_0' := \alpha_{k_0}' - \pi \geq r_2$, and such that

$$\sum_{j=1}^{n_0} \binom{n_0}{j} \left[\frac{\pi + d_0}{\alpha_k'} + \frac{|\beta_k'|}{\alpha_k'}\right]^j \left[1 + \frac{\gamma_1}{r_2}\right] \leq \frac{m_0}{4} \quad \text{for all } k \geq k_0.$$

Then for $k \geq k_0$ and for $\rho \in \Omega_k'$, we have

$$|g_k(\rho)| \leq \frac{m_0}{2} < m_0 \leq |f_k(\rho)| \tag{8.35}$$

and

$$|\Delta_0(\rho)| \geq \frac{m_0 |a_p| |\mu_0|}{2(2\xi)^{n_0}} |\rho|^q > 0. \tag{8.36}$$

The estimate (8.35) is local in character in that it depends on k: it is valid only on the region Ω_k'. In contrast, the estimate (8.36) is global because the constant on the right is independent of k. If we introduce the *punctured logarithmic strip*

$$\Omega_*' := \bigcup_{k=k_0}^{\infty} \Omega_k',$$

then we see that Ω_*' consists of all points $\rho = a + ib \in \Omega_0$ with $a \geq y_0'$ which do not lie inside any of the circles Γ_k' for $k \geq k_0$, and from the above

$$|\Delta_0(\rho)| \geq \frac{m_0 |a_p| |\mu_0|}{2(2\xi)^{n_0}} |\rho|^q > 0 \tag{8.37}$$

for all $\rho \in \Omega_*'$.

With the basic estimates (8.35) and (8.37) in place, consider one of the circles Γ_k' for $k \geq k_0$. Since (8.35) is valid for each point ρ on Γ_k', it follows by Rouché's Theorem that Δ_0 and $f_k + g_k$ have the same number of zeros as f_k inside Γ_k'. But μ_k' is the only zero of f_k inside Γ_k', μ_k' being a zero of order 1. Consequently, the characteristic determinant Δ_0 has a unique zero ρ_k' inside Γ_k' with ρ_k' having order 1 for $k \geq k_0$. Setting

$$\lambda_k' := (\rho_k')^n, \qquad k = k_0, k_0 + 1, \ldots,$$

the complex numbers λ_k' are eigenvalues of L, and by our earlier work the corresponding algebraic multiplicities and ascents are given by

$$\nu(\lambda_k') = m(\lambda_k') = 1, \qquad k = k_0, k_0 + 1, \ldots. \tag{8.38}$$

For the zeros ρ_k' we obtain the asymptotic formulas

$$|\rho_k' - \mu_k'| \leq \frac{\gamma \ln k}{k}, \qquad k = k_0, k_0 + 1, \ldots. \tag{8.39}$$

To determine the zeros of Δ_1 in the logarithmic strip Ω_1, we adopt an approach similar to the above. For the constant $\eta := 1 + n_0/\alpha_1 = \xi$, we see that

$$|\rho| \leq \eta |a| \quad \text{for } \rho = a + ib \in \Omega_1. \tag{8.40}$$

Relative to the strip Ω_1 let Δ_1 be expressed in the form

$$\Delta_1(\rho) = a_q' \rho^q \Big\{ e^{-i\rho} - \frac{1}{\mu_1 \rho^{n_0}} \left[1 + h_1(\rho)\right] \Big\} \tag{8.41}$$

for $\rho \in G_1$, where h_1 is the analytic function given by

$$h_1(\rho) := -\mu_1 \rho^{n_0} e^{-i\rho} \phi_{11}(\rho) + \phi_{10}(\rho)$$

for $\rho \in G_1$. Let $\delta > 0$ be defined as above, and for $k = 1, 2, \ldots$ define

$$\begin{cases} \alpha''_k := -(2\pi k - \operatorname{Arg} \mu_1 + \pi n_0), & \beta''_k := -n_0 \ln[-|\mu_1|^{1/n_0} \alpha''_k], \\ \mu''_k := \alpha''_k + i\beta''_k, \end{cases}$$

and introduce the circles

$$\Gamma''_k := \{\rho \in \mathbb{C} \mid |\rho - \mu''_k| = \delta\}.$$

Choose an integer $k_1 \geq 2$ such that the real number $y''_1 := \alpha''_{k_1} + \pi$ satisfies the condition $y''_1 \leq -r_2$. Also, introduce the *logarithmic rectangles*

$$R''_k := \{\rho = a + ib \in \mathbb{C} \mid \alpha''_k - \pi \leq a \leq \alpha''_k + \pi, \\ - n_0 \ln[-a/\alpha_1] \leq b \leq -n_0 \ln[-a/\beta_1]\}$$

for $k = k_1, k_1 + 1, \ldots$. We can assume that this new k_1 is identical to the k_1 introduced earlier, and that k_1 is sufficiently large to guarantee that each R''_k is contained in the sector S_1, and hence, for $k = k_1, k_1 + 1, \ldots$ the point μ''_k lies in the interior of the logarithmic rectangle R''_k with R''_k a subset of the logarithmic strip Ω_1. The circle Γ''_k also lies in the interior of the logarithmic rectangle R''_k for $k = k_1, k_1 + 1, \ldots$. Lastly, for $k = k_1, k_1 + 1, \ldots$ let Ω''_k be the *punctured logarithmic rectangle* formed from R''_k by removing all the points inside Γ''_k.

Next, we establish the growth rate of Δ_1 on each of the regions Ω''_k. Observe that

$$e^{i\mu''_k} = \mu_1(\alpha''_k)^{n_0} \quad \text{for } k = k_1, k_1 + 1, \ldots.$$

Introducing the analytic functions F_k, $k = k_1, k_1 + 1, \ldots$, and G_k, $k = k_1, k_1 + 1, \ldots$, defined by

$$F_k(\rho) := e^{i(\mu''_k - \rho)} - 1 \quad \text{for } \rho \in \mathbb{C},$$

$$G_k(\rho) := -h_1(\rho) - \sum_{j=1}^{n_0} \binom{n_0}{j} \left[\frac{1}{\rho}(\mu''_k - \rho) - \frac{i\beta''_k}{\rho}\right]^j [1 + h_1(\rho)] \quad \text{for } \rho \in G_1,$$

we can use (8.41) to write Δ_1 in its final form:

$$\Delta_1(\rho) = \frac{a'_q \rho^q}{\mu_1(\alpha''_k)^{n_0}} [F_k(\rho) + G_k(\rho)] \tag{8.42}$$

for $\rho \in G_1$ and for $k = k_1, k_1 + 1, \ldots$. Here we have a family of representations for Δ_1 which depends on the integer k. We use the kth representation to determine the growth rate of Δ_1 on the kth region Ω''_k.

In terms of the constants d_0 and m_0 introduced earlier, we can determine an integer $k_0 \geq k_1$ and the corresponding real number $y''_0 := \alpha''_{k_0} + \pi \leq -r_2$ and proceed to derive the estimates

$$|G_k(\rho)| \leq \frac{m_0}{2} < m_0 \leq |F_k(\rho)| \tag{8.43}$$

and

$$|\Delta_1(\rho)| \geq \frac{m_0 |a'_q|}{2|\mu_1|(2^{n_0})} |\rho|^p > 0 \tag{8.44}$$

for $k \geq k_0$ and for $\rho \in \Omega''_k$. If we introduce the *punctured logarithmic strip*

$$\Omega''_* := \bigcup_{k=k_0}^{\infty} \Omega''_k,$$

then we see that Ω''_* consists of all points $\rho = a + ib \in \Omega_1$ with $a \le y''_0$ and with ρ not inside any of the circles Γ''_k for $k \ge k_0$, and from the above

$$|\Delta_1(\rho)| \ge \frac{m_0|a'_q|}{2|\mu_1|(2^{n_0})}|\rho|^p > 0 \tag{8.45}$$

for all $\rho \in \Omega''_*$.

Consider one of the circles Γ''_k for $k \ge k_0$. Since (8.43) is valid for each point ρ on Γ''_k, it follows by Rouché's Theorem that Δ_1 and $F_k + G_k$ have the same number of zeros as F_k inside Γ''_k. But μ''_k is the only zero of F_k inside Γ''_k, μ''_k being a zero of order 1. Consequently, the characteristic determinant Δ_1 has a unique zero ρ''_k inside Γ''_k with ρ''_k having order 1 for $k \ge k_0$. Setting

$$\lambda''_k := (\rho''_k)^n, \qquad k = k_0, k_0 + 1, \ldots,$$

the λ''_k are eigenvalues of L with corresponding algebraic multiplicities and ascents

$$\nu(\lambda''_k) = m(\lambda''_k) = 1, \qquad k = k_0, k_0 + 1, \ldots. \tag{8.46}$$

For the zeros ρ''_k we have the asymptotic formulas

$$|\rho''_k - \mu''_k| \le \frac{\gamma \ln k}{k}, \qquad k = k_0, k_0 + 1, \ldots. \tag{8.47}$$

The λ'_k, λ''_k, $k = k_0, k_0 + 1, \ldots$, account for all but a finite number of the eigenvalues of the differential operator L.

We summarize the results for this logarithmic case as a theorem. Previously this theorem was stated in Chapter 1 as Theorem 1.7.

THEOREM 8.4. *Let the differential operator L belong to Case 2, a logarithmic case, where the integers p and q satisfy the conditions* $-\infty < p < q \le p_0$, *and let* $n_0 = q - p$, $\mu_0 = -b_q/a_p \ne 0$, *and* $\mu_1 = -a'_q/b'_p \ne 0$ *(so* $|\mu_1| = |\mu_0|$ *and* $\arg \mu_1 = \arg \mu_0 + 2\pi(n_0\nu + p)/n$*). Then the elements of the spectrum* $\sigma(L)$ *can be listed as two sequences*

$$\lambda'_k = (\rho'_k)^n, \quad k = k_0, k_0 + 1, \ldots, \qquad \lambda''_k = (\rho''_k)^n, \quad k = k_0, k_0 + 1, \ldots,$$

plus a finite number of additional points, where k_0 *is a positive integer and*

$$\rho'_k = (2\pi k + \operatorname{Arg} \mu_0) - in_0 \ln[|\mu_0|^{1/n_0}(2\pi k + \operatorname{Arg} \mu_0)] + \epsilon'_k, \qquad k = k_0, k_0 + 1, \ldots,$$

$$\rho''_k = -(2\pi k - \operatorname{Arg} \mu_1 + \pi n_0) - in_0 \ln[|\mu_0|^{1/n_0}(2\pi k - \operatorname{Arg} \mu_1 + \pi n_0)] + \epsilon''_k, \qquad k = k_0, k_0 + 1, \ldots,$$

with $|\epsilon'_k| \le \gamma \ln k/k$ *and* $|\epsilon''_k| \le \gamma \ln k/k$ *for* $k = k_0, k_0 + 1, \ldots$. *In addition, the corresponding algebraic multiplicities and ascents are*

$$\nu(\lambda'_k) = m(\lambda'_k) = 1, \quad k = k_0, k_0 + 1, \ldots,$$
$$\nu(\lambda''_k) = m(\lambda''_k) = 1, \quad k = k_0, k_0 + 1, \ldots.$$

8.3. Case 3. $p > q$

Assume that $p > q$. The differential operators belonging to this case are always simply irregular. It is also a logarithmic case. As in Case 2 we will assume that the constant R_0 has been chosen sufficiently large to guarantee that $\operatorname{Re}\rho \geq 0$ for all $\rho \in G_0$ and $\operatorname{Re}\rho \leq 0$ for all $\rho \in G_1$. Set $n_0 = p - q > 0$, and let us begin work in the sector T_0 and the open set G_0. By (5.46) or Theorem 5.3 and (8.1) we can write the characteristic determinant Δ_0 in the form

$$\Delta_0(\rho) = \rho^p \Big\{ a_p e^{i\rho}[1 + \phi_{01}(\rho)] + \frac{b_q}{\rho^{n_0}}\,[1 + \phi_{00}(\rho)] \Big\} \tag{8.48}$$

for $\rho \in G_0$, where

$$\phi_{01}(\rho) := \sum_{\kappa=-(m-p_0-1)}^{p-1} \frac{a_\kappa}{a_p \rho^{p-\kappa}} + \frac{1}{a_p \rho^p}\,\Phi_{01}(\rho),$$

$$\phi_{00}(\rho) := \sum_{\kappa=-(m-p_0-1)}^{q-1} \frac{b_\kappa}{b_q \rho^{q-\kappa}} + \frac{1}{b_q \rho^q}\,\Phi_{00}(\rho)$$

for $\rho \in G_0$. The functions ϕ_{01}, ϕ_{00} are analytic on the open set G_0.

Set $\mu_0 := -a_p/b_q \neq 0$. Choosing a constant $r_1 > R_0$ sufficiently large, we rewrite (8.48) in the form

$$\Delta_0(\rho) = a_p \rho^p e^{i\rho} \Big\{ [1 + \phi_{01}(\rho)] + \frac{b_q}{a_p \rho^{n_0}}\,[1 + \phi_{00}(\rho)] e^{-i\rho} \Big\}$$

for $\rho \in G_0$, and then derive the growth rate

$$|\Delta_0(\rho)| \geq \frac{1}{4}\,|a_p||\rho|^p e^{-b} > 0 \tag{8.49}$$

for $\rho = a + ib \in G_0$ with $|\rho| \geq r_1$ and $b \leq 0$. Consequently, as we search for the zeros of Δ_0 in the open set G_0, we will concentrate our search in Quadrant I.

Let ω be the real number defined by the equation $1/\omega := 1/|\mu_0|^{1/n_0} + n_0$. Choose real numbers α and β with $0 < \alpha < [1/(2|\mu_0|)]^{1/n_0}$, $\beta > [2/|\mu_0|]^{1/n_0}$, and

$$2|\mu_0|(2\alpha)^{n_0} \leq \frac{1}{4}, \qquad \frac{2}{|\mu_0|\beta^{n_0}} \leq \frac{1}{4}.$$

Clearly $1/\beta < |\mu_0|^{1/n_0} < 1/\alpha$. In terms of these constants we form the *logarithmic strip*

$$\begin{aligned} \Omega_0 &:= \big\{ \rho = a + ib \in S_0 \;\big|\; b \geq 0,\ \alpha e^{b/n_0} \leq a \leq \beta e^{b/n_0} \big\} \\ &= \big\{ \rho = a + ib \in S_0 \;\big|\; a \geq \alpha,\ b \geq 0,\ n_0 \ln[a/\beta] \leq b \leq n_0 \ln[a/\alpha] \big\} \end{aligned}$$

and the two complementary regions

$$\Omega_{0\infty} := \big\{ \rho = a + ib \in S_0 \;\big|\; b \geq 0,\ a \leq \alpha e^{b/n_0} \big\},$$

$$\Omega_{0\square} := \big\{ \rho = a + ib \in S_0 \;\big|\; b \geq 0,\ \beta e^{b/n_0} \leq a \big\}.$$

These three regions are contained in Quadrant I.

Let us begin by calculating the growth rates of Δ_0 on the regions $\Omega_{0\infty}$ and $\Omega_{0\square}$. Choose a real number r_2 sufficiently large with $r_2 \geq r_1$ and $r_2 \geq \beta$ and such that

$$|\rho| \leq 2\alpha e^{|b|/n_0} \tag{8.50}$$

for $\rho = a + ib \in \Omega_{0\infty}$ with $|\rho| \geq r_2$. Then rewriting (8.48) in the form

$$\Delta_0(\rho) = b_q\rho^q\{[1 + \phi_{00}(\rho)] + (a_p/b_q)\rho^{n_0}[1 + \phi_{01}(\rho)]e^{i\rho}\}$$

for $\rho \in G_0$, we proceed to establish the growth rate

$$|\Delta_0(\rho)| \geq \frac{1}{4}\,|b_q||\rho|^q > 0 \tag{8.51}$$

for all $\rho = a + ib \in \Omega_{0\infty}$ with $|\rho| \geq r_2$. Similarly, for the region $\Omega_{0\square}$, we again express the characteristic determinant in the form

$$\Delta_0(\rho) = a_p\rho^p e^{i\rho}\Big\{[1 + \phi_{01}(\rho)] + \frac{b_q}{a_p\rho^{n_0}}\,[1 + \phi_{00}(\rho)]e^{-i\rho}\Big\}$$

for $\rho \in G_0$, and then derive the growth rate

$$|\Delta_0(\rho)| \geq \frac{1}{4}\,|a_p||\rho|^p e^{-b} \geq \frac{1}{4}\,|a_p|(\beta)^{n_0}|\rho|^q > 0 \tag{8.52}$$

for all $\rho = a + ib \in \Omega_{0\square}$.

Relative to the sector T_1 and the open set G_1, we use (5.53) or Theorem 5.4 and (8.2) to write the characteristic determinant Δ_1 in the form

$$\Delta_1(\rho) = \rho^q\Big\{a'_q e^{-i\rho}\big[1 + \phi_{11}(\rho)\big] + b'_p\rho^{n_0}\big[1 + \phi_{10}(\rho)\big]\Big\} \tag{8.53}$$

for $\rho \in G_1$, where

$$\phi_{11}(\rho) := \sum_{\kappa=-(m-p_0-1)}^{q-1} \frac{a'_\kappa}{a'_q\rho^{q-\kappa}} + \frac{1}{a'_q\rho^q}\,\Phi_{11}(\rho),$$

$$\phi_{10}(\rho) := \sum_{\kappa=-(m-p_0-1)}^{p-1} \frac{b'_\kappa}{b'_p\rho^{p-\kappa}} + \frac{1}{b'_p\rho^p}\,\Phi_{10}(\rho)$$

for $\rho \in G_1$. The functions ϕ_{11}, ϕ_{10} are analytic on the open set G_1.

Set $\mu_1 := -b'_p/a'_q \neq 0$. From (5.56) we have $b'_p = \omega^p_{\nu-1}a_p$ and $a'_q = \omega^q_\nu b_q$, and hence,

$$\mu_1 = -\omega^p_{\nu-1}a_p/(\omega^q_\nu b_q) = \omega_{n_0\nu-p}\mu_0. \tag{8.54}$$

It follows that $|\mu_1| = |\mu_0|$ and $\arg\mu_1 = \arg\mu_0 + 2\pi(n_0\nu - p)/n$. Choosing a constant $r'_1 > R_0$ sufficiently large, we rewrite (8.53) in the form

$$\Delta_1(\rho) = b'_p\rho^p\Big\{[1 + \phi_{10}(\rho)] + \frac{a'_q}{b'_p\rho^{n_0}}\,[1 + \phi_{11}(\rho)]e^{-i\rho}\Big\}$$

for $\rho \in G_1$, and then derive the growth rate

$$|\Delta_1(\rho)| \geq \frac{1}{4}\,|b'_p||\rho|^p > 0 \tag{8.55}$$

for $\rho = a + ib \in G_1$ with $|\rho| \geq r'_1$ and $b \leq 0$. In view of this result, our search for the zeros of the characteristic determinant Δ_1 in the open set G_1 will be concentrated in Quadrant II.

Set $\omega_1 := \omega$, $\alpha_1 := \alpha$, and $\beta_1 := \beta$. Then these constants satisfy the conditions $1/\omega_1 = 1/|\mu_1|^{1/n_0} + n_0$, $0 < \alpha_1 < [1/(2|\mu_1|)]^{1/n_0}$, $\beta_1 > [2/|\mu_1|]^{1/n_0}$, and

$$2|\mu_1|(2\alpha_1)^{n_0} \leq \frac{1}{4}, \qquad \frac{2}{|\mu_1|(\beta_1)^{n_0}} \leq \frac{1}{4}.$$

In terms of them we form the *logarithmic strip*

$$\begin{aligned}\Omega_1 &:= \{\rho = a + ib \in S_1 \mid b \geq 0,\ -\beta_1 e^{b/n_0} \leq a \leq -\alpha_1 e^{b/n_0}\} \\ &= \{\rho = a + ib \in S_1 \mid a \leq -\alpha_1,\ b \geq 0,\ n_0 \ln[-a/\beta_1] \leq b \leq n_0 \ln[-a/\alpha_1]\}\end{aligned}$$

and the two complementary regions

$$\Omega_{1\infty} := \{\rho = a + ib \in S_1 \mid b \geq 0,\ -\alpha_1 e^{b/n_0} \leq a\},$$

$$\Omega_{1\square} := \{\rho = a + ib \in S_1 \mid b \geq 0,\ a \leq -\beta_1 e^{b/n_0}\}.$$

These three regions are contained in Quadrant II.

Let us determine the growth rates of Δ_1 on the regions $\Omega_{1\infty}$ and $\Omega_{1\square}$. Choose a real number r_2' sufficiently large with $r_2' \geq r_1'$ and $r_2' \geq \beta_1$ and such that

$$|\rho| \leq 2\alpha_1 e^{|b|/n_0} \tag{8.56}$$

for $\rho = a + ib \in \Omega_{1\infty}$ with $|\rho| \geq r_2'$. Without loss of generality we can assume that $r_2 = r_2'$. Rewriting (8.53) in the form

$$\Delta_1(\rho) = a_q' \rho^q e^{-i\rho} \{[1 + \phi_{11}(\rho)] + (b_p'/a_q')\rho^{n_0}[1 + \phi_{10}(\rho)]e^{i\rho}\}$$

for $\rho \in G_1$, we obtain the growth rate

$$|\Delta_1(\rho)| \geq \frac{1}{4}\,|a_q'||\rho|^q e^b \geq \frac{|a_q'|}{4(2\alpha_1)^{n_0}}\,|\rho|^p > 0 \tag{8.57}$$

for all $\rho = a + ib \in \Omega_{1\infty}$ with $|\rho| \geq r_2$. Similarly, for the region $\Omega_{1\square}$ we again express the characteristic determinant Δ_1 in the form

$$\Delta_1(\rho) = b_p' \rho^p \Big\{[1 + \phi_{10}(\rho)] + \frac{a_q'}{b_p' \rho^{n_0}}\,[1 + \phi_{11}(\rho)]e^{-i\rho}\Big\}$$

for $\rho \in G_1$, and then establish the growth rate

$$|\Delta_1(\rho)| \geq \frac{1}{4}\,|b_p'||\rho|^p > 0 \tag{8.58}$$

for all $\rho = a + ib \in \Omega_{1\square}$.

The growth rates on the regions $\Omega_{0\infty}$, $\Omega_{0\square}$ and $\Omega_{1\infty}$, $\Omega_{1\square}$ combine to give the following *apriori estimates* for the eigenvalues of L.

THEOREM 8.5. *Assume that $p > q$. Let $\lambda = \rho^n \in \mathbb{C}$ with $\rho = a + ib \in S_0$ and with $|\rho| \geq r_2$.*

(a) *If $\rho \in \Omega_{0\infty}$ or $\rho \in \Omega_{0\square}$, then $\Delta_0(\rho) \neq 0$ and $\lambda \in \rho(L)$.*

(b) *If λ is an eigenvalue of L, then $\Delta_0(\rho) = 0$ and ρ lies in the interior of the logarithmic strip Ω_0.*

In addition, let $\lambda = \rho^n \in \mathbb{C}$ with $\rho = a + ib \in S_1$ and with $|\rho| \geq r_2$.

(c) *If $\rho \in \Omega_{1\infty}$ or $\rho \in \Omega_{1\square}$, then $\Delta_1(\rho) \neq 0$ and $\lambda \in \rho(L)$.*

(d) *If λ is an eigenvalue of L, then $\Delta_1(\rho) = 0$ and ρ lies in the interior of the logarithmic strip Ω_1.*

The theorem shows that the resolvent set $\rho(L)$ is nonempty, and hence, the spectrum $\sigma(L)$ is a countable set having no limit points in $\mathbb{C}$.

Next, let us compute the actual zeros of Δ_0 in the logarithmic strip Ω_0. Setting $\xi := 1 + n_0/\alpha$, we see that

$$|\rho| \leq \xi |a| \quad \text{for all } \rho = a + ib \in \Omega_0. \tag{8.59}$$

Relative to the strip Ω_0 the characteristic determinant can be written in the form

$$\Delta_0(\rho) = a_p\rho^p\Big\{e^{i\rho} - \frac{1}{\mu_0\rho^{n_0}}\,[1+h_0(\rho)]\Big\} \tag{8.60}$$

for $\rho \in G_0$, where h_0 is the analytic function given by

$$h_0(\rho) := -\mu_0\rho^{n_0}e^{i\rho}\phi_{01}(\rho) + \phi_{00}(\rho)$$

for $\rho \in G_0$. The function h_0 satisfies the growth rate

$$|h_0(\rho)| \le \frac{\gamma_1}{|\rho|} \tag{8.61}$$

for all $\rho = a + ib \in \Omega_0$ with $|\rho| \ge r_2$.

Fix a real number δ with $0 < \delta \le \pi/4$ and $0 < \delta < (\ln 2)/(1+n_0)$, and then for the integers $k = 1, 2, \dots$ define

$$\begin{cases} \alpha'_k := 2\pi k - \operatorname{Arg}\mu_0, & \beta'_k := n_0\ln[|\mu_0|^{1/n_0}\alpha'_k], \\ \mu'_k := \alpha'_k + i\beta'_k, \end{cases}$$

and introduce the circles

$$\Gamma'_k := \{\rho \in \mathbb{C} \mid |\rho - \mu'_k| = \delta\}.$$

Choose an integer $k_1 \ge 2$ such that the real number $y'_1 := \alpha'_{k_1} - \pi$ satisfies the condition $y'_1 \ge r_2$. Also, introduce the *logarithmic rectangles*

$$R'_k := \big\{\rho = a + ib \in \mathbb{C} \;\big|\; \alpha'_k - \pi \le a \le \alpha'_k + \pi,\ n_0\ln[a/\beta] \le b \le n_0\ln[a/\alpha]\big\}$$

for $k = k_1, k_1{+}1, \dots$. Without loss of generality we can assume that k_1 is sufficiently large to guarantee that each R'_k is contained in the sector S_0, and hence, for $k = k_1, k_1{+}1, \dots$ the point μ'_k lies in the interior of R'_k with R'_k a subset of Ω_0. The circle Γ'_k also lies in the interior of the logarithmic rectangle R'_k for $k = k_1, k_1+1, \dots$. To complete the setup of the geometry, for $k = k_1, k_1+1, \dots$ let Ω'_k be the *punctured logarithmic rectangle* formed from R'_k by removing all the points inside Γ'_k.

To establish the growth rate of Δ_0 on the region Ω'_k, first note that

$$e^{i\mu'_k} = \frac{1}{\mu_0(\alpha'_k)^{n_0}}$$

for $k = k_1, k_1+1, \dots$. Let f_k, $k = k_1, k_1+1, \dots$, and g_k, $k = k_1, k_1+1, \dots$, be the sequences of functions defined by

$$f_k(\rho) := e^{i(\rho-\mu'_k)} - 1 \quad \text{for } \rho \in \mathbb{C},$$

$$g_k(\rho) := -h_0(\rho) - \sum_{j=1}^{n_0}\binom{n_0}{j}\Big[\frac{1}{\rho}(\mu'_k - \rho) - \frac{i\beta'_k}{\rho}\Big]^j[1+h_0(\rho)] \quad \text{for } \rho \in G_0.$$

Then we use (8.60) to write Δ_0 in its final form:

$$\Delta_0(\rho) = \frac{a_p\rho^p}{\mu_0(\alpha'_k)^{n_0}}\,[f_k(\rho) + g_k(\rho)] \tag{8.62}$$

for $\rho \in G_0$ and for $k = k_1, k_1+1, \dots$. We will use the kth representation in this family of equations to compute the growth rate of Δ_0 on the kth region Ω'_k.

Choose a constant $d_0 > 0$ such that

$$n_0\ln[2/(|\mu_0|^{1/n_0}\alpha)] \le d_0, \qquad n_0\ln[2|\mu_0|^{1/n_0}\beta] \le d_0,$$

and $\delta < d_0$, and then form the punctured rectangle

$$R_* := \{\rho = a + ib \in \mathbb{C} \mid -\pi \leq a \leq \pi,\ -d_0 \leq b \leq d_0, \text{ and } |\rho| \geq \delta\}.$$

Set $m_0 := \min\{\, |e^{i\rho} - 1| \mid \rho \in R_*\} > 0$. Then for any index $k \geq k_1$ and for any point $\rho = a + ib \in \Omega'_k$, the translate $\rho - \mu'_k$ belongs to the punctured rectangle R_* with $|\rho - \mu'_k| \leq \pi + d_0$, and hence,

$$|f_k(\rho)| \geq m_0 \tag{8.63}$$

for $k \geq k_1$ and for $\rho \in \Omega'_k$. Note that the constant m_0 is independent of the index k.

Clearly $\lim_{k\to\infty} 1/\alpha'_k = \lim_{k\to\infty} \beta'_k/\alpha'_k = 0$. In terms of the constant γ_1 that appears in the inequality (8.61), select an integer $k_0 \geq k_1$ such that

$$\frac{\gamma_1}{|\rho|} \leq \frac{m_0}{4} \quad \text{for all } \rho \in \mathbb{C} \text{ with } |\rho| \geq y'_0,$$

where $y'_0 := \alpha'_{k_0} - \pi \geq r_2$, and such that

$$\sum_{j=1}^{n_0} \binom{n_0}{j} \left[\frac{2}{|\alpha'_k|}(\pi + d_0) + \frac{2|\beta'_k|}{|\alpha'_k|}\right]^j \left[1 + \frac{\gamma_1}{r_2}\right] \leq \frac{m_0}{4} \quad \text{for all } k \geq k_0.$$

Then for $k \geq k_0$ and for $\rho \in \Omega'_k$, we have

$$|g_k(\rho)| \leq \frac{m_0}{2} < m_0 \leq |f_k(\rho)| \tag{8.64}$$

and

$$|\Delta_0(\rho)| \geq \frac{m_0|a_p|}{2|\mu_0|(2^{n_0})}\,|\rho|^q > 0. \tag{8.65}$$

Introducing the *punctured logarithmic strip*

$$\Omega'_* = \bigcup_{k=k_0}^{\infty} \Omega'_k,$$

we see that Ω'_* consists of all points $\rho = a + ib \in \Omega_0$ with $a \geq y'_0$ which do not lie inside any of the circles Γ'_k for $k \geq k_0$, and from the above

$$|\Delta_0(\rho)| \geq \frac{m_0|a_p|}{2|\mu_0|(2^{n_0})}\,|\rho|^q > 0 \tag{8.66}$$

for all $\rho \in \Omega'_*$.

Now consider one of the circles Γ'_k for $k \geq k_0$. The inequality (8.64) is valid for each point ρ on Γ'_k, and hence, by Rouché's Theorem Δ_0 and $f_k + g_k$ have the same number of zeros as f_k inside Γ'_k. But the only zero of f_k inside the circle Γ'_k occurs at the center μ'_k, with μ'_k being a zero of order 1. It follows that the characteristic determinant Δ_0 has a unique zero ρ'_k inside Γ'_k, and ρ'_k is a zero of order 1 for $k \geq k_0$. If we set

$$\lambda'_k := (\rho'_k)^n, \qquad k = k_0, k_0 + 1, \ldots,$$

then the λ'_k are eigenvalues of L with algebraic multiplicities and ascents

$$\nu(\lambda'_k) = m(\lambda'_k) = 1, \qquad k = k_0, k_0 + 1, \ldots. \tag{8.67}$$

For the zeros ρ'_k we have the asymptotic formulas

$$|\rho'_k - \mu'_k| \leq \frac{\gamma \ln k}{k}, \qquad k = k_0, k_0 + 1, \ldots. \tag{8.68}$$

The calculation of the zeros of Δ_1 in the logarithmic strip Ω_1 is similar to the above. Indeed, for the constant $\eta := 1 + n_0/\alpha_1 = \xi$, we see that

$$|\rho| \leq \eta |a| \quad \text{for } \rho = a + ib \in \Omega_1. \tag{8.69}$$

Relative to Ω_1 let Δ_1 be expressed in the form

$$\Delta_1(\rho) = a'_q \rho^q \{e^{-i\rho} - \mu_1 \rho^{n_0}[1 + h_1(\rho)]\} \tag{8.70}$$

for $\rho \in G_1$, where h_1 is the analytic function given by

$$h_1(\rho) := -\frac{e^{-i\rho}}{\mu_1 \rho^{n_0}} \phi_{11}(\rho) + \phi_{10}(\rho)$$

for $\rho \in G_1$. Let $\delta > 0$ be defined as above, and for $k = 1, 2, \ldots$ set

$$\begin{cases} \alpha''_k := -(2\pi k + \operatorname{Arg} \mu_1 + \pi n_0), & \beta''_k := n_0 \ln[-|\mu_1|^{1/n_0} \alpha''_k], \\ \mu''_k := \alpha''_k + i\beta''_k, \end{cases}$$

and then form the circles

$$\Gamma''_k := \{\rho \in \mathbb{C} \mid |\rho - \mu''_k| = \delta\}.$$

Choose an integer $k_1 \geq 2$ such that the real number $y''_1 := \alpha''_{k_1} + \pi$ satisfies the condition $y''_1 \leq -r_2$. Also, form the *logarithmic rectangles*

$$R''_k := \{\rho = a + ib \in \mathbb{C} \mid \alpha''_k - \pi \leq a \leq \alpha''_k + \pi,\ n_0 \ln[-a/\beta_1] \leq b \leq n_0 \ln[-a/\alpha_1]\}$$

for $k = k_1, k_1 + 1, \ldots$. We can assume that this new index k_1 is identical to the k_1 introduced earlier, and that k_1 is sufficiently large to guarantee that each R''_k is contained in the sector S_1. Consequently, for $k = k_1, k_1 + 1, \ldots$ the point μ''_k lies in the interior of the logarithmic rectangle R''_k, and R''_k is a subset of the logarithmic strip Ω_1. The circle Γ''_k also lies in the interior of the logarithmic rectangle R''_k for $k = k_1, k_1 + 1, \ldots$. Lastly, for $k = k_1, k_1 + 1, \ldots$ let Ω''_k be the *punctured logarithmic rectangle* formed from R''_k by removing all the points inside Γ''_k.

Next, let us calculate the growth rate of Δ_1 on each of the regions Ω''_k. From the definition of μ''_k it follows that

$$e^{i\mu''_k} = \frac{1}{\mu_1 (\alpha''_k)^{n_0}} \quad \text{for } k = k_1, k_1 + 1, \ldots.$$

Let F_k, $k = k_1, k_1 + 1, \ldots$, and G_k, $k = k_1, k_1 + 1, \ldots$, be the sequences of functions defined by

$$F_k(\rho) := e^{i(\mu''_k - \rho)} - 1 \quad \text{for } \rho \in \mathbb{C},$$

$$G_k(\rho) := -h_1(\rho) - \sum_{j=1}^{n_0} \binom{n_0}{j} \left[\frac{1}{\alpha''_k}(\rho - \mu''_k) + \frac{i\beta''_k}{\alpha''_k}\right]^j [1 + h_1(\rho)] \quad \text{for } \rho \in G_1.$$

Then (8.70) can be used to express Δ_1 in its final form:

$$\Delta_1(\rho) = a'_q \mu_1 (\alpha''_k)^{n_0} \rho^q [F_k(\rho) + G_k(\rho)] \tag{8.71}$$

for $\rho \in G_1$ and for $k = k_1, k_1 + 1, \ldots$. We will use (8.71) to calculate the growth rate of Δ_1 on the region Ω''_k.

In terms of the constants d_0 and m_0 introduced earlier, we can determine an integer $k_0 \geq k_1$ and the corresponding real number $y''_0 := \alpha''_{k_0} + \pi \leq -r_2$ and proceed to derive the estimates

$$|G_k(\rho)| \leq \frac{m_0}{2} < m_0 \leq |F_k(\rho)| \tag{8.72}$$

and

$$(8.73) \qquad |\Delta_1(\rho)| \geq \frac{m_0|a_q'||\mu_1|}{2(2\eta)^{n_0}}\,|\rho|^p > 0$$

for $k \geq k_0$ and for $\rho \in \Omega_k''$. If we introduce the *punctured logarithmic strip*

$$\Omega_*'' = \bigcup_{k=k_0}^{\infty} \Omega_k'',$$

then Ω_*'' consists of all points $\rho = a + ib \in \Omega_1$ with $a \leq y_0''$ and with ρ not inside any of the circles Γ_k'' for $k \geq k_0$, and from the above

$$(8.74) \qquad |\Delta_1(\rho)| \geq \frac{m_0|a_q'||\mu_1|}{2(2\eta)^{n_0}}\,|\rho|^p > 0$$

for all $\rho \in \Omega_*''$.

Let us examine one of the circles Γ_k'' for $k \geq k_0$. Since (8.72) is valid for each point ρ on Γ_k'', by Rouché's Theorem Δ_1 and $F_k + G_k$ must have the same number of zeros as F_k inside Γ_k''. Clearly μ_k'' is the only zero of F_k inside Γ_k'', μ_k'' being a zero of order 1. Therefore, Δ_1 has a unique zero ρ_k'' of order 1 inside the circle Γ_k'' for $k \geq k_0$. Corresponding to these zeros, the complex numbers

$$\lambda_k'' := (\rho_k'')^n, \qquad k = k_0, k_0+1, \ldots,$$

are eigenvalues of L with algebraic multiplicities and ascents

$$(8.75) \qquad \nu(\lambda_k'') = m(\lambda_k'') = 1, \qquad k = k_0, k_0+1, \ldots.$$

For the zeros ρ_k'' we have the asymptotic formulas

$$(8.76) \qquad |\rho_k'' - \mu_k''| \leq \frac{\gamma \ln k}{k}, \qquad k = k_0, k_0+1, \ldots.$$

Again the λ_k', λ_k'', $k = k_0, k_0+1, \ldots$, account for all but a finite number of the eigenvalues of L.

The results for this logarithmic case are summarized below in a theorem. This theorem was stated previously in Chapter 1 as Theorem 1.8.

THEOREM 8.6. *Let the differential operator L belong to Case 3, a logarithmic case, where the integers p and q satisfy the conditions* $-\infty < q < p \leq p_0$, *and let* $n_0 = p - q$, $\mu_0 = -a_p/b_q \neq 0$, *and* $\mu_1 = -b_p'/a_q' \neq 0$ *(so* $|\mu_1| = |\mu_0|$ *and* $\arg \mu_1 = \arg \mu_0 + 2\pi(n_0\nu - p)/n$ *). Then the elements of the spectrum* $\sigma(L)$ *can be listed as two sequences*

$$\lambda_k' = (\rho_k')^n, \quad k = k_0, k_0+1, \ldots, \qquad \lambda_k'' = (\rho_k'')^n, \quad k = k_0, k_0+1, \ldots,$$

plus a finite number of additional points, where k_0 *is a positive integer and*

$$\rho_k' = (2\pi k - \operatorname{Arg}\mu_0) + in_0 \ln[|\mu_0|^{1/n_0}(2\pi k - \operatorname{Arg}\mu_0)] + \epsilon_k', \qquad k = k_0, k_0+1, \ldots,$$

$$\rho_k'' = -(2\pi k + \operatorname{Arg}\mu_1 + \pi n_0) + in_0 \ln[|\mu_0|^{1/n_0}(2\pi k + \operatorname{Arg}\mu_1 + \pi n_0)] + \epsilon_k'', \qquad k = k_0, k_0+1, \ldots,$$

with $|\epsilon'_k| \leq \gamma \ln k/k$ *and* $|\epsilon''_k| \leq \gamma \ln k/k$ *for* $k = k_0, k_0+1, \ldots$. *In addition, the corresponding algebraic multiplicities and ascents are*

$$\nu(\lambda'_k) = m(\lambda'_k) = 1, \quad k = k_0, k_0+1, \ldots,$$
$$\nu(\lambda''_k) = m(\lambda''_k) = 1, \quad k = k_0, k_0+1, \ldots.$$

CHAPTER 9

Completeness of the Generalized Eigenfunctions

To demonstrate the completeness of the generalized eigenfunctions of L, we will determine growth rates for the Green's function $G(t, s; \lambda)$ and resolvent $R_\lambda(L)$ along various rays in the λ plane, and then appeal to a fundamental completeness theorem: Corollary XI.6.31 of [**8**] or Theorem 2.6.2 of [**34**].

From the results of Chapters 7 and 8, we know that the resolvent set $\rho(L)$ is nonempty, and hence, the differential operator L is a Hilbert-Schmidt discrete linear operator in $L^2[0,1]$. The spectrum $\sigma(L)$ is a countable set having no finite limit points in $\mathbb{C}$, and in Chapters 7 and 8 we have given a detailed description of $\sigma(L)$. Let

$$\sigma(L) = \{\lambda_i\}_{i=1}^{\infty}$$

be any enumeration of $\sigma(L)$, let m_i ($0 < m_i < \infty$) denote the ascent of the operator $\lambda_i I - L$ for $i = 1, 2, \ldots$, and let P_i, $i = 1, 2, \ldots$, denote the projection of $L^2[0,1]$ onto the generalized eigenspace $\mathcal{N}((\lambda_i I - L)^{m_i})$ along the range $\mathcal{R}((\lambda_i I - L)^{m_i})$. These projections have the property $P_i P_j = \delta_{ij} P_i$ for all i, j (see Theorems V.10.1 and V.10.2 in [**49**] and [**34**, p. 60]). Let $\operatorname{sp}(L)$ denote the subspace of $L^2[0,1]$ spanned by the generalized eigenfunctions of L, and let us introduce the subspaces

$$S_\infty(L) = \Big\{u \in L^2[0,1] \;\Big|\; u = \sum_{i=1}^{\infty} P_i u\Big\}$$

and

$$M_\infty(L) = \{u \in L^2[0,1] \mid P_i u = 0 \text{ for } i = 1, 2, \ldots\}.$$

Clearly $M_\infty(L)$ is closed, $\operatorname{sp}(L)$ is a subset of $S_\infty(L)$, and it is easy to check that $\overline{\operatorname{sp}(L)} = \overline{S_\infty(L)}$. Our goal in this chapter is to prove that $\overline{\operatorname{sp}(L)} = \overline{S_\infty(L)} = L^2[0,1]$ and $M_\infty(L) = \{0\}$.

9.1. Completeness for n Even

Assume that n is even, $n = 2\nu \geq 2$. For the Green's function $G(t, s; \lambda)$ of the differential operator $\lambda I - L$, we have established growth rates for it in equations (6.37) and (6.53). These growth rates depend on the sectors S_0 and S_1, but they are independent of whether L belongs to Case 1, Case 2, or Case 3. On the other hand, growth rates for the characteristic determinants Δ_0 and Δ_1 are case dependent.

Suppose the differential operator L belongs to either Case 1 or Case 2 where $p = q \leq p_0$. Combining (6.37) with (7.9), we obtain the estimate

$$|G(t, s; \lambda)| \leq \gamma|\rho|^{p_0 - p - n + 1} = \gamma|\lambda|^{(p_0 - p - n + 1)/n} \quad \text{for } t \neq s \text{ in } [0,1], \tag{9.1}$$

for $\lambda = \rho^n$ in $\mathbb{C}$ with $\rho = a + ib \in S_0$ and with $|\rho| \geq r_1$ and $b \geq d$. Similarly, combining (6.53) with (7.10), we get

$$|G(t, s; \lambda)| \leq \gamma|\rho|^{p_0 - p - n + 1} = \gamma|\lambda|^{(p_0 - p - n + 1)/n} \quad \text{for } t \neq s \text{ in } [0,1], \tag{9.2}$$

for $\lambda = \rho^n$ in $\mathbb{C}$ with $\rho = a+ib \in S_1$ and with $|\rho| \geq r_1$ and $b \leq -d$. Thus, in Case 1 and Case 2 the resolvent satisfies the growth rate

$$\|R_\lambda(L)\| \leq \gamma|\rho|^{p_0-p-n+1} = \gamma|\lambda|^{(p_0-p-n+1)/n} \tag{9.3}$$

for $\lambda = \rho^n$ in $\mathbb{C}$ with $\rho = a+ib \in S_0$, $|\rho| \geq r_1$, and $b \geq d$, or with $\rho = a+ib \in S_1$, $|\rho| \geq r_1$, and $b \leq -d$.

Suppose the differential operator L belongs to Case 3 where $p < q \leq p_0$, the logarithmic case. Combining (6.37) with (7.31) and (6.53) with (7.36), we obtain the estimates

$$|G(t,s;\lambda)| \leq \gamma|\rho|^{p_0-p-n+1} = \gamma|\lambda|^{(p_0-p-n+1)/n} \quad \text{for } t \neq s \text{ in } [0,1], \tag{9.4}$$

for $\lambda = \rho^n$ in $\mathbb{C}$ with $\rho = a+ib \in \Omega_{0\infty}$ and $|\rho| \geq r_2$, and

$$|G(t,s;\lambda)| \leq \gamma|\rho|^{p_0-p-n+1} = \gamma|\lambda|^{(p_0-p-n+1)/n} \quad \text{for } t \neq s \text{ in } [0,1], \tag{9.5}$$

for $\lambda = \rho^n$ in $\mathbb{C}$ with $\rho = a+ib \in \Omega_{1\infty}$ and $|\rho| \geq r_2$. Therefore, the resolvent in Case 3 satisfies the growth rate

$$\|R_\lambda(L)\| \leq \gamma|\rho|^{p_0-p-n+1} = \gamma|\lambda|^{(p_0-p-n+1)/n} \tag{9.6}$$

for $\lambda = \rho^n$ in $\mathbb{C}$ with $\rho \in \Omega_{0\infty} \cup \Omega_{1\infty}$ and $|\rho| \geq r_2$.

With our estimates (9.3) and (9.6) for the resolvent in place, we are now ready to establish the completeness of the generalized eigenfunctions of the differential operator L. Let N be a positive integer that satisfies the condition

$$N \geq (p_0 - p - n + 1)/n.$$

Fix any real number θ_0 with $\sigma_0 \leq \theta_0 \leq \pi$ or $-\pi \leq \theta_0 \leq -\sigma_0$, and let us consider the ray

$$\mathcal{R}_{\theta_0}\colon \ \lambda = |\lambda|e^{i\theta_0}, \quad 0 < |\lambda| < \infty,$$

in the λ plane. If $\sigma_0 \leq \theta_0 \leq \pi$, then for $\lambda \in \mathcal{R}_{\theta_0}$ we form the nth root

$$\rho = \sqrt[n]{\lambda} = |\lambda|^{1/n}e^{i\theta_0/n},$$

with $\sigma_0/n \leq \theta_0/n \leq \pi/n$ and with $\rho \in S_0$. For this nth root we have

$$\begin{cases} \rho = a+ib = |\rho|\cos(\theta_0/n) + i|\rho|\sin(\theta_0/n), \\ a = |\rho|\cos(\theta_0/n) \geq 0, \qquad b = |\rho|\sin(\theta_0/n) > 0. \end{cases}$$

Similarly, if $-\pi \leq \theta_0 \leq -\sigma_0$, then for $\lambda \in \mathcal{R}_{\theta_0}$ we form the nth root

$$\rho = \sqrt[n]{\lambda} = |\lambda|^{1/n}e^{i\theta_0/n},$$

with $-\pi/n \leq \theta_0/n \leq -\sigma_0/n$ and with $\rho \in S_1$. Here the nth root has the simple properties

$$\begin{cases} \rho = a+ib = |\rho|\cos(\theta_0/n) + i|\rho|\sin(\theta_0/n), \\ a = |\rho|\cos(\theta_0/n) \geq 0, \qquad b = |\rho|\sin(\theta_0/n) < 0. \end{cases}$$

The geometry is slightly different for Case 1 and Case 2 where $p = q \leq p_0$ and for Case 3 where $p < q \leq p_0$. First, suppose that L belongs to Case 1 or Case 2 where $p = q$. Assume that $\sigma_0 \leq \theta_0 \leq \pi$, and set

$$r(\theta_0) := \max\{r_1, d/\sin(\theta_0/n)\}.$$

If $\lambda \in \mathcal{R}_{\theta_0}$ with $|\lambda| \geq r(\theta_0)^n$, then the point $\rho = \sqrt[n]{\lambda} = a + ib$ belongs to the sector S_0 with $|\rho| \geq r(\theta_0) \geq r_1$, $|\rho| \geq r(\theta_0) \geq d/\sin(\theta_0/n)$, and $b = |\rho|\sin(\theta_0/n) \geq d$. Thus, by (9.3)

$$\|R_\lambda(L)\| \leq \gamma|\lambda|^{(p_0-p-n+1)/n} \leq \gamma|\lambda|^N \tag{9.7}$$

for $\lambda \in \mathcal{R}_{\theta_0}$ with $|\lambda| \geq r(\theta_0)^n$. Similarly, for $-\pi \leq \theta_0 \leq -\sigma_0$, we set

$$r(\theta_0) := \max\{r_1, d/|\sin(\theta_0/n)|\}$$

and then derive the result

$$\|R_\lambda(L)\| \leq \gamma|\lambda|^{(p_0-p-n+1)/n} \leq \gamma|\lambda|^N \tag{9.8}$$

for $\lambda \in \mathcal{R}_{\theta_0}$ with $|\lambda| \geq r(\theta_0)^n$.

Second, suppose L belongs to Case 3 where $p < q$. Assume that $\sigma_0 \leq \theta_0 \leq \pi$, and set

$$r(\theta_0) := \max\{r_2, (2n_0^2/\alpha)\cot(\theta_0/n)/\sin(\theta_0/n)\}.$$

Take any point $\lambda \in \mathcal{R}_{\theta_0}$ with $|\lambda| \geq r(\theta_0)^n$. Then the point $\rho = \sqrt[n]{\lambda} = a+ib$ belongs to S_0, and $|\rho| \geq r(\theta_0) \geq r_2$ and $b = |\rho|\sin(\theta_0/n) > 0$. Also,

$$b = |\rho|\sin(\theta_0/n) \geq (2n_0^2/\alpha)\cot(\theta_0/n),$$

$$e^{b/n_0} \geq b \cdot \frac{b}{2n_0^2} \geq \frac{b}{\alpha}\cot(\theta_0/n),$$

and

$$a = |\rho|\cos(\theta_0/n) = b\cot(\theta_0/n) \leq \alpha e^{b/n_0},$$

and hence, $\rho = a + ib \in \Omega_{0\infty}$ with $|\rho| \geq r_2$. It follows from (9.6) that

$$\|R_\lambda(L)\| \leq \gamma|\lambda|^{(p_0-p-n+1)/n} \leq \gamma|\lambda|^N \tag{9.9}$$

for $\lambda \in \mathcal{R}_{\theta_0}$ with $|\lambda| \geq r(\theta_0)^n$. Similarly, for $-\pi \leq \theta_0 \leq -\sigma_0$, we set

$$r(\theta_0) := \max\{r_2, (2n_0^2/\alpha)|\cot(\theta_0/n)|/|\sin(\theta_0/n)|\}$$

and proceed to derive the result

$$\|R_\lambda(L)\| \leq \gamma|\lambda|^{(p_0-p-n+1)/n} \leq \gamma|\lambda|^N \tag{9.10}$$

for $\lambda \in \mathcal{R}_{\theta_0}$ with $|\lambda| \geq r(\theta_0)^n$.

Recall that the constant σ_0 was selected in Chapter 5 with $0 < \sigma_0 < \pi/10$. The rays $\mathcal{R}_{\theta_0}$, with $\sigma_0 \leq \theta_0 \leq \pi$ or $-\pi \leq \theta_0 \leq -\sigma_0$, clearly cover the sector

$$\Sigma_0\colon\ \text{all } \lambda = |\lambda|e^{i\theta} \in \mathbb{C} \text{ with } \sigma_0 \leq \theta \leq 2\pi - \sigma_0$$

in the λ plane. Observe that the five equally spaced rays

$$\mathcal{R}_1\colon\ \arg\lambda = \theta_1 := \frac{\pi}{5} = 36^\circ,$$

$$\mathcal{R}_2\colon\ \arg\lambda = \theta_2 := \frac{\pi}{5} + \frac{2\pi}{5} = \frac{3\pi}{5} = 108^\circ,$$

$$\mathcal{R}_3\colon\ \arg\lambda = \theta_3 := \frac{3\pi}{5} + \frac{2\pi}{5} = \pi = 180^\circ,$$

$$\mathcal{R}_4\colon\ \arg\lambda = \theta_4 := \pi + \frac{2\pi}{5} = \frac{7\pi}{5} = 252^\circ,$$

$$\mathcal{R}_5\colon\ \arg\lambda = \theta_5 := \frac{7\pi}{5} + \frac{2\pi}{5} = \frac{9\pi}{5} = 324^\circ$$

all lie in Σ_0, and the angle between adjacent rays is $2\pi/5 = 72^\circ < \pi/2$. From the estimates in equations (9.7)–(9.10), we obtain the growth rate

$$\|R_\lambda(L)\| = O\big(|\lambda|^N\big) \quad \text{as } \lambda \to \infty \text{ along each ray } \mathcal{R}_j,$$

$j = 1, \ldots, 5$, valid for all three cases, Case 1, Case 2, Case 3. By the completeness theorem, Theorem 2.6.2 of [**34**], we conclude that

$$\overline{\mathrm{sp}(L)} = \overline{S_\infty(L)} = L^2[0,1] \quad \text{and} \quad M_\infty(L) = \{0\}. \tag{9.11}$$

We summarize the above results for the even order case in the following theorem. This theorem was stated previously as a part of Theorem 1.9 in Chapter 1.

THEOREM 9.1. *Let the differential operator L be of even order $n = 2\nu$, and let L be either regular or simply irregular. Then the spectrum $\sigma(L)$ is an infinite countable subset of $\mathbb{C}$ having no limit points in $\mathbb{C}$, and if $\sigma(L) = \{\lambda_i\}_{i=1}^\infty$ is any enumeration of $\sigma(L)$ and if $S_\infty(L)$ and $M_\infty(L)$ are the corresponding subspaces, then*

$$\overline{\mathrm{sp}(L)} = \overline{S_\infty(L)} = L^2[0,1] \quad \textit{and} \quad M_\infty(L) = \{0\}.$$

9.2. Completeness for n Odd

Assume that n is odd, $n = 2\nu - 1 \geq 3$. For the Green's function $G(t, s; \lambda)$ of the differential operator $\lambda I - L$, we have established growth rates for it in equations (6.69) and (6.82) relative to the sector S_0 and in equations (6.99) and (6.112) relative to the sector S_1. These growth rates depend on the sectors S_0 and S_1, but they are independent of whether L belongs to Case 1, Case 2, or Case 3. On the other hand, growth rates for the characteristic determinants Δ_0 and Δ_1 are case dependent.

Suppose the differential operator L belongs to Case 1 where $p = q$. Combining (6.69) and (8.6), we obtain the estimate

$$|G(t,s;\lambda)| \leq \gamma|\rho|^{p_0 - p - n + 1} \quad \text{for } t \neq s \text{ in } [0,1], \tag{9.12}$$

for $\lambda = \rho^n$ in $\mathbb{C}$ with $\rho = a + ib \in S_0$ and with $|\rho| \geq r_1$ and $b \geq d$; and combining (6.82) and (8.7), we obtain the estimate

$$|G(t,s;\lambda)| \leq \gamma|\rho|^{p_0 - p - n + 1} \quad \text{for } t \neq s \text{ in } [0,1], \tag{9.13}$$

for $\lambda = \rho^n$ in $\mathbb{C}$ with $\rho = a + ib \in S_0$ and with $|\rho| \geq r_1$ and $b \leq -d$. Similarly, combining (6.99) with (8.9), we get

$$|G(t,s;\lambda)| \leq \gamma|\rho|^{p_0 - p - n + 1} \quad \text{for } t \neq s \text{ in } [0,1], \tag{9.14}$$

for $\lambda = \rho^n$ in $\mathbb{C}$ with $\rho = a + ib \in S_1$ and with $|\rho| \geq r_1$ and $b \leq -d$; and combining (6.112) with (8.10), we get

$$|G(t,s;\lambda)| \leq \gamma|\rho|^{p_0 - p - n + 1} \quad \text{for } t \neq s \text{ in } [0,1], \tag{9.15}$$

for $\lambda = \rho^n$ in $\mathbb{C}$ with $\rho = a + ib \in S_1$ and with $|\rho| \geq r_1$ and $b \geq d$. Thus, in Case 1 the resolvent satisfies the growth rate

$$\|R_\lambda(L)\| \leq \gamma|\rho|^{p_0 - p - n + 1} = \gamma|\lambda|^{(p_0 - p - n + 1)/n} \tag{9.16}$$

for $\lambda = \rho^n$ in $\mathbb{C}$ with $\rho = a + ib \in S_0$ and $|\rho| \geq r_1$ and $|b| \geq d$, or with $\rho = a + ib \in S_1$ and $|\rho| \geq r_1$ and $|b| \geq d$.

Next, suppose that L belongs to Case 2 where $p < q$, a logarithmic case. Combining (6.69) and (8.20), (6.82) and (8.22), (6.99) and (8.28), and (6.112) and (8.26), we obtain the following estimates:

$$|G(t,s;\lambda)| \le \gamma|\rho|^{p_0-q-n+1} \quad \text{for } t \ne s \text{ in } [0,1], \tag{9.17}$$

for $\lambda = \rho^n$ in $\mathbb{C}$ with $\rho = a+ib \in S_0$ and with $|\rho| \ge r_1$ and $b \ge 0$;

$$|G(t,s;\lambda)| \le \gamma|\rho|^{p_0-p-n+1} \quad \text{for } t \ne s \text{ in } [0,1], \tag{9.18}$$

for $\lambda = \rho^n$ in $\mathbb{C}$ with $\rho = a+ib \in \Omega_{0\infty}$ and $|\rho| \ge r_2$;

$$|G(t,s;\lambda)| \le \gamma|\rho|^{p_0-p-n+1} \quad \text{for } t \ne s \text{ in } [0,1], \tag{9.19}$$

for $\lambda = \rho^n$ in $\mathbb{C}$ with $\rho = a+ib \in \Omega_{1\infty}$ and $|\rho| \ge r_2' = r_2$; and

$$|G(t,s;\lambda)| \le \gamma|\rho|^{p_0-q-n+1} \quad \text{for } t \ne s \text{ in } [0,1], \tag{9.20}$$

for $\lambda = \rho^n$ in $\mathbb{C}$ with $\rho = a+ib \in S_1$ and with $|\rho| \ge r_1'$ and $b \ge 0$. Therefore, the resolvent in Case 2 satisfies the growth rate

$$\|R_\lambda(L)\| \le \gamma|\rho|^{p_0-p-n+1} = \gamma|\lambda|^{(p_0-p-n+1)/n} \tag{9.21}$$

for $\lambda = \rho^n$ in $\mathbb{C}$ with $\rho = a+ib \in S_0$ and $|\rho| \ge r_1$ and $b \ge 0$, or with $\rho = a+ib \in \Omega_{0\infty}$ and $|\rho| \ge r_2$, or with $\rho = a+ib \in \Omega_{1\infty}$ and $|\rho| \ge r_2' = r_2$, or with $\rho = a+ib \in S_1$ and $|\rho| \ge r_1'$ and $b \ge 0$.

Finally, suppose that L belongs to Case 3 where $p > q$, another logarithmic case. Combining (6.69) and (8.51), (6.82) and (8.49), (6.99) and (8.55), and (6.112) and (8.57), we obtain the following estimates:

$$|G(t,s;\lambda)| \le \gamma|\rho|^{p_0-q-n+1} \quad \text{for } t \ne s \text{ in } [0,1], \tag{9.22}$$

for $\lambda = \rho^n$ in $\mathbb{C}$ with $\rho = a+ib \in \Omega_{0\infty}$ and $|\rho| \ge r_2$;

$$|G(t,s;\lambda)| \le \gamma|\rho|^{p_0-p-n+1} \quad \text{for } t \ne s \text{ in } [0,1], \tag{9.23}$$

for $\lambda = \rho^n$ in $\mathbb{C}$ with $\rho = a+ib \in S_0$ and with $|\rho| \ge r_1$ and $b \le 0$;

$$|G(t,s;\lambda)| \le \gamma|\rho|^{p_0-p-n+1} \quad \text{for } t \ne s \text{ in } [0,1], \tag{9.24}$$

for $\lambda = \rho^n$ in $\mathbb{C}$ with $\rho = a+ib \in S_1$ and with $|\rho| \ge r_1'$ and $b \le 0$; and

$$|G(t,s;\lambda)| \le \gamma|\rho|^{p_0-q-n+1} \quad \text{for } t \ne s \text{ in } [0,1], \tag{9.25}$$

for $\lambda = \rho^n$ in $\mathbb{C}$ with $\rho = a+ib \in \Omega_{1\infty}$ and $|\rho| \ge r_2' = r_2$. Therefore, the resolvent in Case 3 satisfies the growth rate

$$\|R_\lambda(L)\| \le \gamma|\rho|^{p_0-q-n+1} = \gamma|\lambda|^{(p_0-q-n+1)/n} \tag{9.26}$$

for $\lambda = \rho^n$ in $\mathbb{C}$ with $\rho = a+ib \in \Omega_{0\infty}$ and $|\rho| \ge r_2$, or with $\rho = a+ib \in S_0$ and $|\rho| \ge r_1$ and $b \le 0$, or with $\rho = a+ib \in S_1$ and $|\rho| \ge r_1'$ and $b \le 0$, or with $\rho = a+ib \in \Omega_{1\infty}$ and $|\rho| \ge r_2' = r_2$.

With our estimates (9.16), (9.21), and (9.26) for the resolvent in place, we are now ready to prove the completeness of the generalized eigenfunctions of the differential operator L. Let N be a positive integer that satisfies the conditions

$$N \ge (p_0 - p - n + 1)/n \quad \text{and} \quad N \ge (p_0 - q - n + 1)/n.$$

Fix any real number θ_0 satisfying one of the following conditions: $\sigma_0 \le \theta_0 \le \pi/2$, $-\pi/2 \le \theta_0 \le -\sigma_0$, $n\pi + \sigma_0 \le \theta_0 \le n\pi + \pi/2$, or $n\pi - \pi/2 \le \theta_0 \le n\pi - \sigma_0$. Consider the ray

$$\mathcal{R}_{\theta_0}:\ \lambda = |\lambda|e^{i\theta_0}, \quad 0 < |\lambda| < \infty,$$

in the λ plane. We assert that for λ on $\mathcal{R}_{\theta_0}$ with $|\lambda|$ sufficiently large, the resolvent $R_\lambda(L)$ exists and satisfies the growth rate $\|R_\lambda(L)\| = O(|\lambda|^N)$.

Assume that $\sigma_0 \le \theta_0 \le \pi/2$. For each $\lambda \in \mathcal{R}_{\theta_0}$ we form the nth root

$$\rho = \sqrt[n]{\lambda} = |\lambda|^{1/n} e^{i\theta_0/n},$$

with $\sigma_0/n \le \theta_0/n \le \pi/2n$ and with $\rho \in S_0$. For this nth root we have

$$\begin{cases} \rho = a + ib = |\rho|\cos(\theta_0/n) + i|\rho|\sin(\theta_0/n), \\ a = |\rho|\cos(\theta_0/n) > 0, \qquad b = |\rho|\sin(\theta_0/n) > 0. \end{cases}$$

Similarly, if $-\pi/2 \le \theta_0 \le -\sigma_0$, then for $\lambda \in \mathcal{R}_{\theta_0}$ we form the nth root

$$\rho = \sqrt[n]{\lambda} = |\lambda|^{1/n} e^{i\theta_0/n},$$

with $-\pi/2n \le \theta_0/n \le -\sigma_0/n$ and with $\rho \in S_0$; if $n\pi + \sigma_0 \le \theta_0 \le n\pi + \pi/2$, then for $\lambda \in \mathcal{R}_{\theta_0}$ we form the nth root

$$\rho = \sqrt[n]{\lambda} = |\lambda|^{1/n} e^{i\theta_0/n},$$

with $\pi + \sigma_0/n \le \theta_0/n \le \pi + \pi/2n$ and with $\rho \in S_1$; and if $n\pi - \pi/2 \le \theta_0 \le n\pi - \sigma_0$, then for $\lambda \in \mathcal{R}_{\theta_0}$ we form the nth root

$$\rho = \sqrt[n]{\lambda} = |\lambda|^{1/n} e^{i\theta_0/n},$$

with $\pi - \pi/2n \le \theta_0/n \le \pi - \sigma_0/n$ and with $\rho \in S_1$.

The geometry is slightly different for Case 1, Case 2, and Case 3. First, suppose that L belongs to Case 1 where $p = q$. Assume that $\sigma_0 \le \theta_0 \le \pi/2$, and set

$$r(\theta_0) := \max\{r_1, d/\sin(\theta_0/n)\}.$$

If $\lambda \in \mathcal{R}_{\theta_0}$ with $|\lambda| \ge r(\theta_0)^n$, then the point $\rho = \sqrt[n]{\lambda} = a + ib$ belongs to the sector S_0 with $|\rho| \ge r(\theta_0) \ge r_1$, $|\rho| \ge r(\theta_0) \ge d/\sin(\theta_0/n)$, and $b = |\rho|\sin(\theta_0/n) \ge d$. Thus, by (9.16)

$$\|R_\lambda(L)\| \le \gamma|\lambda|^{(p_0-p-n+1)/n} \le \gamma|\lambda|^N \tag{9.27}$$

for $\lambda \in \mathcal{R}_{\theta_0}$ with $|\lambda| \ge r(\theta_0)^n$. Similarly, for $-\pi/2 \le \theta_0 \le -\sigma_0$, we set

$$r(\theta_0) := \max\{r_1, d/|\sin(\theta_0/n)|\};$$

for $n\pi + \sigma_0 \le \theta_0 \le n\pi + \pi/2$, we set

$$r(\theta_0) := \max\{r_1, d/|\sin(\theta_0/n)|\};$$

and for $n\pi - \pi/2 \le \theta_0 \le n\pi - \sigma_0$, we set

$$r(\theta_0) := \max\{r_1, d/\sin(\theta_0/n)\}.$$

Then for each of these latter cases we obtain the result

$$\|R_\lambda(L)\| \le \gamma|\lambda|^{(p_0-p-n+1)/n} \le \gamma|\lambda|^N \tag{9.28}$$

for $\lambda \in \mathcal{R}_{\theta_0}$ with $|\lambda| \ge r(\theta_0)^n$.

Second, suppose that L belongs to Case 2 where $p < q$. Assume $\sigma_0 \le \theta_0 \le \pi/2$, and set $r(\theta_0) := r_1$. Take any point $\lambda \in \mathcal{R}_{\theta_0}$ with $|\lambda| \ge r(\theta_0)^n$. Then the point $\rho = \sqrt[n]{\lambda} = a + ib$ belongs to S_0, $|\rho| \ge r_1$, and $b > 0$. It follows from (9.21) that

$$\|R_\lambda(L)\| \le \gamma|\lambda|^{(p_0-p-n+1)/n} \le \gamma|\lambda|^N \tag{9.29}$$

for $\lambda \in \mathcal{R}_{\theta_0}$ with $|\lambda| \ge r(\theta_0)^n$. Similarly, for $-\pi/2 \le \theta_0 \le -\sigma_0$, we set

$$r(\theta_0) := \max\{r_2, (2n_0^2/\alpha)|\cot(\theta_0/n)|/|\sin(\theta_0/n)|\};$$

for $n\pi + \sigma_0 \le \theta_0 \le n\pi + \pi/2$, we set

$$r(\theta_0) := \max\{r_2', (2n_0^2/\alpha_1)\cot(\theta_0/n)/|\sin(\theta_0/n)|\};$$

and for $n\pi - \pi/2 \le \theta_0 \le n\pi - \sigma_0$, we set

$$r(\theta_0) := r_1'.$$

Then for each of these latter cases we obtain the result

$$\|R_\lambda(L)\| \le \gamma|\lambda|^{(p_0-p-n+1)/n} \le \gamma|\lambda|^N \tag{9.30}$$

for $\lambda \in \mathcal{R}_{\theta_0}$ with $|\lambda| \ge r(\theta_0)^n$.

Third, suppose that L belongs to Case 3 where $p > q$. Assume $\sigma_0 \le \theta_0 \le \pi/2$, and set

$$r(\theta_0) := \max\{r_2, (2n_0^2/\alpha)\cot(\theta_0/n)/\sin(\theta_0/n)\}.$$

Take any point $\lambda \in \mathcal{R}_{\theta_0}$ with $|\lambda| \ge r(\theta_0)^n$. Then the point $\rho = \sqrt[n]{\lambda} = a + ib$ belongs to S_0, and $|\rho| \ge r(\theta_0) \ge r_2$ and $b = |\rho|\sin(\theta_0/n) > 0$. Also,

$$b = |\rho|\sin(\theta_0/n) \ge (2n_0^2/\alpha)\cot(\theta_0/n),$$

$$e^{b/n_0} \ge b \cdot \frac{b}{2n_0^2} \ge \frac{b}{\alpha}\cot(\theta_0/n),$$

and

$$a = |\rho|\cos(\theta_0/n) = b\cot(\theta_0/n) \le \alpha e^{b/n_0},$$

and hence, $\rho = a + ib \in \Omega_{0\infty}$ with $|\rho| \ge r_2$. It follows from (9.26) that

$$\|R_\lambda(L)\| \le \gamma|\lambda|^{(p_0-q-n+1)/n} \le \gamma|\lambda|^N \tag{9.31}$$

for $\lambda \in \mathcal{R}_{\theta_0}$ with $|\lambda| \ge r(\theta_0)^n$. Similarly, for $-\pi/2 \le \theta_0 \le -\sigma_0$, set

$$r(\theta_0) := r_1;$$

for $n\pi + \sigma_0 \le \theta_0 \le n\pi + \pi/2$, set

$$r(\theta_0) := r_1';$$

and for $n\pi - \pi/2 \le \theta_0 \le n\pi - \sigma_0$, set

$$r(\theta_0) := \max\{r_2', (2n_0^2/\alpha_1)|\cot(\theta_0/n)|/\sin(\theta_0/n)\}.$$

Then for each of these latter cases we obtain the result

$$\|R_\lambda(L)\| \le \gamma|\lambda|^{(p_0-q-n+1)/n} \le \gamma|\lambda|^N \tag{9.32}$$

for $\lambda \in \mathcal{R}_{\theta_0}$ with $|\lambda| \ge r(\theta_0)^n$.

Recall that the constant σ_0 was selected in Chapter 5 with $0 < \sigma_0 < \pi/10$. The family of rays $\mathcal{R}_{\theta_0}$, with either $\sigma_0 \le \theta_0 \le \pi/2$, or $-\pi/2 \le \theta_0 \le -\sigma_0$, or $n\pi + \sigma_0 \le \theta_0 \le n\pi + \pi/2$, or $n\pi - \pi/2 \le \theta_0 \le n\pi - \sigma_0$, clearly covers the union of the two sectors

$$\Sigma_0\colon \text{ all } \lambda = |\lambda|e^{i\theta} \in \mathbb{C} \text{ with } \sigma_0 \le \theta \le \pi - \sigma_0,$$

$$\Sigma_1\colon \text{ all } \lambda = |\lambda|e^{i\theta} \in \mathbb{C} \text{ with } -\pi + \sigma_0 \le \theta \le -\sigma_0$$

in the λ plane. Note that the five equally spaced rays

$$\begin{aligned}
\mathcal{R}_1\colon\ \arg\lambda = \theta_1 &:= \frac{\pi}{10} = 18^\circ,\\
\mathcal{R}_2\colon\ \arg\lambda = \theta_2 &:= \frac{\pi}{10} + \frac{2\pi}{5} = \frac{\pi}{2} = 90^\circ,\\
\mathcal{R}_3\colon\ \arg\lambda = \theta_3 &:= \frac{\pi}{2} + \frac{2\pi}{5} = \frac{9\pi}{10} = 162^\circ,\\
\mathcal{R}_4\colon\ \arg\lambda = \theta_4 &:= \frac{9\pi}{10} + \frac{2\pi}{5} = \frac{13\pi}{10} = 234^\circ,\\
\mathcal{R}_5\colon\ \arg\lambda = \theta_5 &:= \frac{13\pi}{10} + \frac{2\pi}{5} = \frac{17\pi}{10} = 306^\circ
\end{aligned}$$

all lie in $\Sigma_0 \cup \Sigma_1$, and the angle between adjacent rays is $2\pi/5 = 72^\circ < \pi/2$. From the estimates (9.27)–(9.32) we have the growth rate

$$\|R_\lambda(L)\| = O\big(|\lambda|^N\big) \quad \text{as } \lambda \to \infty \text{ along each ray } \mathcal{R}_j,$$

$j = 1, \ldots, 5$, valid for all three cases, Case 1, Case 2, Case 3. By the completeness theorem, Theorem 2.6.2 of [**34**], we conclude that

$$\overline{\mathrm{sp}(L)} = \overline{S_\infty(L)} = L^2[0,1] \quad \text{and} \quad M_\infty(L) = \{0\}. \tag{9.33}$$

We summarize the above results for the odd order case in the following theorem. This theorem was stated previously as a part of Theorem 1.9 in Chapter 1.

THEOREM 9.2. *Let the differential operator L be of odd order $n = 2\nu - 1$, and let L be either regular or simply irregular. Then the spectrum $\sigma(L)$ is an infinite countable subset of $\mathbb{C}$ having no limit points in $\mathbb{C}$, and if $\sigma(L) = \{\lambda_i\}_{i=1}^\infty$ is any enumeration of $\sigma(L)$ and if $S_\infty(L)$ and $M_\infty(L)$ are the corresponding subspaces, then*

$$\overline{\mathrm{sp}(L)} = \overline{S_\infty(L)} = L^2[0,1] \quad \textit{and} \quad M_\infty(L) = \{0\}.$$

CHAPTER 10

The Case $L = T$, Degenerate Irregular Examples

In the previous chapters we have developed the spectral theory of the differential operator L for the cases in which L is either *regular* or *simply irregular*. The differential operators L that are neither regular nor simply irregular have been grouped together in the *degenerate irregular* class, for lack of a better name. This degenerate irregular class contains many strange differential operators, and has never been studied. In the future when this class is better understood, we envision it being subdivided into various subclasses, and having better names assigned to these subclasses.

10.1. The Case $L = T$

To illustrate some of the unusual features of the degenerate irregular differential operators, we present in this chapter some examples for the special case when the differential operator L is equal to its principal part T. Assume that $\ell = \tau$ and $\sigma = 0$, so $L = T$ and

$$\mathcal{D}(L) = \mathcal{D}(T) = \{u \in H^n[0,1] \mid B_i(u) = 0,\ i = 1, \dots, n\},$$
$$Lu = Tu = i^{-n}u^{(n)}.$$

This important special case has been studied previously in Chapter 4 of [**34**] under the assumption that L is either regular or simply irregular, but there has been no discussion of the degenerate irregular case. Let us begin by observing some of the important simplifications that occur in Chapters 2–5 when $L = T$.

Fix any integer m with $m > n$ and $m > p_0$. Consider the approximate solutions and the approximate characteristic determinant defined in Chapter 2 and Chapter 3. First, for the special case $L = T$, the functions $c_{0\ell p}(t)$ simplify to $c_{0\ell 0}(t) = \binom{n}{\ell} i^{n-\ell}$, $\ell = 1, \dots, n$, with all other $c_{0\ell p}(t)$ identically zero on $[0, 1]$, and hence, the differential operators ℓ_q, $q = 2, \dots, n$, are given by

$$\ell_q u(t) = c_{0q0}(t)u^{(q)}(t) = \binom{n}{q} i^{n-q}u^{(q)}(t).$$

If $u(t)$ is a constant function, then clearly $\ell_q u(t) = 0$ for $q = 2, \dots, n$, and from equations (2.15), (2.16), and (2.8) it then follows that $z_{0j}(t) = 0$ for $j = 1, \dots, m-1$. Thus, $z_0(t, \rho) = e^{i\rho t}$, and by (2.11) the approximate solutions are simply

$$z_k(t, \rho) = e^{i\rho\omega_k t}, \qquad k = 0, 1, \dots, n-1,$$

for $\rho \neq 0$ in $\mathbb{C}$. This is the expected result for this special case.

The approximate solutions $z_k(t, \rho)$, $k = 0, 1, \dots, n-1$, are actual solutions of the differential equation (2.1),

$$\rho^n u(t) - i^{-n}u^{(n)}(t) = 0,$$

and they are independent of the integer m. The associated residual functions are simply

$$\eta_k(t,\rho) = e^{-i\rho\omega_k t}(\rho^n I - \ell)z_k(t,\rho) = 0, \qquad k = 0,1,\dots,n-1,$$

for $\rho \neq 0$ in $\mathbb{C}$.

Second, the modified approximate solutions are given by

$$\begin{aligned} y_k(t,\rho) &= z_k(t,\rho) = e^{i\rho\omega_k t}, & k &= 0,1,\dots,\nu-1, \\ y_k(t,\rho) &= e^{-i\rho\omega_k} z_k(t,\rho) = e^{i\rho\omega_k(t-1)}, & k &= \nu,\dots,n-1, \end{aligned}$$

for $\rho \neq 0$ in $\mathbb{C}$. These functions are also actual solutions of the differential equation (2.1). In terms of the boundary values $B_1,\dots,B_n$, we introduce the polynomials

$$P_i(\rho) := \sum_{p=0}^{m_i} \alpha_{ip}\rho^p, \qquad Q_i(\rho) := \sum_{p=0}^{m_i} \beta_{ip}\rho^p, \qquad i = 1,\dots,n.$$

Then for $i = 1,\dots,n$ and $k = 0,1,\dots,\nu-1$ we have

$$\begin{aligned} B_i(y_k(\,\cdot\,,\rho)) &= \sum_{p=0}^{m_i} \alpha_{ip}(i\rho\omega_k)^p + \sum_{p=0}^{m_i} \beta_{ip}(i\rho\omega_k)^p e^{i\rho\omega_k} \\ &= P_i(i\rho\omega_k) + Q_i(i\rho\omega_k)e^{i\rho\omega_k} \\ &= \widehat{P}_{ik}(\rho) + \widehat{Q}_{ik}(\rho)e^{i\rho\omega_k} \end{aligned} \tag{10.1}$$

for $\rho \neq 0$ in $\mathbb{C}$, while for $i = 1,\dots,n$ and $k = \nu,\dots,n-1$

$$\begin{aligned} B_i(y_k(\,\cdot\,,\rho)) &= \sum_{p=0}^{m_i} \alpha_{ip}(i\rho\omega_k)^p e^{-i\rho\omega_k} + \sum_{p=0}^{m_i} \beta_{ip}(i\rho\omega_k)^p \\ &= Q_i(i\rho\omega_k) + P_i(i\rho\omega_k)e^{-i\rho\omega_k} \\ &= \widehat{P}_{ik}(\rho) + \widehat{Q}_{ik}(\rho)e^{-i\rho\omega_k} \end{aligned} \tag{10.2}$$

for $\rho \neq 0$ in $\mathbb{C}$. Thus, the functions $\widehat{P}_{ik}(\rho)$, $\widehat{Q}_{ik}(\rho)$ defined in equations (3.1), (3.2) are polynomials in ρ that are independent of the integer m. Equations (3.7) and (3.9) now simplify to

$$\widehat{P}_{ik}(\rho) = \sum_{s=0}^{m_i} p_{iks}\rho^s, \qquad \widehat{Q}_{ik}(\rho) = \sum_{s=0}^{m_i} q_{iks}\rho^s \tag{10.3}$$

for $\rho \neq 0$ in $\mathbb{C}$ and for $i = 1,\dots,n$ and $k = 0,1,\dots,n-1$. In terms of these polynomials, the approximate characteristic determinant is given by

$$\begin{aligned} \widehat{\Delta}(\rho) &= \det(B_i(y_k(\,\cdot\,,\rho))) \\ &= \det\Big(\underbrace{P_i(i\rho\omega_k) + Q_i(i\rho\omega_k)e^{i\rho\omega_k}}_{0\le k\le \nu-1} \;\; \underbrace{Q_i(i\rho\omega_k) + P_i(i\rho\omega_k)e^{-i\rho\omega_k}}_{\nu\le k\le n-1} \Big) \\ &= \det\Big(\underbrace{\widehat{P}_{ik}(\rho) + \widehat{Q}_{ik}(\rho)e^{i\rho\omega_k}}_{0\le k\le \nu-1} \;\; \underbrace{\widehat{P}_{ik}(\rho) + \widehat{Q}_{ik}(\rho)e^{-i\rho\omega_k}}_{\nu\le k\le n-1} \Big) \end{aligned} \tag{10.4}$$

for $\rho \neq 0$ in $\mathbb{C}$. This shows that the approximate characteristic determinant $\widehat{\Delta}(\rho)$ is independent of the integer m.

We can also construct the approximate characteristic determinant using the functions $z_k(t, \rho)$, $k = 0, 1, \ldots, n-1$. Indeed, these functions form a basis for the solution space of the differential equation (2.1), and for $i = 1, \ldots, n$ and $k = 0, 1, \ldots, n-1$ we have

$$B_i(z_k(\,\cdot\,, \rho)) = P_i(i\rho\omega_k) + Q_i(i\rho\omega_k)e^{i\rho\omega_k}$$

for $\rho \neq 0$ in $\mathbb{C}$. An alternate approximate characteristic determinant for the differential operator $L = T$ is defined by

$$\Delta_*(\rho) := \det(B_i(z_k(\,\cdot\,, \rho))) = \det\big(P_i(i\rho\omega_k) + Q_i(i\rho\omega_k)e^{i\rho\omega_k}\big)$$

for $\rho \neq 0$ in $\mathbb{C}$. The approximate characteristic determinants are related by the relation

$$\widehat{\Delta}(\rho) = e^{i\rho\eta}\Delta_*(\rho) \tag{10.5}$$

for $\rho \neq 0$ in $\mathbb{C}$, where the constant η is defined by $\eta := -\omega_\nu - \omega_{\nu+1} - \cdots - \omega_{n-1}$. Later in this chapter it will be convenient to utilize $\Delta_*(\rho)$ in place of $\widehat{\Delta}(\rho)$.

Assume n is even: $n = 2\nu \geq 2$. From equation (10.3) it follows that the functions $\widehat{\pi}_i(\rho)$, $i = 0, 1, 2$, introduced in Chapter 3, are polynomials of degree $\leq p_0$. Also, from (3.24), (3.27), and (3.28) these polynomials have the representations

$$\widehat{\pi}_2(\rho) = \sum_{\kappa=0}^{p_0} a_\kappa \rho^\kappa, \qquad \widehat{\pi}_1(\rho) = \sum_{\kappa=0}^{p_0} b_\kappa \rho^\kappa, \qquad \widehat{\pi}_0(\rho) = \sum_{\kappa=0}^{p_0} c_\kappa \rho^\kappa$$

for $\rho \neq 0$ in $\mathbb{C}$, with $a_\kappa = b_\kappa = c_\kappa = 0$ for $\kappa = -1, -2, \ldots, -(m - p_0 - 1)$. Clearly these polynomials are independent of the integer m. Since m can be chosen arbitrarily large, we conclude that

$$a_\kappa = b_\kappa = c_\kappa = 0 \quad \text{for } \kappa = -1, -2, \ldots. \tag{10.6}$$

It is immediate that the functions $\pi_i(\rho)$, $i = 0, 1, 2$, introduced in Chapter 3 are polynomials that coincide with the polynomials $\widehat{\pi}_i(\rho)$, $i = 0, 1, 2$:

$$\pi_2(\rho) = \widehat{\pi}_2(\rho) = \sum_{\kappa=0}^{p_0} a_\kappa \rho^\kappa,$$
$$\pi_1(\rho) = \widehat{\pi}_1(\rho) = \sum_{\kappa=0}^{p_0} b_\kappa \rho^\kappa,$$
$$\pi_0(\rho) = \widehat{\pi}_0(\rho) = \sum_{\kappa=0}^{p_0} c_\kappa \rho^\kappa$$

for $\rho \neq 0$ in $\mathbb{C}$, and hence, the polynomials $\pi_i(\rho)$, $i = 0, 1, 2$, are independent of the integer m.

Assume n is odd: $n = 2\nu - 1 \geq 3$. Again from (10.3) we see that the functions $\widehat{\pi}_i(\rho)$, $i = 0, 1$, and the functions $\pi_i(\rho)$, $i = 0, 1$, introduced in Chapter 3, are polynomials of degree $\leq p_0$ which are identical and independent of the integer m:

$$\pi_1(\rho) = \widehat{\pi}_1(\rho) = \sum_{\kappa=0}^{p_0} a_\kappa \rho^\kappa, \qquad \pi_0(\rho) = \widehat{\pi}_0(\rho) = \sum_{\kappa=0}^{p_0} b_\kappa \rho^\kappa$$

for $\rho \neq 0$ in $\mathbb{C}$. For this case we have

$$a_\kappa = b_\kappa = 0 \quad \text{for } \kappa = -1, -2, \ldots. \tag{10.7}$$

These results for the special case $L = T$ show that the classification scheme defined in our current work (see Definition 3.2 and Definition 3.3) is consistent with the classification scheme defined in the monograph [**34**] (see Definition 4.5.1 and Definition 4.6.1 of [**34**]): regular $\equiv$ regular, simply irregular $\equiv$ irregular, and degenerate irregular $\equiv$ degenerate. Also, the classification of the differential operator $L = T$ is greatly simplified because of equations (10.6) and (10.7): the relevant constants a_κ, b_κ, c_κ, $\kappa = p_0, \ldots, 1, 0$, can be calculated in one step by expanding the determinants that define the polynomials $\widehat{\pi}_i(\rho)$.

In Chapter 4 we developed asymptotic expansions for solutions of the differential equation (2.1). Let us examine the form of these expansions in the special case $L = T$. Indeed, fix any integer m with $m > n$ and $m > p_0$, and take any integer k with $0 \leq k \leq n-1$ and any point $\rho \neq 0$ in $\mathbb{C}$. Then for $|\rho|$ sufficiently large the integral equations (4.13), (4.23) and (4.34), (4.38) have only the trivial solutions $\psi_{0k}(t, \rho) = 0$ and $\psi_{1k}(t, \rho) = 0$. Applying (4.15), (4.25) and (4.36), (4.40), we obtain the functions $v_{0k}(t, \rho) = z_k(t, \rho)$ and $v_{1k}(t, \rho) = z_k(t, \rho)$, which are solutions of the differential equation (2.1). Setting $v_k(t, \rho) := z_k(t, \rho)$ for $k = 0, 1, \ldots, n-1$ and for $\rho \neq 0$ in $\mathbb{C}$, we see that the functions $v_k(\,\cdot\,, \rho)$ are solutions of the differential equation (2.1), the solutions $v_{0k}(\,\cdot\,, \rho)$ and $v_{1k}(\,\cdot\,, \rho)$ are restrictions of $v_k(\,\cdot\,, \rho)$ to the sectors T_0 and T_1, and Theorems 4.3, 4.4 and Theorems 4.6, 4.7 can be replaced by the stronger statement that the functions

$$v_k(t, \rho) = z_k(t, \rho) = e^{i\rho\omega_k t}, \qquad k = 0, 1, \ldots, n-1,$$

are n linearly independent solutions of the differential equation (2.1) for all $\rho \neq 0$ in $\mathbb{C}$.

Finally, in Chapter 5 the solutions $v_k(\,\cdot\,, \rho)$, $k = 0, 1, \ldots, n-1$, are used to form the characteristic determinant of the differential operator $L = T$. Introducing the modified solutions

$$\begin{aligned} u_k(t, \rho) &:= v_k(t, \rho) = y_k(t, \rho), & k &= 0, 1, \ldots, \nu - 1, \\ u_k(t, \rho) &:= e^{-i\rho\omega_k} v_k(t, \rho) = y_k(t, \rho), & k &= \nu, \ldots, n-1, \end{aligned}$$

for $0 \leq t \leq 1$ and for $\rho \neq 0$ in $\mathbb{C}$, and the characteristic determinant

$$\Delta(\rho) := \det(B_i(u_k(\,\cdot\,, \rho)))$$

for $\rho \neq 0$ in $\mathbb{C}$, we see that the modified solutions $u_{0k}(\,\cdot\,, \rho)$, $k = 0, 1, \ldots, n-1$, in Chapter 5 are the restrictions of the modified solutions $u_k(\,\cdot\,, \rho)$, $k = 0, 1, \ldots, n-1$, to the sector T_0, and the characteristic determinant $\Delta_0(\rho)$ in Chapter 5 is the restriction of the characteristic determinant $\Delta(\rho)$ to the sector T_0. It is immediate that

$$\Delta(\rho) = \det(B_i(u_k(\,\cdot\,, \rho))) = \det(B_i(y_k(\,\cdot\,, \rho))) = \widehat{\Delta}(\rho) \tag{10.8}$$

for $\rho \neq 0$ in $\mathbb{C}$. The characteristic determinant $\Delta(\rho)$ that appears in (10.8) is identical to the characteristic determinant $\Delta(\rho)$ defined for the differential operator $L = T$ in [**34**, p. 100]; $\Delta(\rho)$ can be formed irregardless of whether $L = T$ is regular, simply irregular, or degenerate irregular, and it can be used to calculate the eigenvalues of $L = T$ in all cases.

10.2. Two Degenerate Irregular Examples

With the above results for the special case $L = T$ as a foundation, we next look at some examples of the degenerate irregular case.

EXAMPLE 10.1. Consider the differential operator $L = T$ determined by initial value conditions at the endpoint $t = 0$:

$$B_i(u) = u^{(n-i)}(0), \qquad i = 1, \ldots, n.$$

For this model we have $p_0 = (n/2)(n-1)$. In equations (10.1) and (10.2), for $i = 1, \ldots, n$ the functions are given by

$$\begin{aligned} B_i(y_k(\cdot\,,\rho)) &= (i\rho\omega_k)^{n-i}, & k &= 0, 1, \ldots, \nu - 1, \\ B_i(y_k(\cdot\,,\rho)) &= (i\rho\omega_k)^{n-i}e^{-i\rho\omega_k}, & k &= \nu, \ldots, n-1, \end{aligned}$$

and hence, for $i = 1, \ldots, n$ we have

$$\begin{aligned} &\widehat{P}_{ik}(\rho) = (i\rho\omega_k)^{n-i}, \quad \widehat{Q}_{ik}(\rho) = 0, & k &= 0, 1, \ldots, \nu - 1, \\ &\widehat{P}_{ik}(\rho) = 0, \quad \widehat{Q}_{ik}(\rho) = (i\rho\omega_k)^{n-i}, & k &= \nu, \ldots, n-1. \end{aligned} \tag{10.9}$$

Setting $\eta := -\omega_\nu - \omega_{\nu+1} - \cdots - \omega_{n-1}$, it follows from equation (10.4) that

$$\begin{aligned} \widehat{\Delta}(\rho) = \Delta(\rho) &= \det\Big(\overset{0\le k\le \nu-1}{(i\rho\omega_k)^{n-i}} \quad \overset{\nu\le k\le n-1}{(i\rho\omega_k)^{n-i}e^{-i\rho\omega_k}} \Big) \\ &= (i\rho)^{\frac{n}{2}(n-1)} e^{i\rho\eta} \det \begin{pmatrix} \omega_0^{n-1} & \omega_1^{n-1} & \cdots & \omega_{n-1}^{n-1} \\ \omega_0^{n-2} & \omega_1^{n-2} & \cdots & \omega_{n-1}^{n-2} \\ \vdots & \vdots & & \vdots \\ 1 & 1 & \cdots & 1 \end{pmatrix}, \end{aligned}$$

or

$$\widehat{\Delta}(\rho) = \Delta(\rho) = \gamma(i\rho)^{\frac{n}{2}(n-1)} e^{i\rho\eta} \quad \text{for } \rho \neq 0 \text{ in } \mathbb{C}, \tag{10.10}$$

where the constant γ is the displayed Vandermonde determinant. Thus, $\Delta(\rho) \neq 0$ for all $\rho \neq 0$ in $\mathbb{C}$, which shows that the differential operator L has no nonzero eigenvalues. Using the basis $1, t, t^2, \ldots, t^{n-1}$ for the solution space of the differential equation $-(i)^{-n}u^{(n)}(t) = 0$, it is easy to check that $\lambda = 0$ is not an eigenvalue of L. We conclude that

$$\sigma(L) = \emptyset \quad \text{and} \quad \rho(L) = \mathbb{C}. \tag{10.11}$$

Of course, this result follows directly from the Existence-Uniqueness Theorem for initial value problems.

Assume that n is even: $n = 2\nu \geq 2$. Using the definitions of the functions $\widehat{\pi}_i(\rho)$, $i = 0, 1, 2$, given in Chapter 3 together with equation (10.9), we see that

$$\widehat{\pi}_2(\rho) = \pi_2(\rho) = 0, \qquad \widehat{\pi}_0(\rho) = \pi_0(\rho) = 0,$$

and

$$\begin{aligned} \widehat{\pi}_1(\rho) = \pi_1(\rho) &= \det \begin{pmatrix} (i\rho\omega_0)^{n-1} & \cdots & (i\rho\omega_{\nu-1})^{n-1} & (i\rho\omega_\nu)^{n-1} & 0 & \cdots & 0 \\ (i\rho\omega_0)^{n-2} & \cdots & (i\rho\omega_{\nu-1})^{n-2} & (i\rho\omega_\nu)^{n-2} & 0 & \cdots & 0 \\ \vdots & & \vdots & \vdots & \vdots & & \vdots \\ 1 & \cdots & 1 & 1 & 0 & \cdots & 0 \end{pmatrix} \\ &= \begin{cases} 2i\rho & \text{for } n = 2, \\ 0 & \text{for } n = 4, 6, 8, \ldots. \end{cases} \end{aligned}$$

Thus, for $n = 2$ we have $p_0 = 1$ and

$$\begin{aligned} a_\kappa &= c_\kappa = 0 && \text{for } \kappa = 1, 0, -1, \dots, \\ b_1 &= 2i, \quad b_\kappa = 0 && \text{for } \kappa = 0, -1, \dots, \end{aligned} \tag{10.12}$$

and for $n = 4, 6, 8, \dots$ we have $p_0 = (n/2)(n-1)$ and

$$a_\kappa = b_\kappa = c_\kappa = 0 \quad \text{for } \kappa = p_0, \dots, 1, 0, -1, \dots. \tag{10.13}$$

According to Definition 3.2, the differential operator L is degenerate irregular.

Assume that n is odd: $n = 2\nu - 1 \geq 3$. Using the definitions of the functions $\widehat{\pi}_1(\rho)$ and $\widehat{\pi}_0(\rho)$ given in Chapter 3 together with equation (10.9), we see immediately that

$$\widehat{\pi}_1(\rho) = \pi_1(\rho) = 0, \qquad \widehat{\pi}_0(\rho) = \pi_0(\rho) = 0,$$

and hence,

$$a_\kappa = b_\kappa = 0 \quad \text{for } \kappa = p_0, \dots, 1, 0, -1, \dots \tag{10.14}$$

where $p_0 = (n/2)(n-1)$. By Definition 3.3 the differential operator L is degenerate irregular.

This example illustrates the fact that the principal exponentials,

$$e^{2i\rho},\ e^{i\rho},\ 1 \quad \text{for } n \text{ even}, \qquad e^{i\rho},\ 1 \quad \text{for } n \text{ odd},$$

may not appear in the characteristic determinant $\Delta(\rho)$. In this situation other exponentials become the principal terms. For example, in equation (10.10) the exponential

$$e^{i\rho\eta} = e^{-i\rho(\omega_\nu + \omega_{\nu+1} + \cdots + \omega_{n-1})}$$

is the principal exponential appearing in $\Delta(\rho)$, and in fact, it is the only exponential that appears.

EXAMPLE 10.2. For n even, $n = 2\nu \geq 2$, let $L = T$ be the differential operator determined by the boundary values

$$B_i(u) = u^{(n-i)}(0) + (-1)^{i+1} u^{(n-i)}(1), \qquad i = 1, \dots, n.$$

Again we have $p_0 = (n/2)(n-1)$, and for $i = 1, \dots, n$ equations (10.1) and (10.2) become

$$\begin{aligned} B_i(y_k(\cdot, \rho)) &= (i\rho\omega_k)^{n-i} + (-1)^{i+1}(i\rho\omega_k)^{n-i} e^{i\rho\omega_k}, && k = 0, 1, \dots, \nu - 1, \\ B_i(y_k(\cdot, \rho)) &= (i\rho\omega_k)^{n-i} e^{-i\rho\omega_k} + (-1)^{i+1}(i\rho\omega_k)^{n-i}, && k = \nu, \dots, n-1. \end{aligned} \tag{10.15}$$

Hence, for $i = 1, \dots, n$

$$\begin{aligned} \widehat{P}_{ik}(\rho) &= (i\rho\omega_k)^{n-i}, \quad \widehat{Q}_{ik}(\rho) = (-1)^{i+1}(i\rho\omega_k)^{n-i}, && k = 0, 1, \dots, \nu - 1, \\ \widehat{P}_{ik}(\rho) &= (-1)^{i+1}(i\rho\omega_k)^{n-i}, \quad \widehat{Q}_{ik}(\rho) = (i\rho\omega_k)^{n-i}, && k = \nu, \dots, n-1. \end{aligned} \tag{10.16}$$

Observe that in (10.15), for $i = 1, \dots, n$ and $k = 0$ we have $\omega_0 = 1$ and

$$B_i(y_0(\cdot, \rho)) = (i\rho)^{n-i}[1 + (-1)^{i+1} e^{i\rho}],$$

while for $i = 1, \dots, n$ and $k = \nu$ we have $\omega_\nu = -1$ and

$$\begin{aligned} B_i(y_\nu(\cdot, \rho)) &= (-i\rho)^{n-i}[e^{i\rho} + (-1)^{i+1}] \\ &= (-1)^{n+1}(i\rho)^{n-i}[(-1)^{i+1} e^{i\rho} + 1] = -B_i(y_0(\cdot, \rho)). \end{aligned}$$

It follows from (10.4) that

$$\widehat{\Delta}(\rho) = \Delta(\rho) = 0 \quad \text{for } \rho \neq 0 \text{ in } \mathbb{C}. \tag{10.17}$$

Thus, each $\lambda = \rho^n \neq 0$ in $\mathbb{C}$ is an eigenvalue of L, and a corresponding eigenfunction is given by

$$\phi(t,\rho) = \cos\rho(t - 1/2) = \frac{1}{2}\, e^{-i\rho/2} \cdot e^{i\rho t} + \frac{1}{2}\, e^{-i\rho/2} \cdot e^{-i\rho(t-1)}.$$

For $\lambda = 0$ we use the basis $1, t, t^2, \ldots, t^{n-1}$, showing that $\lambda = 0$ is also an eigenvalue of L with eigenfunctions $\phi_j(t) = (t - 1/2)^{2j}$, $j = 0, 1, \ldots, \nu - 1$. We conclude that

$$\sigma(L) = \mathbb{C} \quad \text{and} \quad \rho(L) = \emptyset. \tag{10.18}$$

In equation (10.16) we have $\widehat{Q}_{i0}(\rho) = (-1)^{i+1}(i\rho)^{n-i}$ for $i = 1, \ldots, n$ and

$$\widehat{Q}_{i\nu}(\rho) = (-i\rho)^{n-i} = -(-1)^{i+1}(i\rho)^{n-i} = -\widehat{Q}_{i0}(\rho) \tag{10.19}$$

for $i = 1, \ldots, n$, and $\widehat{P}_{i0}(\rho) = (i\rho)^{n-i}$ for $i = 1, \ldots, n$ and

$$\widehat{P}_{i\nu}(\rho) = (-1)^{i+1}(-i\rho)^{n-i} = -(i\rho)^{n-i} = -\widehat{P}_{i0}(\rho) \tag{10.20}$$

for $i = 1, \ldots, n$. Therefore, from the definitions of the functions $\widehat{\pi}_2(\rho)$ and $\widehat{\pi}_0(\rho)$ given in Chapter 3, it follows that

$$\widehat{\pi}_2(\rho) = \pi_2(\rho) = 0, \qquad \widehat{\pi}_0(\rho) = \pi_0(\rho) = 0,$$

and hence,

$$a_\kappa = c_\kappa = 0 \quad \text{for } \kappa = p_0, \ldots, 1, 0, -1, \ldots. \tag{10.21}$$

In the definition of the function $\widehat{\pi}_1(\rho)$, if we substitute (10.19) and (10.20) into the νth columns of the two determinants appearing there, then $\widehat{\pi}_1(\rho)$ becomes the sum of two determinants that are the negatives of each other, and hence,

$$\widehat{\pi}_1(\rho) = \pi_1(\rho) = 0$$

and

$$b\kappa = 0 \quad \text{for } \kappa = p_0, \ldots, 1, 0, -1, \ldots. \tag{10.22}$$

According to Definition 3.2, the differential operator L is degenerate irregular.

The differential operators L appearing in Example 10.1 and Example 10.2 are certainly degenerate, but for very different reasons. In the first example, the spectrum $\sigma(L)$ is empty, there are no eigenvalues nor any eigenfunctions, and the resolvent $R_\lambda(L)$ exists on all of $\mathbb{C}$; in the second, the spectrum $\sigma(L)$ is equal to all of $\mathbb{C}$, there are uncountably many eigenvalues and eigenfunctions, and the resolvent $R_\lambda(L)$ fails to exist at any point of $\mathbb{C}$. What kind of spectral theory can one develop in these extreme cases?

10.3. The Case $n = 4,\ L = T$

Next, we give some additional examples for the special case $n = 4$ and $L = T$. The integer $n = 4$ is small enough that calculations can still be made with pencil and paper, while it is large enough to exhibit some of the more subtle features of these fourth order differential operators. In this special case $\ell = \tau = (d/dt)^4$, the boundary values have the form

$$B_i(u) = \sum_{p=0}^{3} \alpha_{ip} u^{(p)}(0) + \sum_{p=0}^{3} \beta_{ip} u^{(p)}(1), \qquad i = 1, \ldots, 4,$$

and the associated boundary coefficient matrix becomes

$$A = \begin{pmatrix} \alpha_{13} & \beta_{13} & \alpha_{12} & \beta_{12} & \alpha_{11} & \beta_{11} & \alpha_{10} & \beta_{10} \\ \alpha_{23} & \beta_{23} & \alpha_{22} & \beta_{22} & \alpha_{21} & \beta_{21} & \alpha_{20} & \beta_{20} \\ \alpha_{33} & \beta_{33} & \alpha_{32} & \beta_{32} & \alpha_{31} & \beta_{31} & \alpha_{30} & \beta_{30} \\ \alpha_{43} & \beta_{43} & \alpha_{42} & \beta_{42} & \alpha_{41} & \beta_{41} & \alpha_{40} & \beta_{40} \end{pmatrix}.$$

Recall that we are assuming that A is in reduced row echelon form with rank 4. Note that the integer $p_0 = \sum_{i=1}^{4} m_i$ satisfies the inequalities $2 \le p_0 \le 10$. Here the fourth roots of unity are simply $\omega_0 = 1$, $\omega_1 = i$, $\omega_2 = -1$, $\omega_3 = -i$; the (approximate) solutions are given by

$$z_0(t,\rho) = e^{i\rho t}, \quad z_1(t,\rho) = e^{-\rho t}, \quad z_2(t,\rho) = e^{-i\rho t}, \quad z_3(t,\rho) = e^{\rho t};$$

and the modified (approximate) solutions are

$$y_0(t,\rho) = e^{i\rho t}, \quad y_1(t,\rho) = e^{-\rho t}, \quad y_2(t,\rho) = e^{-i\rho(t-1)}, \quad y_3(t,\rho) = e^{\rho(t-1)}.$$

The characteristic determinant becomes

$$\Delta(\rho) = \widehat{\Delta}(\rho) = \det(B_i(y_k(\cdot\,,\rho)))$$

$$= \det\begin{pmatrix} P_1(i\rho){+}Q_1(i\rho)e^{i\rho} & P_1(-\rho){+}Q_1(-\rho)e^{-\rho} & P_1(-i\rho)e^{i\rho}{+}Q_1(-i\rho) & P_1(\rho)e^{-\rho}{+}Q_1(\rho) \\ P_2(i\rho){+}Q_2(i\rho)e^{i\rho} & P_2(-\rho){+}Q_2(-\rho)e^{-\rho} & P_2(-i\rho)e^{i\rho}{+}Q_2(-i\rho) & P_2(\rho)e^{-\rho}{+}Q_2(\rho) \\ P_3(i\rho){+}Q_3(i\rho)e^{i\rho} & P_3(-\rho){+}Q_3(-\rho)e^{-\rho} & P_3(-i\rho)e^{i\rho}{+}Q_3(-i\rho) & P_3(\rho)e^{-\rho}{+}Q_3(\rho) \\ P_4(i\rho){+}Q_4(i\rho)e^{i\rho} & P_4(-\rho){+}Q_4(-\rho)e^{-\rho} & P_4(-i\rho)e^{i\rho}{+}Q_4(-i\rho) & P_4(\rho)e^{-\rho}{+}Q_4(\rho) \end{pmatrix}$$

for $\rho \neq 0$ in $\mathbb{C}$, while the alternate characteristic determinant is

$$\Delta_*(\rho) = \det(B_i(z_k(\cdot\,,\rho)))$$

$$= \det\begin{pmatrix} P_1(i\rho){+}Q_1(i\rho)e^{i\rho} & P_1(-\rho){+}Q_1(-\rho)e^{-\rho} & P_1(-i\rho){+}Q_1(-i\rho)e^{-i\rho} & P_1(\rho){+}Q_1(\rho)e^{\rho} \\ P_2(i\rho){+}Q_2(i\rho)e^{i\rho} & P_2(-\rho){+}Q_2(-\rho)e^{-\rho} & P_2(-i\rho){+}Q_2(-i\rho)e^{-i\rho} & P_2(\rho){+}Q_2(\rho)e^{\rho} \\ P_3(i\rho){+}Q_3(i\rho)e^{i\rho} & P_3(-\rho){+}Q_3(-\rho)e^{-\rho} & P_3(-i\rho){+}Q_3(-i\rho)e^{-i\rho} & P_3(\rho){+}Q_3(\rho)e^{\rho} \\ P_4(i\rho){+}Q_4(i\rho)e^{i\rho} & P_4(-\rho){+}Q_4(-\rho)e^{-\rho} & P_4(-i\rho){+}Q_4(-i\rho)e^{-i\rho} & P_4(\rho){+}Q_4(\rho)e^{\rho} \end{pmatrix}$$

for $\rho \neq 0$ in $\mathbb{C}$. The two characteristic determinants are related by the equation

$$\Delta(\rho) = e^{i\rho} e^{-\rho} \Delta_*(\rho) \tag{10.23}$$

for $\rho \neq 0$ in $\mathbb{C}$. Observe that

$$\Delta_*(i\rho) = -\Delta_*(\rho), \qquad \Delta_*(-\rho) = \Delta_*(\rho), \qquad \Delta_*(-i\rho) = -\Delta_*(\rho) \tag{10.24}$$

for $\rho \neq 0$ in $\mathbb{C}$.

Suppose we expand $\Delta_*(\rho)$ using linearity in all four columns: $\Delta_*(\rho)$ becomes the sum of sixteen terms; each term is the product of a polynomial times an exponential, where the polynomial is the determinant of a 4×4 matrix with polynomial entries. The sixteen exponentials that appear are listed below:

$$e^{0\rho} = 1,$$

$$e^{i\rho} = e^{i\rho}, \quad e^{-\rho} = e^{-\rho}, \quad e^{-i\rho} = e^{-i\rho}, \quad e^{\rho} = e^{\rho},$$

$$e^{i\rho}e^{-\rho} = e^{i\rho}e^{-\rho}, \quad e^{i\rho}e^{-i\rho} = 1, \quad e^{i\rho}e^{\rho} = e^{i\rho}e^{\rho},$$

$$e^{-\rho}e^{-i\rho} = e^{-\rho}e^{-i\rho}, \quad e^{-\rho}e^{\rho} = 1, \quad e^{-i\rho}e^{\rho} = e^{-i\rho}e^{\rho},$$

$$e^{i\rho}e^{-\rho}e^{-i\rho} = e^{-\rho}, \quad e^{i\rho}e^{-\rho}e^{\rho} = e^{i\rho}, \quad e^{i\rho}e^{-i\rho}e^{\rho} = e^{\rho}, \quad e^{-\rho}e^{-i\rho}e^{\rho} = e^{-i\rho},$$

$$e^{i\rho}e^{-\rho}e^{-i\rho}e^{\rho} = 1.$$

We note that there are only nine distinct exponentials in this list. The exponentials $e^{-i\rho}e^{\rho}$, $e^{i\rho}e^{\rho}$, $e^{i\rho}e^{-\rho}$, $e^{-\rho}e^{-i\rho}$ each occurs once in the list; the exponentials e^{ρ}, $e^{i\rho}$, $e^{-\rho}$, $e^{-i\rho}$ each occurs twice in the list; and the exponential $e^{0\rho} = 1$ occurs four times in the list. Thus, upon expansion $\Delta_*(\rho)$ has the form

$$\begin{aligned} \Delta_*(\rho) = {} & \mathbb{P}_0(\rho)e^{-i\rho}e^{\rho} + \mathbb{P}_1(\rho)e^{i\rho}e^{\rho} + \mathbb{P}_2(\rho)e^{i\rho}e^{-\rho} + \mathbb{P}_3(\rho)e^{-\rho}e^{-i\rho} \\ & + \mathbb{Q}_0(\rho)e^{\rho} + \mathbb{Q}_1(\rho)e^{i\rho} + \mathbb{Q}_2(\rho)e^{-\rho} + \mathbb{Q}_3(\rho)e^{-i\rho} + \mathbb{D}(\rho) \end{aligned} \tag{10.25}$$

for $\rho \neq 0$ in $\mathbb{C}$, where the polynomials $\mathbb{P}_i(\rho)$ are each 4×4 determinants, the polynomials $\mathbb{Q}_i(\rho)$ are each the sum of two 4×4 determinants, and the polynomial $\mathbb{D}(\rho)$ is the sum of four 4×4 determinants.

Specifically, the polynomials $\mathbb{P}_0(\rho)$, $\mathbb{Q}_0(\rho)$, and $\mathbb{D}(\rho)$ are given by the equations

$$\mathbb{P}_0(\rho) = \det\begin{pmatrix} P_1(i\rho) & P_1(-\rho) & Q_1(-i\rho) & Q_1(\rho) \\ P_2(i\rho) & P_2(-\rho) & Q_2(-i\rho) & Q_2(\rho) \\ P_3(i\rho) & P_3(-\rho) & Q_3(-i\rho) & Q_3(\rho) \\ P_4(i\rho) & P_4(-\rho) & Q_4(-i\rho) & Q_4(\rho) \end{pmatrix}, \tag{10.26}$$

$$\begin{aligned} \mathbb{Q}_0(\rho) = {} & \det\begin{pmatrix} P_1(i\rho) & P_1(-\rho) & P_1(-i\rho) & Q_1(\rho) \\ P_2(i\rho) & P_2(-\rho) & P_2(-i\rho) & Q_2(\rho) \\ P_3(i\rho) & P_3(-\rho) & P_3(-i\rho) & Q_3(\rho) \\ P_4(i\rho) & P_4(-\rho) & P_4(-i\rho) & Q_4(\rho) \end{pmatrix} \\ & + \det\begin{pmatrix} Q_1(i\rho) & P_1(-\rho) & Q_1(-i\rho) & Q_1(\rho) \\ Q_2(i\rho) & P_2(-\rho) & Q_2(-i\rho) & Q_2(\rho) \\ Q_3(i\rho) & P_3(-\rho) & Q_3(-i\rho) & Q_3(\rho) \\ Q_4(i\rho) & P_4(-\rho) & Q_4(-i\rho) & Q_4(\rho) \end{pmatrix}, \end{aligned} \tag{10.27}$$

$$\mathbb{D}(\rho) = \det\begin{pmatrix} P_1(i\rho) & P_1(-\rho) & P_1(-i\rho) & P_1(\rho) \\ P_2(i\rho) & P_2(-\rho) & P_2(-i\rho) & P_2(\rho) \\ P_3(i\rho) & P_3(-\rho) & P_3(-i\rho) & P_3(\rho) \\ P_4(i\rho) & P_4(-\rho) & P_4(-i\rho) & P_4(\rho) \end{pmatrix}$$

$$+\det\begin{pmatrix} Q_1(i\rho) & P_1(-\rho) & Q_1(-i\rho) & P_1(\rho) \\ Q_2(i\rho) & P_2(-\rho) & Q_2(-i\rho) & P_2(\rho) \\ Q_3(i\rho) & P_3(-\rho) & Q_3(-i\rho) & P_3(\rho) \\ Q_4(i\rho) & P_4(-\rho) & Q_4(-i\rho) & P_4(\rho) \end{pmatrix}$$

(10.28)

$$+\det\begin{pmatrix} P_1(i\rho) & Q_1(-\rho) & P_1(-i\rho) & Q_1(\rho) \\ P_2(i\rho) & Q_2(-\rho) & P_2(-i\rho) & Q_2(\rho) \\ P_3(i\rho) & Q_3(-\rho) & P_3(-i\rho) & Q_3(\rho) \\ P_4(i\rho) & Q_4(-\rho) & P_4(-i\rho) & Q_4(\rho) \end{pmatrix}$$

$$+\det\begin{pmatrix} Q_1(i\rho) & Q_1(-\rho) & Q_1(-i\rho) & Q_1(\rho) \\ Q_2(i\rho) & Q_2(-\rho) & Q_2(-i\rho) & Q_2(\rho) \\ Q_3(i\rho) & Q_3(-\rho) & Q_3(-i\rho) & Q_3(\rho) \\ Q_4(i\rho) & Q_4(-\rho) & Q_4(-i\rho) & Q_4(\rho) \end{pmatrix}.$$

From the definitions of the various polynomials we can show that

$$\mathbb{P}_1(\rho) = -\mathbb{P}_0(i\rho), \qquad \mathbb{P}_2(\rho) = \mathbb{P}_0(-\rho), \qquad \mathbb{P}_3(\rho) = -\mathbb{P}_0(-i\rho), \tag{10.29}$$

$$\mathbb{Q}_1(\rho) = -\mathbb{Q}_0(i\rho), \qquad \mathbb{Q}_2(\rho) = \mathbb{Q}_0(-\rho), \qquad \mathbb{Q}_3(\rho) = -\mathbb{Q}_0(-i\rho). \tag{10.30}$$

Indeed, to prove that $\mathbb{P}_1(\rho) = -\mathbb{P}_0(i\rho)$, we see from (10.25) that $\mathbb{P}_1(\rho)$ is the coefficient of the exponential $e^{i\rho}e^{\rho}$, and hence,

$$\mathbb{P}_1(\rho) = \det\begin{pmatrix} Q_1(i\rho) & P_1(-\rho) & P_1(-i\rho) & Q_1(\rho) \\ Q_2(i\rho) & P_2(-\rho) & P_2(-i\rho) & Q_2(\rho) \\ Q_3(i\rho) & P_3(-\rho) & P_3(-i\rho) & Q_3(\rho) \\ Q_4(i\rho) & P_4(-\rho) & P_4(-i\rho) & Q_4(\rho) \end{pmatrix}.$$

But from (10.26)

$$-\mathbb{P}_0(i\rho) = -\det\begin{pmatrix} P_1(-\rho) & P_1(-i\rho) & Q_1(\rho) & Q_1(i\rho) \\ P_2(-\rho) & P_2(-i\rho) & Q_2(\rho) & Q_2(i\rho) \\ P_3(-\rho) & P_3(-i\rho) & Q_3(\rho) & Q_3(i\rho) \\ P_4(-\rho) & P_4(-i\rho) & Q_4(\rho) & Q_4(i\rho) \end{pmatrix},$$

and the result is now clear. In view of (10.29) and (10.30), to calculate the polynomial coefficients in equation (10.25), it is sufficient to calculate the polynomials $\mathbb{P}_0(\rho)$, $\mathbb{Q}_0(\rho)$, and $\mathbb{D}(\rho)$. Let us proceed with this calculation.

For the 4×8 boundary coefficient matrix A, denote the eight columns of A by α_3, β_3, α_2, β_2, α_1, β_1, α_0, β_0, and let $\langle a, b, c, d\rangle$ denote the determinant of the 4×4

submatrix of A formed by using columns a, b, c, d, e.g.,

$$\langle \alpha_2, \beta_3, \alpha_0, \beta_2 \rangle = \det \begin{pmatrix} \alpha_{12} & \beta_{13} & \alpha_{10} & \beta_{12} \\ \alpha_{22} & \beta_{23} & \alpha_{20} & \beta_{22} \\ \alpha_{32} & \beta_{33} & \alpha_{30} & \beta_{32} \\ \alpha_{42} & \beta_{43} & \alpha_{40} & \beta_{42} \end{pmatrix}.$$

Then using a straightforward but lengthy calculation, we get

$$\begin{aligned} \mathbb{P}_0(\rho) &= 2i\langle \alpha_3, \beta_3, \alpha_2, \beta_2 \rangle \rho^{10} \\ &\quad - (2-2i)[\langle \alpha_3, \beta_3, \alpha_2, \beta_1 \rangle + \langle \alpha_3, \beta_3, \beta_2, \alpha_1 \rangle]\rho^9 \\ &\quad + 2[-\langle \alpha_3, \beta_3, \alpha_2, \beta_0 \rangle + \langle \alpha_3, \beta_3, \beta_2, \alpha_0 \rangle + 2\langle \alpha_3, \beta_3, \alpha_1, \beta_1 \rangle \\ &\quad + \langle \alpha_3, \alpha_2, \beta_2, \beta_1 \rangle - \langle \beta_3, \alpha_2, \beta_2, \alpha_1 \rangle]\rho^8 \\ &\quad + (2+2i)[\langle \alpha_3, \beta_3, \alpha_1, \beta_0 \rangle + \langle \alpha_3, \beta_3, \beta_1, \alpha_0 \rangle + \langle \alpha_3, \alpha_2, \beta_2, \beta_0 \rangle \\ &\quad + \langle \alpha_3, \beta_2, \alpha_1, \beta_1 \rangle + \langle \beta_3, \alpha_2, \beta_2, \alpha_0 \rangle + \langle \beta_3, \alpha_2, \alpha_1, \beta_1 \rangle]\rho^7 \\ &\quad + 2i[-\langle \alpha_3, \beta_3, \alpha_0, \beta_0 \rangle + \langle \alpha_3, \alpha_2, \beta_1, \beta_0 \rangle + 2\langle \alpha_3, \beta_2, \alpha_1, \beta_0 \rangle \\ &\quad + \langle \alpha_3, \beta_2, \beta_1, \alpha_0 \rangle + \langle \beta_3, \alpha_2, \alpha_1, \beta_0 \rangle + 2\langle \beta_3, \alpha_2, \beta_1, \alpha_0 \rangle \\ &\quad + \langle \beta_3, \beta_2, \alpha_1, \alpha_0 \rangle - \langle \alpha_2, \beta_2, \alpha_1, \beta_1 \rangle]\rho^6 \\ &\quad + (2-2i)[\langle \alpha_3, \beta_2, \alpha_0, \beta_0 \rangle + \langle \alpha_3, \alpha_1, \beta_1, \beta_0 \rangle + \langle \beta_3, \alpha_2, \alpha_0, \beta_0 \rangle \\ &\quad + \langle \beta_3, \alpha_1, \beta_1, \alpha_0 \rangle + \langle \alpha_2, \beta_2, \alpha_1, \beta_0 \rangle + \langle \alpha_2, \beta_2, \beta_1, \alpha_0 \rangle]\rho^5 \\ &\quad + 2[\langle \alpha_3, \beta_1, \alpha_0, \beta_0 \rangle - \langle \beta_3, \alpha_1, \alpha_0, \beta_0 \rangle - 2\langle \alpha_2, \beta_2, \alpha_0, \beta_0 \rangle \\ &\quad - \langle \alpha_2, \alpha_1, \beta_1, \beta_0 \rangle + \langle \beta_2, \alpha_1, \beta_1, \alpha_0 \rangle]\rho^4 \\ &\quad - (2+2i)[\langle \alpha_2, \beta_1, \alpha_0, \beta_0 \rangle + \langle \beta_2, \alpha_1, \alpha_0, \beta_0 \rangle]\rho^3 \\ &\quad + 2i\langle \alpha_1, \beta_1, \alpha_0, \beta_0 \rangle \rho^2, \end{aligned} \tag{10.31}$$

$$\begin{aligned} \mathbb{Q}_0(\rho) &= 4i[\langle \alpha_3, \beta_3, \alpha_2, \alpha_1 \rangle + \langle \alpha_3, \beta_3, \beta_2, \beta_1 \rangle]\rho^9 \\ &\quad - 4i[\langle \alpha_3, \beta_3, \alpha_2, \alpha_0 \rangle - \langle \alpha_3, \beta_3, \beta_2, \beta_0 \rangle \\ &\quad + \langle \alpha_3, \alpha_2, \beta_2, \alpha_1 \rangle - \langle \beta_3, \alpha_2, \beta_2, \beta_1 \rangle]\rho^8 \\ &\quad + 4i[\langle \alpha_3, \beta_3, \alpha_1, \alpha_0 \rangle + \langle \alpha_3, \beta_3, \beta_1, \beta_0 \rangle + \langle \alpha_3, \alpha_2, \beta_2, \alpha_0 \rangle \\ &\quad + \langle \alpha_3, \alpha_2, \alpha_1, \beta_1 \rangle + \langle \beta_3, \alpha_2, \beta_2, \beta_0 \rangle + \langle \beta_3, \beta_2, \alpha_1, \beta_1 \rangle]\rho^7 \\ &\quad + 4i[\langle \alpha_3, \alpha_2, \alpha_1, \beta_0 \rangle + \langle \alpha_3, \alpha_2, \beta_1, \alpha_0 \rangle + \langle \alpha_3, \beta_2, \alpha_1, \alpha_0 \rangle \\ &\quad + \langle \alpha_3, \beta_2, \beta_1, \beta_0 \rangle + \langle \beta_3, \alpha_2, \alpha_1, \alpha_0 \rangle + \langle \beta_3, \alpha_2, \beta_1, \beta_0 \rangle \\ &\quad + \langle \beta_3, \beta_2, \alpha_1, \beta_0 \rangle + \langle \beta_3, \beta_2, \beta_1, \alpha_0 \rangle]\rho^6 \\ &\quad - 4i[\langle \alpha_3, \alpha_2, \alpha_0, \beta_0 \rangle + \langle \beta_3, \beta_2, \alpha_0, \beta_0 \rangle + \langle \alpha_3, \alpha_1, \beta_1, \alpha_0 \rangle \\ &\quad + \langle \beta_3, \alpha_1, \beta_1, \beta_0 \rangle + \langle \alpha_2, \beta_2, \alpha_1, \alpha_0 \rangle + \langle \alpha_2, \beta_2, \beta_1, \beta_0 \rangle]\rho^5 \\ &\quad + 4i[\langle \alpha_3, \alpha_1, \alpha_0, \beta_0 \rangle - \langle \beta_3, \beta_1, \alpha_0, \beta_0 \rangle \\ &\quad + \langle \alpha_2, \alpha_1, \beta_1, \alpha_0 \rangle - \langle \beta_2, \alpha_1, \beta_1, \beta_0 \rangle]\rho^4 \\ &\quad - 4i[\langle \alpha_2, \alpha_1, \alpha_0, \beta_0 \rangle + \langle \beta_2, \beta_1, \alpha_0, \beta_0 \rangle]\rho^3, \end{aligned} \tag{10.32}$$

and

$$\begin{aligned}\mathbb{D}(\rho) = & -8i\langle\alpha_3,\beta_3,\alpha_2,\beta_2\rangle\rho^{10} \\ & + 8i[\langle\alpha_3,\beta_3,\alpha_0,\beta_0\rangle + 2\langle\alpha_3,\alpha_2,\alpha_1,\alpha_0\rangle + \langle\alpha_3,\alpha_2,\beta_1,\beta_0\rangle \\ & + \langle\alpha_3,\beta_2,\beta_1,\alpha_0\rangle + \langle\beta_3,\alpha_2,\alpha_1,\beta_0\rangle + \langle\beta_3,\beta_2,\alpha_1,\alpha_0\rangle \\ & + 2\langle\beta_3,\beta_2,\beta_1,\beta_0\rangle + \langle\alpha_2,\beta_2,\alpha_1,\beta_1\rangle]\rho^6 \\ & - 8i\langle\alpha_1,\beta_1,\alpha_0,\beta_0\rangle\rho^2.\end{aligned} \tag{10.33}$$

From equations (10.23) and (10.25) we see that

$$\begin{aligned}\Delta(\rho) &= e^{i\rho}e^{-\rho}\Delta_*(\rho) \\ &= \mathbb{P}_0(\rho) + \mathbb{P}_1(\rho)e^{2i\rho} + \mathbb{P}_2(\rho)e^{2i\rho}e^{-2\rho} + \mathbb{P}_3(\rho)e^{-2\rho} \\ &\quad + \mathbb{Q}_0(\rho)e^{i\rho} + \mathbb{Q}_1(\rho)e^{2i\rho}e^{-\rho} + \mathbb{Q}_2(\rho)e^{i\rho}e^{-2\rho} + \mathbb{Q}_3(\rho)e^{-\rho} + \mathbb{D}(\rho)e^{i\rho}e^{-\rho},\end{aligned}$$

or

$$\begin{aligned}\Delta(\rho) = & \mathbb{P}_1(\rho)e^{2i\rho} + \mathbb{Q}_0(\rho)e^{i\rho} + \mathbb{P}_0(\rho) \\ & + [\mathbb{P}_2(\rho)e^{-2\rho} + \mathbb{Q}_1(\rho)e^{-\rho}]e^{2i\rho} \\ & + [\mathbb{Q}_2(\rho)e^{-2\rho} + \mathbb{D}(\rho)e^{-\rho}]e^{i\rho} \\ & + [\mathbb{P}_3(\rho)e^{-2\rho} + \mathbb{Q}_3(\rho)e^{-\rho}]\end{aligned} \tag{10.34}$$

for $\rho \neq 0$ in $\mathbb{C}$. This is equation (3.32) or equation (5.31) for the special case $n = 4$ and $L = T$ with

$$\pi_2(\rho) = \mathbb{P}_1(\rho) = -\mathbb{P}_0(i\rho), \qquad \pi_1(\rho) = \mathbb{Q}_0(\rho), \qquad \pi_0(\rho) = \mathbb{P}_0(\rho) \tag{10.35}$$

for $\rho \neq 0$ in $\mathbb{C}$. Cf. equation (3.18).

We know that a nonzero complex number $\lambda = \rho^4$ is an eigenvalue of L if and only if $\Delta(\rho) = 0$, or equivalently, if and only if $\Delta_*(\rho) = 0$. To determine if $\lambda = 0$ is an eigenvalue for L, we use the basis $\phi_0(t) = 1$, $\phi_1(t) = t$, $\phi_2(t) = t^2$, $\phi_3(t) = t^3$ for the solution space of the differential equation $-u^{(4)}(t) = 0$. Clearly

$$\begin{aligned}B_i(\phi_0(\,\cdot\,)) &= \alpha_{i0} + \beta_{i0}, \\ B_i(\phi_1(\,\cdot\,)) &= \alpha_{i1} + \beta_{i0} + \beta_{i1}, \\ B_i(\phi_2(\,\cdot\,)) &= 2\alpha_{i2} + \beta_{i0} + 2\beta_{i1} + 2\beta_{i2}, \\ B_i(\phi_3(\,\cdot\,)) &= 6\alpha_{i3} + \beta_{i0} + 3\beta_{i1} + 6\beta_{i2} + 6\beta_{i3}\end{aligned}$$

for $i = 1, \dots, 4$. In terms of these quantities we form the determinant

$$\Delta_0 := \det(B_i(\phi_k(\,\cdot\,))).$$

Then $\lambda = 0$ is an eigenvalue for L if and only if $\Delta_0 = 0$. Upon expansion, we obtain the following expression for the constant Δ_0:

$$
\begin{aligned}
\Delta_0 = 12[&\langle\alpha_3,\alpha_2,\alpha_1,\alpha_0\rangle + \langle\alpha_3,\alpha_2,\alpha_1,\beta_0\rangle + \langle\alpha_3,\alpha_2,\beta_1,\alpha_0\rangle \\
&+ \langle\alpha_3,\beta_2,\alpha_1,\alpha_0\rangle + \langle\beta_3,\alpha_2,\alpha_1,\alpha_0\rangle + \langle\alpha_3,\alpha_2,\beta_1,\beta_0\rangle \\
&+ \langle\alpha_3,\beta_2,\alpha_1,\beta_0\rangle + \langle\alpha_3,\beta_2,\beta_1,\alpha_0\rangle + \langle\beta_3,\alpha_2,\alpha_1,\beta_0\rangle \\
&+ \langle\beta_3,\alpha_2,\beta_1,\alpha_0\rangle + \langle\beta_3,\beta_2,\alpha_1,\alpha_0\rangle + \langle\alpha_3,\beta_2,\beta_1,\beta_0\rangle \\
&+ \langle\beta_3,\alpha_2,\beta_1,\beta_0\rangle + \langle\beta_3,\beta_2,\alpha_1,\beta_0\rangle + \langle\beta_3,\beta_2,\beta_1,\alpha_0\rangle \\
&+ \langle\beta_3,\beta_2,\beta_1,\beta_0\rangle] \\
&- 12[\langle\alpha_3,\alpha_2,\alpha_0,\beta_0\rangle + \langle\alpha_3,\alpha_1,\beta_1,\alpha_0\rangle + \langle\alpha_2,\beta_2,\alpha_1,\alpha_0\rangle \\
&+ \langle\alpha_3,\alpha_1,\beta_1,\beta_0\rangle + \langle\alpha_3,\beta_2,\alpha_0,\beta_0\rangle + \langle\alpha_2,\beta_2,\alpha_1,\beta_0\rangle \\
&+ \langle\beta_3,\alpha_2,\alpha_0,\beta_0\rangle + \langle\alpha_2,\beta_2,\beta_1,\alpha_0\rangle + \langle\beta_3,\alpha_1,\beta_1,\alpha_0\rangle \\
&+ \langle\alpha_2,\beta_2,\beta_1,\beta_0\rangle + \langle\beta_3,\alpha_1,\beta_1,\beta_0\rangle + \langle\beta_3,\beta_2,\alpha_0,\beta_0\rangle] \\
&+ 6[\langle\alpha_3,\alpha_1,\alpha_0,\beta_0\rangle + \langle\alpha_2,\alpha_1,\beta_1,\alpha_0\rangle - \langle\alpha_3,\beta_1,\alpha_0,\beta_0\rangle \\
&+ \langle\alpha_2,\alpha_1,\beta_1,\beta_0\rangle + 2\langle\alpha_2,\beta_2,\alpha_0,\beta_0\rangle + \langle\beta_3,\alpha_1,\alpha_0,\beta_0\rangle \\
&- \langle\beta_2,\alpha_1,\beta_1,\alpha_0\rangle - \langle\beta_2,\alpha_1,\beta_1,\beta_0\rangle - \langle\beta_3,\beta_1,\alpha_0,\beta_0\rangle] \\
&- 2[\langle\alpha_2,\alpha_1,\alpha_0,\beta_0\rangle - 2\langle\alpha_2,\beta_1,\alpha_0,\beta_0\rangle \\
&- 2\langle\beta_2,\alpha_1,\alpha_0,\beta_0\rangle + \langle\beta_2,\beta_1,\alpha_0,\beta_0\rangle] \\
&- \langle\alpha_1,\beta_1,\alpha_0,\beta_0\rangle.
\end{aligned}
\tag{10.36}
$$

EXAMPLE 10.3. Consider the 4th order differential operator $L = T$ determined by $\ell = \tau = (d/dt)^4$ and the boundary values

$$B_1(u) = u'''(0) + 6u(0), \quad B_2(u) = u''(0), \quad B_3(u) = u'(0), \quad B_4(u) = u(1),$$

so the boundary coefficient matrix is

$$A = \begin{pmatrix} 1 & 0 & 0 & 0 & 0 & 0 & 6 & 0 \\ 0 & 0 & 1 & 0 & 0 & 0 & 0 & 0 \\ 0 & 0 & 0 & 0 & 1 & 0 & 0 & 0 \\ 0 & 0 & 0 & 0 & 0 & 0 & 0 & 1 \end{pmatrix}.$$

Clearly $p_0 = 6$, and $\langle\alpha_3,\alpha_2,\alpha_1,\beta_0\rangle = 1$ and $\langle\alpha_2,\alpha_1,\alpha_0,\beta_0\rangle = 6$, with all the other determinants $\langle a,b,c,d\rangle$ equal to 0. Thus, from equations (10.31)–(10.33) we have

$$\mathbb{P}_0(\rho) = 0, \qquad \mathbb{Q}_0(\rho) = 4i\rho^6 - 24i\rho^3, \qquad \mathbb{D}(\rho) = 0,$$

and then equations (10.34) and (10.29), (10.30) produce the characteristic determinant

$$
\begin{aligned}
\Delta(\rho) = (4i\rho^6 - 24i\rho^3)e^{i\rho} + (4i\rho^6 + 24\rho^3)e^{-\rho}e^{2i\rho} \\
+ (4i\rho^6 + 24i\rho^3)e^{-2\rho}e^{i\rho} + (4i\rho^6 - 24\rho^3)e^{-\rho}
\end{aligned}
$$

for $\rho \neq 0$ in $\mathbb{C}$. Since $\pi_2(\rho) = -\mathbb{P}_0(i\rho) = 0$ and $\pi_0(\rho) = \mathbb{P}_0(\rho) = 0$, the differential operator L belongs to the degenerate irregular class. Later in this section we will study this differential operator as a model for a more general class (Case II below), showing that it has a countably infinite spectrum that lies near the negative real axis. This is the first time that we have encountered a differential operator L of even order $n = 2\nu$ with its spectrum lying near the *negative* real axis.

Next, we reexamine our classification scheme of Chapter 3 by exploiting the explicit forms of the polynomials $\mathbb{P}_0(\rho)$, $\mathbb{Q}_0(\rho)$, and $\mathbb{D}(\rho)$. From (10.31)–(10.33) we see immediately that

$$\begin{aligned} 2 \le \text{degree } \mathbb{P}_0(\rho) \le 10 \quad &\text{or} \quad \mathbb{P}_0(\rho) \equiv 0, \\ 3 \le \text{degree } \mathbb{Q}_0(\rho) \le 9 \quad &\text{or} \quad \mathbb{Q}_0(\rho) \equiv 0, \\ 2 \le \text{degree } \mathbb{D}_0(\rho) \le 10 \quad &\text{or} \quad \mathbb{D}_0(\rho) \equiv 0. \end{aligned}$$

Consider the following cases.

Case I. $n = 4$, $\mathbb{P}_0(\rho) \not\equiv 0$. From the relations (10.35) we see that the polynomials $\pi_0(\rho) = \mathbb{P}_0(\rho)$ and $\pi_2(\rho) = -\mathbb{P}_0(i\rho)$ are both of degree p with $2 \le p \le 10$. According to Definition 3.2, the fourth order differential operator $L = T$ is either regular or simply irregular. This case has been studied extensively in the previous chapters.

Case II. $n = 4$, $\mathbb{P}_0(\rho) \equiv 0$, $\mathbb{Q}_0(\rho) \not\equiv 0$. For this case the polynomials $\pi_0(\rho) = \mathbb{P}_0(\rho)$ and $\pi_2(\rho) = -\mathbb{P}_0(i\rho)$ are identically zero, and hence, by Definition 3.2 the differential operator $L = T$ is degenerate irregular. Example 10.3 provides a model for this case. We are going to show that the differential operators belonging to Case II have some very unusual properties.

From equation (10.34) the characteristic determinant now takes the form

$$\begin{aligned} \Delta(\rho) = \mathbb{Q}_0(\rho)e^{i\rho} + \mathbb{Q}_1(\rho)e^{-\rho}e^{2i\rho} \\ + [\mathbb{Q}_2(\rho)e^{-2\rho} + \mathbb{D}(\rho)e^{-\rho}]e^{i\rho} + \mathbb{Q}_3(\rho)e^{-\rho} \end{aligned}$$

for $\rho \neq 0$ in $\mathbb{C}$. Let q denote the degree of the polynomial $\mathbb{Q}_0(\rho)$, so $3 \le q \le 9$ and

$$\mathbb{Q}_0(\rho) = \pi_1(\rho) = \sum_{\kappa=3}^{q} b_\kappa \rho^\kappa, \qquad b_q \neq 0.$$

We know that $\mathbb{Q}_1(\rho) = -\mathbb{Q}_0(i\rho)$, $\mathbb{Q}_2(\rho) = \mathbb{Q}_0(-\rho)$, and $\mathbb{Q}_3(\rho) = -\mathbb{Q}_0(-i\rho)$. Also, from (10.31) we have $\langle \alpha_3, \beta_3, \alpha_2, \beta_2 \rangle = 0$ and $\langle \alpha_1, \beta_1, \alpha_0, \beta_0 \rangle = 0$, and hence, setting

$$\begin{aligned} \gamma_0 := \langle \alpha_3, \beta_3, \alpha_0, \beta_0 \rangle + 2\langle \alpha_3, \alpha_2, \alpha_1, \alpha_0 \rangle + \langle \alpha_3, \alpha_2, \beta_1, \beta_0 \rangle \\ + \langle \alpha_3, \beta_2, \beta_1, \alpha_0 \rangle + \langle \beta_3, \alpha_2, \alpha_1, \beta_0 \rangle + \langle \beta_3, \beta_2, \alpha_1, \alpha_0 \rangle \\ + 2\langle \beta_3, \beta_2, \beta_1, \beta_0 \rangle + \langle \alpha_2, \beta_2, \alpha_1, \beta_1 \rangle, \end{aligned}$$

we see that $\mathbb{D}(\rho) = 8i\gamma_0\rho^6$. Fix a real number σ_0 with $0 < \sigma_0 < \pi/10$.

First, we introduce the sector

$$\Sigma_0\colon \text{ all } \rho = |\rho|e^{i\theta} \in \mathbb{C} \text{ with } -\frac{\pi}{4} + \frac{\sigma_0}{4} \le \theta \le \frac{\pi}{4} - \frac{\sigma_0}{4},$$

and then proceed to show that $\Delta(\rho)$ has no zeros in the sector Σ_0 for $|\rho|$ sufficiently large. To study the behavior of $\Delta(\rho)$ on the sector Σ_0, we first rewrite it in the alternate form

$$\begin{aligned} \Delta(\rho) = \rho^q e^{i\rho} \Big\{ \frac{\mathbb{Q}_0(\rho)}{\rho^q} + \frac{\mathbb{Q}_1(\rho)}{\rho^q} e^{-\rho}e^{i\rho} \\ + \frac{\mathbb{Q}_2(\rho)}{\rho^q} e^{-2\rho} + \frac{\mathbb{D}(\rho)}{\rho^q} e^{-\rho} + \frac{\mathbb{Q}_3(\rho)}{\rho^q} e^{-\rho}e^{-i\rho} \Big\} \end{aligned}$$

for $\rho \neq 0$ in $\mathbb{C}$. Set $\alpha := \sin(\pi/4 + \sigma_0/4) > 0$ and $\beta := \sin(\pi/4 - \sigma_0/4) > 0$, and take any point $\rho = a + ib = |\rho|e^{i\theta} \neq 0$ in $\mathbb{C}$ with $-\pi/4 + \sigma_0/4 \le \theta \le \pi/4 - \sigma_0/4$.

Clearly

$$\begin{aligned} a = |a| &= |\rho|\cos\theta \geq |\rho|\cos(\pi/4 - \sigma_0/4) = |\rho|\sin(\pi/4 + \sigma_0/4) = \alpha|\rho|, \\ |b| &= |\rho||\sin\theta| \leq |\rho|\sin(\pi/4 - \sigma_0/4) = \beta|\rho|, \end{aligned}$$

and

$$\begin{aligned} &|e^{-\rho}e^{i\rho}| = e^{-a}e^{-b} \leq e^{-(\alpha-\beta)|\rho|}, && |e^{-2\rho}| = e^{-2a} \leq e^{-2\alpha|\rho|} \leq e^{-(\alpha-\beta)|\rho|}, \\ &|e^{-\rho}| = e^{-a} \leq e^{-\alpha|\rho|} \leq e^{-(\alpha-\beta)|\rho|}, && |e^{-\rho}e^{-i\rho}| = e^{-a}e^{b} \leq e^{-(\alpha-\beta)|\rho|}. \end{aligned}$$

Thus,

$$\begin{aligned} |\Delta(\rho)| &\geq |\rho|^q e^{-b}\left\{|b_q| - \frac{\gamma}{|\rho|} - 4\gamma e^{-\frac{1}{2}(\alpha-\beta)|\rho|}\right\} \\ &\geq \frac{|b_q|}{2}|\rho|^q e^{-b} > 0 \end{aligned} \tag{10.37}$$

for all ρ in the sector Σ_0 with $|\rho|$ sufficiently large.

Second, we introduce the sector

$$\Sigma_1\colon\ \text{all } \rho = |\rho|e^{i\theta} \in \mathbb{C} \text{ with } \frac{\pi}{4} - \frac{\sigma_0}{4} \leq \theta \leq \frac{\pi}{4} + \frac{\sigma_0}{4}.$$

We claim that $\Delta(\rho)$ has an infinite sequence of zeros in the sector Σ_1. In treating the sector Σ_1, we express $\Delta(\rho)$ in the form

$$\begin{aligned} \Delta(\rho) = e^{-\rho}\{&\mathbb{Q}_0(\rho)e^{\rho}e^{i\rho} + \mathbb{Q}_3(\rho) \\ &+ \mathbb{Q}_1(\rho)e^{2i\rho} + \mathbb{Q}_2(\rho)e^{-\rho}e^{i\rho} + \mathbb{D}(\rho)e^{i\rho}\} \end{aligned} \tag{10.38}$$

for $\rho \neq 0$ in $\mathbb{C}$. Take any point $\rho = a + ib = |\rho|e^{i\theta} \neq 0$ in the sector Σ_1, so $\pi/4 - \sigma_0/4 \leq \theta \leq \pi/4 + \sigma_0/4$. Then

$$\begin{aligned} a = |a| &= |\rho|\cos\theta \geq |\rho|\cos(\pi/4 + \sigma_0/4) = |\rho|\sin(\pi/4 - \sigma_0/4) = \beta|\rho|, \\ b = |b| &= |\rho|\sin\theta \geq |\rho|\sin(\pi/4 - \sigma_0/4) = \beta|\rho|, \end{aligned}$$

and

$$\begin{gathered} |e^{\rho}e^{i\rho}| = e^{a}e^{-b}, \quad |e^{0\rho}| = 1, \\ |e^{2i\rho}| = e^{-2b} \leq e^{-2\beta|\rho|} \leq e^{-\beta|\rho|}, \quad |e^{-\rho}e^{i\rho}| = e^{-a}e^{-b} \leq e^{-2\beta|\rho|} \leq e^{-\beta|\rho|}, \\ |e^{i\rho}| = e^{-b} \leq e^{-\beta|\rho|}. \end{gathered}$$

On the ray $\arg\rho = \pi/4$ the exponentials $e^{\rho}e^{i\rho}$ and $e^{0\rho} = 1$ have modulus 1.

Next, we introduce the sector

$$\Sigma_{\oplus}\colon\ \text{all } z = |z|e^{i\phi} \in \mathbb{C} \text{ with } -\frac{\sigma_0}{4} \leq \phi \leq \frac{\sigma_0}{4}.$$

Set $\omega := (1+i)/2$, and let us make the change of variable

$$\rho = \omega z, \qquad |\rho| = \frac{|z|}{\sqrt{2}}.$$

Clearly $\rho \in \Sigma_1$ if and only if $z \in \Sigma_\oplus$. Now

$$(1+i)\rho = (1+i)\left(\frac{1+i}{2}\right)z = iz, \quad 0\rho = 0z = 0,$$
$$2i\rho = 2i\left(\frac{1+i}{2}\right)z = (-1+i)z, \quad (-1+i)\rho = (-1+i)\left(\frac{1+i}{2}\right)z = -z,$$
$$i\rho = i\left(\frac{1+i}{2}\right)z = \frac{1}{2}(-1+i)z,$$

so the above exponentials transform to

$$e^{iz},\ |e^{iz}| = e^{-\operatorname{Im} z}, \qquad e^{0z} = 1,\ |e^{0z}| = 1,$$
$$e^{-z}e^{iz},\ |e^{-z}e^{iz}| \le e^{-\frac{\beta}{\sqrt{2}}|z|}, \qquad e^{-z},\ |e^{-z}| \le e^{-\frac{\beta}{\sqrt{2}}|z|},$$
$$e^{-\frac{1}{2}z}e^{\frac{i}{2}z},\ |e^{-\frac{1}{2}z}e^{\frac{i}{2}z}| \le e^{-\frac{\beta}{\sqrt{2}}|z|},$$

the estimates being valid for all z in the sector $\Sigma_\oplus$.

At this point we introduce a modified form of the characteristic determinant based on the representation (10.38):

$$\begin{aligned} \Delta_\oplus(z) &:= \Delta(\omega z) \\ &= e^{-\omega z}\{\mathbb{Q}_0(\omega z)e^{iz} + \mathbb{Q}_3(\omega z) \\ &\quad + \mathbb{Q}_1(\omega z)e^{-z}e^{iz} + \mathbb{Q}_2(\omega z)e^{-z} + \mathbb{D}(\omega z)e^{-\frac{1}{2}z}e^{\frac{i}{2}z}\} \end{aligned} \tag{10.39}$$

for $z \ne 0$ in $\mathbb{C}$. In this last equation we have

$$\mathbb{Q}_0(\omega z) = \sum_{\kappa=3}^{q} b_\kappa \omega^\kappa z^\kappa, \quad \mathbb{Q}_3(\omega z) = -\mathbb{Q}_0(-i\omega z) = -\sum_{\kappa=3}^{q} b_\kappa(-i)^\kappa \omega^\kappa z^\kappa.$$

Set $\xi_0 := (-i)^q$,

$$f(z) := b_q\omega^q e^{iz} - b_q(-i)^q\omega^q = b_q\omega^q\left[e^{iz} - \xi_0\right]$$

for $z \in \mathbb{C}$, and

$$\begin{aligned} g(z) :=& \sum_{\kappa=3}^{q-1} \frac{b_\kappa\omega^\kappa}{z^{q-\kappa}} e^{iz} - \sum_{\kappa=3}^{q-1} \frac{b_\kappa(-i)^\kappa\omega^\kappa}{z^{q-k}} \\ &+ \frac{1}{z^q}\left[\mathbb{Q}_1(\omega z)e^{-z}e^{iz} + \mathbb{Q}_2(\omega z)e^{-z} + \mathbb{D}(\omega z)e^{-\frac{1}{2}z}e^{\frac{i}{2}z}\right] \end{aligned}$$

for $z \ne 0$ in $\mathbb{C}$. Then equation (10.39) can be rewritten in the simpler form

$$\Delta_\oplus(z) = z^q e^{-\omega z}\left[f(z) + g(z)\right] \tag{10.40}$$

for $z \ne 0$ in $\mathbb{C}$, where

$$|e^{-z}e^{iz}| \le e^{-\frac{\beta}{\sqrt{2}}|z|}, \qquad |e^{-z}| \le e^{-\frac{\beta}{\sqrt{2}}|z|}, \qquad |e^{-\frac{1}{2}z}e^{\frac{i}{2}z}| \le e^{-\frac{\beta}{\sqrt{2}}|z|}$$

for z in the sector $\Sigma_\oplus$. Clearly the zeros of $f(z)$ are given by the sequence

$$\mu_k = 2\pi k + \operatorname{Arg}\xi_0, \qquad k = 0, \pm 1, \pm 2, \ldots.$$

Each μ_k is a zero of order 1 of f, and all of these zeros are real.

Proceeding as in Chapter 8, Case 1, it follows that $\Delta_{\oplus}(z)$ has a sequence of zeros z_k, $k = k_0, k_0 + 1, \ldots$, in the sector $\Sigma_{\oplus}$ with each z_k being a zero of order 1 of $\Delta_{\oplus}(z)$. These zeros satisfy the asymptotic formulas

$$|z_k - \mu_k| \leq \frac{\gamma}{k}, \qquad k = k_0, k_0 + 1, \ldots, \tag{10.41}$$

and are approaching the positive real axis as $k \to \infty$. It is immediate that the characteristic determinant $\Delta(\rho)$ has a sequence of zeros

$$\rho_k = \omega z_k, \qquad k = k_0, k_0 + 1, \ldots,$$

in the sector Σ_1, each ρ_k being a zero of order 1 of $\Delta(\rho)$, with the ρ_k approaching the ray $\arg \rho = \pi/4$ as $k \to \infty$. The sequence

$$\lambda_k = (\rho_k)^4, \qquad k = k_0, k_0 + 1, \ldots,$$

is a sequence of eigenvalues for the differential operator $L = T$, with the λ_k approaching the *negative* real axis as $k \to \infty$. These eigenvalues account for all but a finite number of the eigenvalues of L. The corresponding algebraic multiplicities and ascents are

$$\nu(\lambda_k) = m(\lambda_k) = 1, \qquad k = k_0, k_0 + 1, \ldots.$$

In Case II we have found only one sequence of eigenvalues, with these eigenvalues approaching in the limit the negative real axis. This behavior is very different from our previous work where L is either regular or simply irregular. Cf. Theorem 7.2, Theorem 7.3, and Theorem 7.5.

Case III. $n = 4$, $\mathbb{P}_0(\rho) \equiv 0$, $\mathbb{Q}_0(\rho) \equiv 0$. For this case equation (10.34) for the characteristic determinant simplifies dramatically to

$$\Delta(\rho) = \mathbb{D}(\rho)e^{-\rho}e^{i\rho} = 8i\gamma_0\rho^6 e^{-\rho}e^{i\rho} \tag{10.42}$$

for $\rho \neq 0$ in $\mathbb{C}$, where γ_0 is the constant defined in Case II. Thus, the nonzero part of the spectrum $\sigma(L)$ is determined by the constant γ_0. Again by Definition 3.2 the differential operator $L = T$ is degenerate irregular.

Now in our previous work we have shown that $\lambda = 0$ is an eigenvalue for L if and only if $\Delta_0 = 0$, where the constant Δ_0 is given by equation (10.36). We assert that $\Delta_0 = 6\gamma_0$, and hence, $\lambda = 0$ is an eigenvalue for L if and only if $\gamma_0 = 0$. First, since the ρ^2 term is missing in (10.31), we must have

$$\langle \alpha_1, \beta_1, \alpha_0, \beta_0 \rangle = 0. \tag{10.43}$$

Second, in (10.31) and (10.32) the ρ^3 coefficient must vanish:

$$\langle \alpha_2, \beta_1, \alpha_0, \beta_0 \rangle + \langle \beta_2, \alpha_1, \alpha_0, \beta_0 \rangle = 0, \quad \langle \alpha_2, \alpha_1, \alpha_0, \beta_0 \rangle + \langle \beta_2, \beta_1, \alpha_0, \beta_0 \rangle = 0,$$

and it follows that

$$\langle \alpha_2, \alpha_1, \alpha_0, \beta_0 \rangle - 2\langle \alpha_2, \beta_1, \alpha_0, \beta_0 \rangle - 2\langle \beta_2, \alpha_1, \alpha_0, \beta_0 \rangle + \langle \beta_2, \beta_1, \alpha_0, \beta_0 \rangle = 0. \tag{10.44}$$

Third, the vanishing of the ρ^4 terms in (10.31) and (10.32) yields the equations

$$\begin{gathered} \langle \alpha_3, \beta_1, \alpha_0, \beta_0 \rangle - \langle \beta_3, \alpha_1, \alpha_0, \beta_0 \rangle - 2\langle \alpha_2, \beta_2, \alpha_0, \beta_0 \rangle \\ -\langle \alpha_2, \alpha_1, \beta_1, \beta_0 \rangle + \langle \beta_2, \alpha_1, \beta_1, \alpha_0 \rangle = 0, \\ \langle \alpha_3, \alpha_1, \alpha_0, \beta_0 \rangle - \langle \beta_3, \beta_1, \alpha_0, \beta_0 \rangle + \langle \alpha_2, \alpha_1, \beta_1, \alpha_0 \rangle - \langle \beta_2, \alpha_1, \beta_1, \beta_0 \rangle = 0, \end{gathered}$$

and hence,

$$\begin{aligned}&\langle\alpha_3,\alpha_1,\alpha_0,\beta_0\rangle+\langle\alpha_2,\alpha_1,\beta_1,\alpha_0\rangle-\langle\alpha_3,\beta_1,\alpha_0,\beta_0\rangle\\ &+\langle\alpha_2,\alpha_1,\beta_1,\beta_0\rangle+2\langle\alpha_2,\beta_2,\alpha_0,\beta_0\rangle+\langle\beta_3,\alpha_1,\alpha_0,\beta_0\rangle\\ &-\langle\beta_2,\alpha_1,\beta_1,\alpha_0\rangle-\langle\beta_2,\alpha_1,\beta_1,\beta_0\rangle-\langle\beta_3,\beta_1,\alpha_0,\beta_0\rangle=0.\end{aligned}\tag{10.45}$$

Fourth, the vanishing of the ρ^5 terms in (10.31) and (10.32) implies that

$$\begin{aligned}&\langle\alpha_3,\beta_2,\alpha_0,\beta_0\rangle+\langle\alpha_3,\alpha_1,\beta_1,\beta_0\rangle+\langle\beta_3,\alpha_2,\alpha_0,\beta_0\rangle\\ &+\langle\beta_3,\alpha_1,\beta_1,\alpha_0\rangle+\langle\alpha_2,\beta_2,\alpha_1,\beta_0\rangle+\langle\alpha_2,\beta_2,\beta_1,\alpha_0\rangle=0,\\ &\langle\alpha_3,\alpha_2,\alpha_0,\beta_0\rangle+\langle\beta_3,\beta_2,\alpha_0,\beta_0\rangle+\langle\alpha_3,\alpha_1,\beta_1,\alpha_0\rangle\\ &+\langle\beta_3,\alpha_1,\beta_1,\beta_0\rangle+\langle\alpha_2,\beta_2,\alpha_1,\alpha_0\rangle+\langle\alpha_2,\beta_2,\beta_1,\beta_0\rangle=0,\end{aligned}$$

and it follows that

$$\begin{aligned}&\langle\alpha_3,\alpha_2,\alpha_0,\beta_0\rangle+\langle\alpha_3,\alpha_1,\beta_1,\alpha_0\rangle+\langle\alpha_2,\beta_2,\alpha_1,\alpha_0\rangle\\ &+\langle\alpha_3,\alpha_1,\beta_1,\beta_0\rangle+\langle\alpha_3,\beta_2,\alpha_0,\beta_0\rangle+\langle\alpha_2,\beta_2,\alpha_1,\beta_0\rangle\\ &+\langle\beta_3,\alpha_2,\alpha_0,\beta_0\rangle+\langle\alpha_2,\beta_2,\beta_1,\alpha_0\rangle+\langle\beta_3,\alpha_1,\beta_1,\alpha_0\rangle\\ &+\langle\alpha_2,\beta_2,\beta_1,\beta_0\rangle+\langle\beta_3,\alpha_1,\beta_1,\beta_0\rangle+\langle\beta_3,\beta_2,\alpha_0,\beta_0\rangle=0.\end{aligned}\tag{10.46}$$

Fifth, the vanishing of the ρ^6 terms in (10.32) and (10.31) gives

$$\begin{aligned}&\langle\alpha_3,\alpha_2,\alpha_1,\beta_0\rangle+\langle\alpha_3,\alpha_2,\beta_1,\alpha_0\rangle+\langle\alpha_3,\beta_2,\alpha_1,\alpha_0\rangle\\ &+\langle\alpha_3,\beta_2,\beta_1,\beta_0\rangle+\langle\beta_3,\alpha_2,\alpha_1,\alpha_0\rangle+\langle\beta_3,\alpha_2,\beta_1,\beta_0\rangle\\ &+\langle\beta_3,\beta_2,\alpha_1,\beta_0\rangle+\langle\beta_3,\beta_2,\beta_1,\alpha_0\rangle=0\end{aligned}\tag{10.47}$$

and

$$\begin{aligned}&2\langle\alpha_3,\beta_2,\alpha_1,\beta_0\rangle+2\langle\beta_3,\alpha_2,\beta_1,\alpha_0\rangle\\ &\quad=\langle\alpha_3,\beta_3,\alpha_0,\beta_0\rangle-\langle\alpha_3,\alpha_2,\beta_1,\beta_0\rangle-\langle\alpha_3,\beta_2,\beta_1,\alpha_0\rangle\\ &\qquad-\langle\beta_3,\alpha_2,\alpha_1,\beta_0\rangle-\langle\beta_3,\beta_2,\alpha_1,\alpha_0\rangle+\langle\alpha_2,\beta_2,\alpha_1,\beta_1\rangle.\end{aligned}\tag{10.48}$$

If we now substitute (10.43)–(10.47) into the expression (10.36) for the constant Δ_0, then it simplifies to give

$$\begin{aligned}\Delta_0=6[&2\langle\alpha_3,\alpha_2,\alpha_1,\alpha_0\rangle+2\langle\alpha_3,\alpha_2,\beta_1,\beta_0\rangle+2\langle\alpha_3,\beta_2,\alpha_1,\beta_0\rangle\\ &+2\langle\alpha_3,\beta_2,\beta_1,\alpha_0\rangle+2\langle\beta_3,\alpha_2,\alpha_1,\beta_0\rangle+2\langle\beta_3,\alpha_2,\beta_1,\alpha_0\rangle\\ &+2\langle\beta_3,\beta_2,\alpha_1,\alpha_0\rangle+2\langle\beta_3,\beta_2,\beta_1,\beta_0\rangle],\end{aligned}\tag{10.49}$$

and if we finally substitute (10.48) into (10.49), then we arrive at the result

$$\begin{aligned}\Delta_0=6[&2\langle\alpha_3,\alpha_2,\alpha_1,\alpha_0\rangle+\langle\alpha_3,\alpha_2,\beta_1,\beta_0\rangle+\langle\alpha_3,\beta_3,\alpha_0,\beta_0\rangle\\ &+\langle\alpha_2,\beta_2,\alpha_1,\beta_1\rangle+\langle\alpha_3,\beta_2,\beta_1,\alpha_0\rangle+\langle\beta_3,\alpha_2,\alpha_1,\beta_0\rangle\\ &+\langle\beta_3,\beta_2,\alpha_1,\alpha_0\rangle+2\langle\beta_3,\beta_2,\beta_1,\beta_0\rangle]=6\gamma_0.\end{aligned}$$

This establishes the assertion.

In view of the above, for Case III we conclude that

$$\sigma(L)=\emptyset,\quad \rho(L)=\mathbb{C}\quad\text{when } \gamma_0\neq 0,\tag{10.50}$$

$$\sigma(L)=\mathbb{C},\quad \rho(L)=\emptyset\quad\text{when } \gamma_0=0.\tag{10.51}$$

Examples 10.1 and 10.2 with $n = 4$ are models for these degenerate irregular cases.

Some of the above results for the case $n = 4$ can be extended to the higher order case $n = 2\nu \geq 6$. We will not discuss these generalizations here.

CHAPTER 11

Unsolved Problems

In the previous chapters we have established important results in the spectral theory of two-point differential operators L that are either regular or simply irregular. For these two cases there still remain many unsolved problems, while for the degenerate irregular differential operators the spectral theory is completely wide open. It seems only appropriate to list here some of the important unsolved problems.

Problem 1. The series

$$Z_k(t,\rho) = e^{i\rho\omega_k t}\sum_{j=0}^{\infty} z_{kj}(t)\rho^{-j}, \qquad k=0,1,\dots,n-1,$$

give formal solutions to the differential equation (2.1). Do these formal solutions converge to actual solutions of (2.1)? See Example 2.8.7 in [**36**].

Problem 2. On an appropriate sector, do the approximate characteristic determinants $\widehat{\Delta}(\rho,m)$ of Chapter 3 converge to a characteristic determinant in the limit as $m \to \infty$?

Problem 3. If the differential operator L is simply irregular, is it true that $S_\infty(L) \neq \overline{S_\infty(L)}$? Are the associated projections P_i unbounded? This is true in certain situations for the case $n=2$, as shown in the two series [**25, 26**] and [**30, 31, 32, 33**].

Problem 4. If the differential operator L is degenerate irregular and the spectrum $\sigma(L)$ is a countably infinite set, can it be shown that $\overline{S_\infty(L)} = L^2[0,1]$? Models for this type of problem are given in Case II of Chapter 10.

Problem 5. What is the subspace $S_\infty(L)$ when the differential operator L is simply irregular or degenerate irregular? What is $S_\infty(L)$ when $n=2$, $L=T$, and L is simply irregular? In general, is the domain $\mathcal{D}(L)$ a subset of $S_\infty(L)$?

Problem 6. Is there a natural subdivision of the degenerate irregular class into disjoint subclasses having different spectral properties? The cases where the spectrum $\sigma(L)$ is countably infinite, or is empty, or is equal to all of $\mathbb{C}$, should go into different subclasses. Is there a spectral theory available for any of these subclasses?

Problem 7. For the case $n = 2\nu$ even, it is easy to verify that Dirichlet and Neumann boundary conditions determine a differential operator L that is regular. This is also true for periodic boundary conditions and boundary conditions of

Sturm Type [**37**, pp. 60–63]. Are all self-adjoint differential operators regular? In [**25, 26**] we have shown this to be true for the case $n = 2$ and $L = T$, because in the irregular cases either the associated projections are unbounded (Case IX and Case XI) or the spectrum is either empty (Case XII) or is equal to all of $\mathbb{C}$ (Case XIII). See [**25**, p. 556].

Problem 8. In general, can the spectrum $\sigma(L)$ be a nonempty *finite* set? This is impossible in the special case $n = 2$, $L = T$ (see [**25**, p. 556]) and in the special case $n = 4$, $L = T$ (see Cases I, II, III in the previous chapter).

Problem 9. In Example 10.2 we gave a model of an even order differential operator $L = T$ with $\sigma(L) = \mathbb{C}$ and $\rho(L) = \emptyset$. Can we find an analogous model of an odd order differential operator $L = T$? Are there more general models with $\sigma(L) = \mathbb{C}$ and $\rho(L) = \emptyset$ where L has variable coefficients?

Problem 10. The rows of the $n \times 2n$ boundary coefficient matrix A,

$$(\alpha_{i\,n-1}, \beta_{i\,n-1}, \alpha_{i\,n-2}, \beta_{i\,n-2}, \ldots, \alpha_{i0}, \beta_{i0}), \qquad i = 1, \ldots, n,$$

span an $(n-1)$-dimensional linear space $\mathbb{L}$ in the projective space $\mathbb{P}^{2n-1}$. Let the columns of A be denoted by $\gamma_1, \ldots, \gamma_{2n}$, and for integers $j_1, \ldots, j_n$ with

$$1 \le j_1 < \cdots < j_n \le 2n,$$

let $\langle \gamma_{j_1}, \ldots, \gamma_{j_n} \rangle$ denote the determinant of the $n \times n$ submatrix of A formed by using the columns $\gamma_{j_1}, \ldots, \gamma_{j_n}$. Then the $\binom{2n}{n}$ determinants $\langle \gamma_{j_1}, \ldots, \gamma_{j_n} \rangle$ are the Plücker coordinates of the linear space $\mathbb{L}$. For the special case $n = 4$, $L = T$, the 70 Plücker coordinates appear very prominently in equations (10.31), (10.32), and (10.33) for the polynomial coefficients $\mathbb{P}_0(\rho)$, $\mathbb{Q}_0(\rho)$, and $\mathbb{D}(\rho)$ that determine the characteristic determinant $\Delta(\rho)$. For the general case $L = T$, can we use the Plücker coordinates associated with the boundary coefficient matrix A to effectively construct the characteristic determinant?

Problem 11. For the special case $L = T$, in forming the alternate characteristic determinant $\Delta_*(\rho) = \det(B_i(z_k(\,\cdot\,, \rho)))$, the exponentials

$$e^{i\rho(\delta_0\omega_0 + \delta_1\omega_1 + \cdots + \delta_{n-1}\omega_{n-1})}$$

appear, where the constants δ_i are either 0 or 1. How many *distinct* exponentials are formed from these 2^n exponentials? We know the following results:

$n = 2:$	4 exponentials,	3 distinct exponentials,
$n = 3:$	8 exponentials,	7 distinct exponentials,
$n = 4:$	16 exponentials,	9 distinct exponentials,
$n = 6:$	64 exponentials,	19 distinct exponentials,
$n = 8:$	256 exponentials,	81 distinct exponentials.

Is there a practical way to construct the distinct exponentials so that they can be ordered according to their moduli relative to the sector S_0? Can we calculate the polynomial coefficient of each distinct exponential? The case $n = 4$ has been treated in detail in Chapter 10.

Problem 12. For the special case $n = 8$, $L = T$, we have done some preliminary work with help from Maple. When the characteristic determinant $\Delta(\rho)$ is expanded using linearity in all eight columns, the 256 exponentials that are produced reduce down to 81 distinct exponentials. First, there are the three primary exponentials $e^{0\rho} = 1$, $e^{i\rho}$, $e^{2i\rho}$. Second, there are the six secondary exponentials

$$e^{[-\frac{\sqrt{2}}{2}-\frac{\sqrt{2}}{2}i]\rho}, \quad e^{[-\frac{\sqrt{2}}{2}+(1-\frac{\sqrt{2}}{2})i]\rho}, \quad e^{[-\frac{\sqrt{2}}{2}+\frac{\sqrt{2}}{2}i]\rho},$$
$$e^{[-\frac{\sqrt{2}}{2}+(2-\frac{\sqrt{2}}{2})i]\rho}, \quad e^{[-\frac{\sqrt{2}}{2}+(1+\frac{\sqrt{2}}{2})i]\rho}, \quad e^{[-\frac{\sqrt{2}}{2}+(2+\frac{\sqrt{2}}{2})i]\rho}.$$

And third, there are 72 additional exponentials of lower order. The ordering of these 81 exponentials is according to their moduli relative to the sector S_0. In terms of these exponentials the characteristic determinant takes the form

$$\begin{aligned}\Delta(\rho) = {} & \mathbb{P}_0(\rho) + \mathbb{Q}_0(\rho)e^{i\rho} + \mathbb{P}_1(\rho)e^{2i\rho} + \mathbb{Q}_1(\rho)e^{[-\frac{\sqrt{2}}{2}-\frac{\sqrt{2}}{2}i]\rho} \\ & + \mathbb{R}_0(\rho)e^{[-\frac{\sqrt{2}}{2}+(1-\frac{\sqrt{2}}{2})i]\rho} + \mathbb{S}_0(\rho)e^{[-\frac{\sqrt{2}}{2}+\frac{\sqrt{2}}{2}i]\rho} + \mathbb{T}_0(\rho)e^{[-\frac{\sqrt{2}}{2}+(2-\frac{\sqrt{2}}{2})i]\rho} \\ & + \mathbb{R}_1(\rho)e^{[-\frac{\sqrt{2}}{2}+(1+\frac{\sqrt{2}}{2})i]\rho} + \mathbb{Q}_2(\rho)e^{[-\frac{\sqrt{2}}{2}+(2+\frac{\sqrt{2}}{2})i]\rho} + \cdots\end{aligned}$$

for $\rho \neq 0$ in $\mathbb{C}$, where the functions $\mathbb{P}_0(\rho), \ldots, \mathbb{Q}_2(\rho), \ldots$ are polynomials of degree ≤ 44. For the special case $\mathbb{P}_0(\rho) = 0$ and $\mathbb{Q}_0(\rho) = 0$, this representation simplifies to

$$\begin{aligned}\Delta(\rho) = {} & \mathbb{R}_0(\rho)e^{[-\frac{\sqrt{2}}{2}+(1-\frac{\sqrt{2}}{2})i]\rho} + \mathbb{S}_0(\rho)e^{[-\frac{\sqrt{2}}{2}+\frac{\sqrt{2}}{2}i]\rho} \\ & + \mathbb{T}_0(\rho)e^{[-\frac{\sqrt{2}}{2}+(2-\frac{\sqrt{2}}{2})i]\rho} + \mathbb{R}_1(\rho)e^{[-\frac{\sqrt{2}}{2}+(1+\frac{\sqrt{2}}{2})i]\rho} + \cdots \\ = {} & e^{[-\frac{\sqrt{2}}{2}+(1-\frac{\sqrt{2}}{2})i]\rho}\{\mathbb{R}_0(\rho) + \mathbb{S}_0(\rho)e^{(-1+\sqrt{2})i\rho} \\ & + \mathbb{T}_0(\rho)e^{i\rho} + \mathbb{R}_1(\rho)e^{\sqrt{2}i\rho} + \cdots\}\end{aligned}$$

for $\rho \neq 0$ in $\mathbb{C}$. Clearly the differential operator $L = T$ becomes degenerate irregular. How does one calculate the zeros of $\Delta(\rho)$ in this special case? Are the generalized eigenfunctions of L complete in $L^2[0,1]$? Is it possible that there may be more than two sequences of eigenvalues? Could there be algebraic multiplicities greater than 2? Can we give a specific example that falls within this special case?

CHAPTER 12

Appendix

The appendix contains two lemmas which are used to develop the regularity properties for our solutions of the differential equation (2.1). These lemmas generalize well-known results from classical analysis. The first is a complex version of Leibniz's rule from advanced calculus. See [**7**, p. 68 and p. 73].

LEMMA 12.1. *Let Δ_1 and Δ_2 be the triangles in $\mathbb{R}^2$ consisting of all points (t, s) satisfying the inequalities $0 \le s \le t \le 1$ and $0 \le t \le s \le 1$, respectively, and let G be an open set in the complex ρ plane; let $\phi_1\colon \Delta_1 \times G \to \mathbb{C}$ and $\phi_2\colon \Delta_2 \times G \to \mathbb{C}$ be continuous functions, and let ϕ be the function defined by $\phi(t, s, \rho) := \phi_1(t, s, \rho)$ for $0 \le s < t \le 1$, $\rho \in G$, and $\phi(t, s, \rho) := \phi_2(t, s, \rho)$ for $0 \le t < s \le 1$, $\rho \in G$; and let $u\colon [0, 1] \times G \to \mathbb{C}$ be the function defined by*

$$u(t, \rho) := \int_0^1 \phi(t, s, \rho)\, ds \quad \textit{for } 0 \le t \le 1,\ \rho \in G.$$

Then u is continuous on $[0, 1] \times G$. Moreover, if $\partial\phi_1/\partial\rho$ exists and is continuous on $\Delta_1 \times G$ and if $\partial\phi_2/\partial\rho$ exists and is continuous on $\Delta_2 \times G$, then $\partial u/\partial\rho$ exists and is continuous on $[0, 1] \times G$ with

$$\frac{\partial u}{\partial \rho}(t, \rho) = \int_0^1 \frac{\partial \phi}{\partial \rho}(t, s, \rho)\, ds \quad \textit{for } 0 \le t \le 1,\ \rho \in G.$$

Note: we do not make any assumptions about the values of $\phi_1(t, s, \rho)$ and $\phi_2(t, s, \rho)$ when $t = s$, so we can have a jump across the diagonal $t = s$.

The second lemma generalizes a well-known result on the analyticity of the limit of a sequence of analytic functions. See [**38**, p. 256].

LEMMA 12.2. *Let G be an open set in the ρ plane, let $u_k\colon [0, 1] \times G \to \mathbb{C}$, $k = 1, 2, \ldots$, be a sequence of functions, and let $u\colon [0, 1] \times G \to \mathbb{C}$ be a function, where for each compact subset K of G, it is assumed that the u_k converge uniformly on $[0, 1] \times K$ to u. Assume that each u_k is continuous on $[0, 1] \times G$ and that $\partial u_k/\partial\rho$ exists and is continuous on $[0, 1] \times G$. Then u is continuous on $[0, 1] \times G$, $\partial u/\partial\rho$ exists and is continuous on $[0, 1] \times G$, and for each compact subset K of G, the functions $\partial u_k/\partial\rho$ converge uniformly on $[0, 1] \times K$ to $\partial u/\partial\rho$.*

Bibliography

[1] Shmuel Agmon, *Lectures on elliptic boundary value problems*, D. Van Nostrand Company, Inc., Princeton, New Jersey, 1965.

[2] Harold E. Benzinger, *Green's function for ordinary differential operators*, Journal of Differential Equations **7** (1970), 478–496.

[3] George D. Birkhoff, *Boundary value and expansion problems of ordinary linear differential equations*, Transactions of the American Mathematical Society **9** (1908), 373–395.

[4] ———, *On the asymptotic character of the solutions of certain linear differential equations containing a parameter*, Transactions of the American Mathematical Society **9** (1908), 219–231.

[5] ———, *Note on the expansion problems of ordinary linear differential equations*, Rendiconti Palermo **36** (1913), 115–126.

[6] Earl A. Coddington and Norman Levinson, *Theory of ordinary differential equations*, McGraw-Hill Book Company, Inc., New York, 1955.

[7] John B. Conway, *Functions of one complex variable*, Springer-Verlag, New York, 1978.

[8] Nelson Dunford and Jacob T. Schwartz, *Linear operators, I, II, III*, Wiley-Interscience, New York, 1958, 1963, 1971.

[9] I. C. Gohberg and M. G. Krein, *The basic propositions on defect numbers, root numbers and indices of linear operators*, Uspekhi Mat. Nauk. **12, 2(74)** (1957), 43–118, Translated in Amer. Math. Soc. Transl., vol. 13, Ser. 2, 1960, pp. 185–264.

[10] ———, *Introduction to the theory of linear nonselfadjoint operators*, Translations of Mathematical Monographs, vol. 18, American Mathematical Society, Providence, Rhode Island, 1969.

[11] Seymour Goldberg, *Unbounded linear operators: Theory and applications*, McGraw-Hill Book Company, New York, 1966.

[12] Paul R. Halmos, *Finite-dimensional vector spaces*, Springer-Verlag, New York, 1974.

[13] G. H. Hardy, J. E. Littlewood, and G. Pólya, *Inequalities*, second ed., Cambridge University Press, Cambridge, 1952.

[14] Stephen Hoffman, *Second-order linear differential operators defined by irregular boundary conditions*, Ph.D. thesis, Yale University, 1957.

[15] J. W. Hopkins, *Some convergent developments associated with irregular boundary conditions*, Transactions of the American Mathematical Society **20** (1919), 245–259.

[16] Roger A. Horn and Charles R. Johnson, *Matrix analysis*, Cambridge University Press, Cambridge, 1985.

[17] Shmuel Kaniel and Martin Schechter, *Spectral theory for Fredholm operators*, Communications on Pure and Applied Mathematics **16** (1963), 423–448.

[18] L. V. Kantorovich and G. P. Akilov, *Functional analysis*, second ed., Pergamon Press, Oxford, 1982.

[19] Tosio Kato, *Perturbation theory for nullity, deficiency and other quantities of linear operators*, Journal d'Analyse Mathématique **6** (1958), 261–322.

[20] ———, *Perturbation theory for linear operators*, second ed., Springer-Verlag, Berlin, 1976.

[21] Patrick Lang and John Locker, *Spectral decomposition of a Hilbert space by a Fredholm operator*, Journal of Functional Analysis **79** (1988), 9–17.

[22] ———, *Spectral representation of the resolvent of a discrete operator*, Journal of Functional Analysis **79** (1988), 18–31.

[23] ———, *Denseness of the generalized eigenvectors of an H-S discrete operator*, Journal of Functional Analysis **82** (1989), 316–329.

[24] ———, *Spectral theory for a differential operator: Characteristic determinant and Green's function*, Journal of Mathematical Analysis and Applications **141** (1989), 405–423.

[25] ———, *Spectral theory of two-point differential operators determined by $-D^2$. I. Spectral properties*, Journal of Mathematical Analysis and Applications **141** (1989), 538–558.

[26] ———, *Spectral theory of two-point differential operators determined by $-D^2$. II. Analysis of cases*, Journal of Mathematical Analysis and Applications **146** (1990), 148–191.

[27] R. E. Langer, *The zeros of exponential sums and integrals*, Bulletin of the American Mathematical Society **37** (1931), 213–239.

[28] John Locker, *Functional analysis and two-point differential operators*, Pitman Research Notes in Mathematics, vol. 144, Longmans, Harlow, Essex, 1986.

[29] ———, *The nonspectral Birkhoff-regular differential operators determined by $-D^2$*, Journal of Mathematical Analysis and Applications **154** (1991), 243–254.

[30] ———, *The spectral theory of second order two-point differential operators: I. A priori estimates for the eigenvalues and completeness*, Proceedings of the Royal Society of Edinburgh **121A** (1992), 279–301.

[31] ———, *The spectral theory of second order two-point differential operators: II. Asymptotic expansions and the characteristic determinant*, Journal of Differential Equations **114** (1994), 272–287.

[32] ———, *The spectral theory of second order two-point differential operators: III. The eigenvalues and their asymptotic formulas*, Rocky Mountain Journal of Mathematics **26** (1996), 679–706.

[33] ———, *The spectral theory of second order two-point differential operators: IV. The associated projections and the subspace $\mathcal{S}_\infty(L)$*, Rocky Mountain Journal of Mathematics **26** (1996), 1473–1498.

[34] ———, *Spectral theory of non-self-adjoint two-point differential operators*, Mathematical Surveys and Monographs, vol. 73, American Mathematical Society, Providence, R. I., 2000.

[35] John Locker and Patrick Lang, *Eigenfunction expansions for the nonspectral differential operators determined by $-D^2$*, Journal of Differential Equations **96** (1992), 318–339.

[36] Reinhard Mennicken and Manfred Möller, *Non-self-adjoint boundary eigenvalue problems*, North-Holland Mathematics Studies 192, Elsevier, Amsterdam, 2003.

[37] M. A. Naimark, *Linear differential operators, I*, GITTL, Moscow, 1954, English transl., Ungar, New York, 1967.

[38] Bruce P. Palka, *An introduction to complex function theory*, Springer-Verlag, New York, 1991.

[39] Walter Rudin, *Real and complex analysis*, third ed., McGraw-Hill, New York, 1987.

[40] ———, *Functional analysis*, second ed., McGraw-Hill, New York, 1991.

[41] Martin Schechter, *Principles of functional analysis*, Academic Press, New York, 1971.

[42] Bernd Schultze, *Strongly irregular boundary value problems*, Proceedings of the Royal Society of Edinburgh **82A** (1979), 291–303.

[43] J. Schwartz, *Perturbations of spectral operators, and applications. I. Bounded perturbations*, Pacific Journal of Mathematics **4** (1954), 415–458.

[44] G. Seifert, *A third order irregular boundary value problem and the associated series*, Pacific Journal of Mathematics **2** (1952), 395–406.

[45] M. H. Stone, *A comparison of the series of Fourier and Birkhoff*, Transactions of the American Mathematical Society **28** (1926), 695–761.

[46] ———, *Irregular differential systems of order two and the related expansion problems*, Transactions of the American Mathematical Society **29** (1927), 23–53.

[47] ———, *Linear transformations in Hilbert space and their applications to analysis*, American Mathematical Society Colloquium Publications, vol. XV, American Mathematical Society, New York, 1932.

[48] J. Tamarkin, *Some general problems of the theory of ordinary linear differential equations and expansion of an arbitrary function in series of fundamental functions*, Mathematische Zeitschrift **27** (1927), 1–54.

[49] Angus E. Taylor and David C. Lay, *Introduction to functional analysis*, second ed., Krieger Publishing Company, Malabar, Florida, 1980.

[50] E. C. Titchmarsh, *The theory of functions*, second ed., Oxford University Press, Oxford, 1939.

[51] Isabelle Titeux and Yakov Yakubov, *Applications of abstract differential equations to some mechanical problems*, Mathematics and Its Applications, vol. 558, Kluwer, 2003.

[52] Philip W. Walker, *A nonspectral Birkhoff-regular differential operator*, Proceedings of the American Mathematical Society **66** (1977), 187–188.

[53] L. E. Ward, *An irregular boundary value and expansion problem*, Annals of Mathematics **26** (1925), 21–36.

[54] ———, *A third order irregular boundary value problem and the associated series*, Transactions of the American Mathematical Society **34** (1932), 417–434.

[55] Sasun Yakubov, *Completeness of root functions of regular differential operators*, Pitman Monographs and Surveys in Pure and Applied Mathematics, vol. 71, Longman Scientific & Technical, Harlow, Essex, 1994.

[56] Sasun Yakubov and Yakov Yakubov, *Differential-operator equations, ordinary and partial differential equations*, Monographs and Surveys in Pure and Applied Mathematics, vol. 103, Chapman & Hall/CRC, Boca Raton, Florida, 2000.

[57] Yakov Yakubov, *Irregular boundary value problems for ordinary differential equations*, Analysis **18** (1998), 359–402.

Index

Editorial Information

To be published in the *Memoirs*, a paper must be correct, new, nontrivial, and significant. Further, it must be well written and of interest to a substantial number of mathematicians. Piecemeal results, such as an inconclusive step toward an unproved major theorem or a minor variation on a known result, are in general not acceptable for publication.

Papers appearing in *Memoirs* are generally at least 80 and not more than 200 published pages in length. Papers less than 80 or more than 200 published pages require the approval of the Managing Editor of the Transactions/Memoirs Editorial Board.

As of May 31, 2008, the backlog for this journal was approximately 17 volumes. This estimate is the result of dividing the number of manuscripts for this journal in the Providence office that have not yet gone to the printer on the above date by the average number of monographs per volume over the previous twelve months, reduced by the number of volumes published in four months (the time necessary for preparing a volume for the printer). (There are 6 volumes per year, each usually containing at least 4 numbers.)

A Consent to Publish and Copyright Agreement is required before a paper will be published in the *Memoirs*. After a paper is accepted for publication, the Providence office will send a Consent to Publish and Copyright Agreement to all authors of the paper. By submitting a paper to the *Memoirs*, authors certify that the results have not been submitted to nor are they under consideration for publication by another journal, conference proceedings, or similar publication.

Information for Authors

Memoirs are printed from camera copy fully prepared by the author. This means that the finished book will look exactly like the copy submitted.

Initial submission. The AMS uses Centralized Manuscript Processing for initial submissions. Authors should submit a PDF file using the Initial Manuscript Submission form found at `www.ams.org/cgi-bin/peertrack/submission.pl`, or send one copy of the manuscript to the following address: Centralized Manuscript Processing, MEMOIRS OF THE AMS, 201 Charles Street, Providence, RI 02904-2294 USA. If a paper copy is being forwarded to the AMS, indicate that it is for it Memoirs and include the name of the corresponding author, contact information such as email address or mailing address, and the name of an appropriate Editor to review the paper (see the list of Editors below).

The paper must contain a *descriptive title* and an *abstract* that summarizes the article in language suitable for workers in the general field (algebra, analysis, etc.). The *descriptive title* should be short, but informative; useless or vague phrases such as "some remarks about" or "concerning" should be avoided. The *abstract* should be at least one complete sentence, and at most 300 words. Included with the footnotes to the paper should be the 2000 *Mathematics Subject Classification* representing the primary and secondary subjects of the article. The classifications are accessible from `www.ams.org/msc/`. The list of classifications is also available in print starting with the 1999 annual index of *Mathematical Reviews*. The Mathematics Subject Classification footnote may be followed by a list of *key words and phrases* describing the subject matter of the article and taken from it. Journal abbreviations used in bibliographies are listed in the latest *Mathematical Reviews* annual index. The series abbreviations are also accessible from `www.ams.org/publications/`. To help in preparing and verifying references, the AMS offers MR Lookup, a Reference Tool for Linking, at `www.ams.org/mrlookup/`.

Electronically prepared manuscripts. The AMS encourages electronically prepared manuscripts, with a strong preference for $\mathcal{AMS}$-LaTeX. To this end, the Society has prepared $\mathcal{AMS}$-LaTeX author packages for each AMS publication. Author packages include instructions for preparing electronic manuscripts, samples, and a style file that generates the particular design specifications of that publication series. Though $\mathcal{A}\mathcal{M}\mathcal{S}$-LaTeX is the highly preferred format of TeX, author packages are also available in $\mathcal{A}\mathcal{M}\mathcal{S}$-TeX.

Authors may retrieve an author package from the AMS website starting from `www.ams.org/tex/` or via FTP to `ftp.ams.org` (login as `anonymous`, enter username as password, and type `cd pub/author-info`). The *AMS Author Handbook* and the *Instruction Manual* are available in PDF format following the author packages link from `www.ams.org/tex/`. The author package can also be obtained free of charge by sending email to `tech-support@ams.org` (Internet) or from the Publication Division, American Mathematical Society, 201 Charles St., Providence, RI 02904-2294, USA. When requesting an author package, please specify $\mathcal{A}\mathcal{M}\mathcal{S}$-LaTeX or $\mathcal{A}\mathcal{M}\mathcal{S}$-TeX and the publication in which your paper will appear. Please be sure to include your complete mailing address.

After acceptance. The final version of the electronic file should be sent to the Providence office (this includes any TeX source file, any graphics files, and the DVI or PostScript file) immediately after the paper has been accepted for publication.

Before sending the source file, be sure you have proofread your paper carefully. The files you send must be the EXACT files used to generate the proof copy that was accepted for publication. For all publications, authors are required to send a printed copy of their paper, which exactly matches the copy approved for publication, along with any graphics that will appear in the paper.

Accepted electronically prepared files can be submitted via the web at `www.ams.org/submit-book-journal/`, sent via FTP, or sent on CD-Rom or diskette to the Electronic Prepress Department, American Mathematical Society, 201 Charles Street, Providence, RI 02904-2294 USA. TeX source files, DVI files, and PostScript files can be transferred over the Internet by FTP to the Internet node `ftp.ams.org` (130.44.1.100). When sending a manuscript electronically via CD-Rom or diskette, please be sure to include a message identifying the paper as a Memoir.

Electronically prepared manuscripts can also be sent via email to `pub-submit@ams.org` (Internet). In order to send files via email, they must be encoded properly. (DVI files are binary and PostScript files tend to be very large.)

Electronic graphics. Comprehensive instructions on preparing graphics are available at `www.ams.org/jourhtml/`. A few of the major requirements are given here.

Submit files for graphics as EPS (Encapsulated PostScript) files. This includes graphics originated via a graphics application as well as scanned photographs or other computer-generated images. If this is not possible, TIFF files are acceptable as long as they can be opened in Adobe Photoshop or Illustrator. No matter what method was used to produce the graphic, it is necessary to provide a paper copy to the AMS.

Authors using graphics packages for the creation of electronic art should also avoid the use of any lines thinner than 0.5 points in width. Many graphics packages allow the user to specify a "hairline" for a very thin line. Hairlines often look acceptable when proofed on a typical laser printer. However, when produced on a high-resolution laser imagesetter, hairlines become nearly invisible and will be lost entirely in the final printing process.

Screens should be set to values between 15% and 85%. Screens which fall outside of this range are too light or too dark to print correctly. Variations of screens within a graphic should be no less than 10%.

Inquiries. Any inquiries concerning a paper that has been accepted for publication should be sent to `memo-query@ams.org` or directly to the Electronic Prepress Department, American Mathematical Society, 201 Charles St., Providence, RI 02904-2294 USA.

Editors

This journal is designed particularly for long research papers, normally at least 80 pages in length, and groups of cognate papers in pure and applied mathematics. Papers intended for publication in the *Memoirs* should be addressed to one of the following editors. The AMS uses Centralized Manuscript Processing for initial submissions to AMS journals. Authors should follow instructions listed on the Initial Submission page found at www.ams.org/memo/memosubmit.html.

Algebra to ALEXANDER KLESHCHEV, Department of Mathematics, University of Oregon, Eugene, OR 97403-1222; email: ams@noether.uoregon.edu

Algebraic geometry and its application to MINA TEICHER, Emmy Noether Research Institute for Mathematics, Bar-Ilan University, Ramat-Gan 52900, Israel; email: teicher@macs.biu.ac.il

Algebraic geometry to DAN ABRAMOVICH, Department of Mathematics, Brown University, Box 1917, Providence, RI 02912; email: amsedit@math.brown.edu

Algebraic topology to ALEJANDRO ADEM, Department of Mathematics, University of British Columbia, Room 121, 1984 Mathematics Road, Vancouver, British Columbia, Canada V6T 1Z2; email: adem@math.ubc.ca

Combinatorics to JOHN R. STEMBRIDGE, Department of Mathematics, University of Michigan, Ann Arbor, Michigan 48109-1109; email: FRS@umich.edu

Complex analysis and harmonic analysis to ALEXANDER NAGEL, Department of Mathematics, University of Wisconsin, 480 Lincoln Drive, Madison, WI 53706-1313; email: nagel@math.wisc.edu

Differential geometry and global analysis to LISA C. JEFFREY, Department of Mathematics, University of Toronto, 100 St. George St., Toronto, ON Canada M5S 3G3; email: jeffrey@math.toronto.edu

Dynamical systems and ergodic theory and complex anaysis to YUNPING JIANG, Department of Mathematics, CUNY Queens College and Graduate Center, 65-30 Kissena Blvd., Flushing, NY 11367; email: Yunping.Jiang@qc.cuny.edu

Functional analysis and operator algebras to DIMITRI SHLYAKHTENKO, Department of Mathematics, University of California, Los Angeles, CA 90095; email: shlyakht@math.ucla.edu

Geometric analysis to WILLIAM P. MINICOZZI II, Department of Mathematics, Johns Hopkins University, 3400 N. Charles St., Baltimore, MD 21218; email: trans@math.jhu.edu

Geometric analysis to MARK FEIGHN, Math Department, Rutgers University, Newark, NJ 07102; email: feighn@andromeda.rutgers.edu

Harmonic analysis, representation theory, and Lie theory to ROBERT J. STANTON, Department of Mathematics, The Ohio State University, 231 West 18th Avenue, Columbus, OH 43210-1174; email: stanton@math.ohio-state.edu

Logic to STEFFEN LEMPP, Department of Mathematics, University of Wisconsin, 480 Lincoln Drive, Madison, Wisconsin 53706-1388; email: lempp@math.wisc.edu

Number theory to JONATHAN ROGAWSKI, Department of Mathematics, University of California, Los Angeles, CA 90095; email: jonr@math.ucla.edu

Partial differential equations to GUSTAVO PONCE, Department of Mathematics, South Hall, Room 6607, University of California, Santa Barbara, CA 93106; email: ponce@math.ucsb.edu

Partial differential equations and dynamical systems to PETER POLACIK, School of Mathematics, University of Minnesota, Minneapolis, MN 55455; email: polacik@math.umn.edu

Probability and statistics to RICHARD BASS, Department of Mathematics, University of Connecticut, Storrs, CT 06269-3009; email: bass@math.uconn.edu

Real analysis and partial differential equations to DANIEL TATARU, Department of Mathematics, University of California, Berkeley, Berkeley, CA 94720; email: tataru@math.berkeley.edu

All other communications to the editors should be addressed to the Managing Editor, ROBERT GURALNICK, Department of Mathematics, University of Southern California, Los Angeles, CA 90089-1113; email: guralnic@math.usc.edu.

Titles in This Series

913 **Ethan Akin, Joseph Auslander, and Eli Glasner,** The topological dynamics of Ellis actions, 2008

912 **Igor Chueshov and Irena Lasiecka,** Long-time behavior of second order evolution equations with nonlinear damping, 2008

911 **John Locker,** Eigenvalues and completeness for regular and simply irregular two-point differential operators, 2008

910 **Joel Friedman,** A proof of Alon's second eigenvalue conjecture and related problems, 2008

909 **Cameron McA. Gordon and Ying-Qing Wu,** Toroidal Dehn fillings on hyperbolic 3-manifolds, 2008

908 **J.-L. Waldspurger,** L'endoscopie tordue n'est pas si tordue, 2008

907 **Yuanhua Wang and Fei Xu,** Spinor genera in characteristic 2, 2008

906 **Raphaël S. Ponge,** Heisenberg calculus and spectral theory of hypoelliptic operators on Heisenberg manifolds, 2008

905 **Dominic Verity,** Complicial sets characterising the simplicial nerves of strict ω-categories, 2008

904 **William M. Goldman and Eugene Z. Xia,** Rank one Higgs bundles and representations of fundamental groups of Riemann surfaces, 2008

903 **Gail Letzter,** Invariant differential operators for quantum symmetric spaces, 2008

902 **Bertrand Toën and Gabriele Vezzosi,** Homotopical algebraic geometry II: Geometric stacks and applications, 2008

901 **Ron Donagi and Tony Pantev (with an appendix by Dmitry Arinkin),** Torus fibrations, gerbes, and duality, 2008

900 **Wolfgang Bertram,** Differential geometry, Lie groups and symmetric spaces over general base fields and rings, 2008

899 **Piotr Hajłasz, Tadeusz Iwaniec, Jan Malý, and Jani Onninen,** Weakly differentiable mappings between manifolds, 2008

898 **John Rognes,** Galois extensions of structured ring spectra/Stably dualizable groups, 2008

897 **Michael I. Ganzburg,** Limit theorems of polynomial approximation with exponential weights, 2008

896 **Michael Kapovich, Bernhard Leeb, and John J. Millson,** The generalized triangle inequalities in symmetric spaces and buildings with applications to algebra, 2008

895 **Steffen Roch,** Finite sections of band-dominated operators, 2008

894 **Martin Dindoš,** Hardy spaces and potential theory on C^1 domains in Riemannian manifolds, 2008

893 **Tadeusz Iwaniec and Gaven Martin,** The Beltrami Equation, 2008

892 **Jim Agler, John Harland, and Benjamin J. Raphael,** Classical function theory, operator dilation theory, and machine computation on multiply-connected domains, 2008

891 **John H. Hubbard and Peter Papadopol,** Newton's method applied to two quadratic equations in $\mathbb{C}^2$ viewed as a global dynamical system, 2008

890 **Steven Dale Cutkosky,** Toroidalization of dominant morphisms of 3-folds, 2007

889 **Michael Sever,** Distribution solutions of nonlinear systems of conservation laws, 2007

888 **Roger Chalkley,** Basic global relative invariants for nonlinear differential equations, 2007

887 **Charlotte Wahl,** Noncommutative Maslov index and eta-forms, 2007

886 **Robert M. Guralnick and John Shareshian,** Symmetric and alternating groups as monodromy groups of Riemann surfaces I: Generic covers and covers with many branch points, 2007

885 **Jae Choon Cha,** The structure of the rational concordance group of knots, 2007

884 **Dan Haran, Moshe Jarden, and Florian Pop,** Projective group structures as absolute Galois structures with block approximation, 2007

883 **Apostolos Beligiannis and Idun Reiten,** Homological and homotopical aspects of torsion theories, 2007

882 **Lars Inge Hedberg and Yuri Netrusov,** An axiomatic approach to function spaces, spec tral synthesis and Luzin approximation, 2007

881 **Tao Mei,** Operator valued Hardy spaces, 2007

880 **Bruce C. Berndt, Geumlan Choi, Youn-Seo Choi, Heekyoung Hahn, Boon Pin Yeap, Ae Ja Yee, Hamza Yesilyurt, and Jinhee Yi,** Ramanujan's forty identities for Rogers-Ramanujan functions, 2007

879 **O. García-Prada, P. B. Gothen, and V. Muñoz,** Betti numbers of the moduli space of rank 3 parabolic Higgs bundles, 2007

878 **Alessandra Celletti and Luigi Chierchia,** KAM stability and celestial mechanics, 2007

877 **María J. Carro, José A. Raposo, and Javier Soria,** Recent developments in the theory of Lorentz spaces and weighted inequalities, 2007

876 **Gabriel Debs and Jean Saint Raymond,** Borel liftings of Borel sets: Some decidable and undecidable statements, 2007

875 **C. Krattenthaler and T. Rivoal,** Hypergéométrie et fonction zêta de Riemann, 2007

874 **Sonia Natale,** Semisolvability of semisimple Hopf algebras of low dimension, 2007

873 **A. J. Duncan,** Exponential genus problems in one-relator products of groups, 2007

872 **Anthony V. Geramita, Tadahito Harima, Juan C. Migliore, and Yong Su Shin,** The Hilbert function of a level algebra, 2007

871 **Pascal Auscher,** On necessary and sufficient conditions for L^p-estimates of Riesz transforms associated to elliptic operators on $\mathbb{R}^n$ and related estimates, 2007

870 **Takuro Mochizuki,** Asymptotic behaviour of tame harmonic bundles and an application to pure twistor D-modules, Part 2, 2007

869 **Takuro Mochizuki,** Asymptotic behaviour of tame harmonic bundles and an application to pure twistor D-modules, Part 1, 2007

868 **Gelu Popescu,** Entropy and multivariable interpolation, 2006

867 **Vilmos Totik,** Metric properties of harmonic measures, 2006

866 **William Craig,** Semigroups underlying first-order logic, 2006

865 **Nathanial P. Brown,** Invariant means and finite representation theory of $C*$-algebras, 2006

864 **John M. Lee,** Fredholm operators and Einstein metrics on conformally compact manifolds, 2006

863 **M. Lübke and A. Teleman,** The Universal Kobayashi-Hitchin correspondence on Hermitian manifolds, 2006

862 **Alberto Canonaco,** The Beilinson complex and canonical rings of irregular surfaces, 2006

861 **Leon A. Takhtajan and Lee-Peng Teo,** Weil-Petersson metric on the universal Teichmüller space, 2006

860 **Thomas M. Fiore,** Pseudo limits, biadjoints and pseudo algebras: Categorical foundations of conformal field theory, 2006

859 **N. Arcozzi, R. Rochberg, and E. Sawyer,** Carleson measures and interpolating sequences for Besov spaces on complex balls, 2006

858 **Enrico Valdinoci, Berardino Sciunzi, and Vasile Ovidiu Savin,** Flat level set regularity of p-Laplace phase transitions, 2006

For a [redacted] of titles in this series, visit the
[redacted] at **www.ams.org/bookstore/.**